CANADA

A NEW GEOGRAPHY

THIRD EDITION

Ralph Krueger

Ray Corder

Holt, Rinehart and Winston of Canada, Limited
Toronto

Copyright © 1974, 1982 by
Holt, Rinehart and Winston of Canada, Limited
Toronto

ISBN 0-03-925223-X

All Rights Reserved

It is illegal to reproduce any portion of this book except by special arrangement with the publishers. Reproduction of this material without authorization by any duplication process whatsoever is a violation of copyright.

Ralph R. Krueger
Professor, Department of Geography
University of Waterloo

Raymond G. Corder
Head of Geography
Bramalea Secondary School, Brampton

Cover Design: **Wycliffe Smith**
Figures and Sketch Maps: **Gus Fantuz**, **Frank Zsigo**, and **Acorn Technical Art**
Relief Sketch Maps: **Hans Stolle**

Canadian Cataloguing in Publication Data
   Krueger, Ralph R., 1927–
     Canada, a new geography
   Includes bibliographies and index.
   ISBN 0-03-925223-X

   1. Canada–Description and travel–1950–   *
I. Corder, Raymond G., 1939–   II. Title.
FC75.K78 1982   917.1   C82-094279-0
F1016.K78 1982

The Metric Commission has granted use of the National Symbol for Metric Conversion.

Printed in Canada

    3 4 5 86 85 84

CONTENTS

INTRODUCTION   1

CHAPTER 1. THE GLOBAL VIEW   5

WORLD LOCATION   5
   THE IMPORTANCE OF CANADA'S
   LOCATION   9
     Transportation   9
     Climate and the Seasons   12
CANADA'S SIZE   13
Problems and Projects   16

CHAPTER 2. CANADA AND THE WORLD   19

CANADA'S NEIGHBOURS   19
   INTERNATIONAL
   ORGANIZATIONS   22
     The Commonwealth of Nations   22
     NATO, NORAD, OAS, UN   24
     FOREIGN AID   26
CANADA'S WORLD TRADE   30
Problems and Projects   36

CHAPTER 3. CANADA FROM SEA TO SEA   39

THE FIRST CANADIANS   39
EUROPEAN EXPLORERS   42
EARLY COLONIZATION AND
SETTLEMENT   46
THE BRITISH-FRENCH STRUGGLE   49
CANADIAN-AMERICAN RELATIONS   51
LATER SETTLEMENT   52
FROM SEA TO SEA   56
CANADA COMES OF AGE   58
NATIONAL UNITY   59
Problems and Projects   63

CHAPTER 4. THE FORMATION OF THE LAND 65

HOW WAS THE LAND FORMED? 65
   Local Study 66
   USING AIR PHOTOS AND MAPS IN LANDFORM AND TERRAIN DESCRIPTION 67
   EARTH MATERIALS 68
     Igneous Rocks 70
     Sedimentary Rocks 70
     Metamorphic Rocks 71
   Local Study 71
   LANDFORM PROCESSES 72
   HOW OLD IS CANADA? 72
THE HISTORY OF CANADA'S ROCK FOUNDATIONS 75
   THE CANADIAN SHIELD 75
   THE APPALACHIAN MOUNTAIN SYSTEM AND THE INNUITIAN MOUNTAINS 77
   THE WESTERN CORDILLERA 79
   THE INTERIOR PLAINS AND THE GREAT LAKES-ST. LAWRENCE LOWLANDS 80
   THE HUDSON BAY AND ARCTIC LOWLANDS 84
   SUMMARY 84
SCULPTURING THE LANDSCAPE 86
   THE WORK OF RUNNING WATER 86
   THE GREAT ICE AGE 88
     Glacial Erosion 90
     Glacial Drift 91
     Glacial Landforms 91
     Alpine Glaciers 95
A SCENIC TRIP 98
   APPALACHIAN REGION 98
   GREAT LAKES-ST. LAWRENCE LOWLANDS 109
   THE CANADIAN SHIELD 118
   THE INTERIOR PLAINS 120
   THE WESTERN CORDILLERA 124
Problems and Projects 131

CHAPTER 5. WEATHER AND CLIMATE 133

WEATHER 133
   Local Study 133
CLIMATE 140
   TEMPERATURE 140

PRECIPITATION 147
CLIMOGRAPHS 148
CLIMATIC REGIONS 150
APPLIED CLIMATOLOGY 151
THE IMPORTANCE OF CLIMATE TO
CANADIANS 159
Problems and Projects 161

CHAPTER 6. NATURAL VEGETATION, WILDLIFE, AND SOILS   165

NATURAL VEGETATION 165
  Local Study 166
  NATURAL VEGETATION AND
  CLIMATE 166
WILDLIFE 167
SOILS 168
NATURAL VEGETATION REGIONS
(INCLUDING WILDLIFE AND
SOILS) 170
  ARCTIC TUNDRA 172
  SUBARCTIC 177
  Fragile Ecosystems 178
  CONIFEROUS FOREST 179
  MIXED FOREST 183
  DECIDUOUS FOREST 186
  GRASSLAND 187
  PARKLAND 192
  INTERIOR MOUNTAIN 193
  WEST COAST FOREST 196
Problems and Projects 197

CHAPTER 7. POPULATION   199

  Local Study 199
THE POPULATION OF CANADA 200
  POPULATION GROWTH 200
  POPULATION DISTRIBUTION 202
  ETHNIC ORIGINS 204
  SOME MINORITY GROUPS 208
  The Indians 209
  The Inuit 213
  The Old Order Mennonites 218
Problems and Projects 223

CHAPTER 8. SETTLEMENT AND LAND-USE PATTERNS   225

RURAL SETTLEMENT 225
  Local Study 225

SOME RURAL SETTLEMENT
PATTERNS 226
  Rural Quebec 227
  Settlement in Southern Ontario 230
  Prairie Settlement Pattern 233
  Variations in Settlement Patterns
  in Manitoba 236
A LAND OF CITIES 238
  URBAN GEOGRAPHY—A STUDY OF
  CITIES 239
  THE ORIGIN AND GROWTH OF
  CITIES 241
  Local Study 244
  History of the Growth of the Cities in
  the Waterloo Region 244
  LOCATIONS OF CITIES 248
  SITES OF TOWNS AND CITIES 252
  Local Study 252
  USES OF LAND IN CITIES 253
  Commercial Land Uses 254
  Industrial Land Uses 255
  Residential Land Uses 255
  GROWTH PATTERNS 258
  URBAN AND REGIONAL
  PLANNING 259
  Local Study 263
  SOME CANADIAN CITIES 263
Problems and Projects 273

CHAPTER 9. AGRICULTURE   275

  Local Study 275
TRENDS IN THE AGRICULTURAL
INDUSTRY 276
PROBLEMS OF THE AGRICULTURAL
INDUSTRY 283
  NATURAL HAZARDS 283
  MARKETING PROBLEMS 283
  PART-TIME FARMING 284
  FARM POVERTY 284
AGRICULTURE REGIONS 286
  TYPES OF FARMING 286
  THE ATLANTIC REGION 287
  THE GREAT LAKES-ST. LAWRENCE
  LOWLANDS 291
  THE PRAIRIES 294
  BRITISH COLUMBIA 299

DETAILED STUDIES OF SOME TYPES OF
FARMING   301
   WHEAT GROWING   301
   Physical Factors   302
   Overcoming Problems   303
   The Farm Operation   304
   Transportation and Marketing   307
   RANCHING IN THE WEST   308
   Ranching in the Prairies   308
   Ranching in Interior British
   Columbia   310
   DAIRY FARMING   311
   A Dairy Farm in Quebec   312
   THE ORCHARD INDUSTRY   315
   The Annapolis Valley   318
   The Orchard Regions of Quebec   320
   The Niagara Fruit Belt   323
   The Okanagan Valley   328
   VEGETABLE FARMING   332
   The Holland Marsh   334
Problems and Projects   336

CHAPTER 10.   MINING, FORESTRY, AND FISHING   339

THE MINING INDUSTRY   340
   Local Study   342
   WHERE MINERALS ARE FOUND   344
   Metals   344
   Nonmetallic Minerals   346
   Mineral Fuels   349
   WHERE MINERALS ARE MINED   349
   THE OIL AND GAS INDUSTRY   350
   Locating Productive Fields   352
   Developing our Resources   355
   MINING IRON ORE   360
   MINING COAL   365
   URANIUM MINING: BOOM AND
   BUST   366
THE FOREST INDUSTRY   368
   Local Study   368
   CANADA'S FOREST RESOURCES   371
   THE LOCATION OF CANADA'S
   FOREST INDUSTRIES   374
   The Forest Industry in British
   Columbia   375

THE FISHING INDUSTRY   377
   Local Study   381
   THE ATLANTIC FISHERIES   381
   Lobster Fishing   384
   PACIFIC COAST FISHERIES   386
   Salmon Fishing   387
   THE INLAND FISHERIES   389
Problems and Projects   390

CHAPTER 11.   MANUFACTURING   393

   Local Study   394
LOCATION FACTORS   395
LOCATION OF MANUFACTURING IN CANADA   397
   THE ATLANTIC PROVINCES   397
   ONTARIO AND QUEBEC   399
   THE PRAIRIE PROVINCES   403
   BRITISH COLUMBIA   404
SOME SELECTED INDUSTRIES   406
   THE IRON AND STEEL INDUSTRY   406
   Hamilton   409
   Sault Ste Marie   411
   Sydney   411
   THE ALUMINUM INDUSTRY   413
   Aluminum Production in Quebec   415
   Aluminum Production in British Columbia   418
   THE AUTOMOBILE INDUSTRY   420
   Development of the Automobile Industry in Canada   420
   Location of the Automobile Industry   422
   THE FURNITURE INDUSTRY   424
Problems and Projects   426

CHAPTER 12.   CONSERVATION AND RESOURCE MANAGEMENT   427

   Local Study   430
AGRICULTURAL RESOURCE MANAGEMENT   431
   SOIL EROSION   432
   URBAN SPRAWL   434
WATER RESOURCE MANAGEMENT   435
   THE UPPER THAMES VALLEY CONSERVATION AUTHORITY   441
WATER TRANSFER   444

USING WATER FOR HYDROELECTRIC
POWER 448
   THE COLUMBIA RIVER PROJECT 448
   THE PEACE RIVER PROJECT 451
   THE JAMES BAY PROJECT 454
Problems and Projects 458

CHAPTER 13. CANADIAN REGIONAL DEVELOPMENT PROBLEMS   459

REGIONAL DISPARITIES 459
CAUSES OF REGIONAL DISPARITIES 461
ATTEMPTS TO SOLVE THE PROBLEM OF
REGIONAL DISPARITIES 465
REGIONAL DEVELOPMENT
PROGRAMS 465
Problems and Projects 471

CHAPTER 14. REGIONS OF CANADA   473

STUDY OF LOCAL REGION 473
GEOGRAPHIC REGIONS OF
CANADA 475
   THE ATLANTIC REGION 478
   SOUTHERN ONTARIO AND
   QUEBEC 480
   THE PRAIRIE REGION 482
   THE WESTERN MOUNTAINS 484
   THE NEAR NORTH 486
   THE FAR NORTH 488

GEOGRAPHERS AT WORK   490

GLOSSARY   491

ACKNOWLEDGEMENTS   498

INDEX   499

INTRODUCTION

We are proud to be Canadians and to know that we live in such a great and interesting country. In this book we are trying to share our enthusiasm about Canada with students all across the nation.

By words, pictures, maps, and diagrams, we try to show what our land is really like, from the barrens of the Arctic to the farmlands of Southern Ontario, and from the mountains of British Columbia to the rocky shores of Newfoundland. We explain the origins of the natural landscape, and the ecological relationships among the climate, soil, vegetation, and wildlife. After looking at the physical geography of the country, we introduce Man and show how he explored and settled the land, made use of the country's resources, and ultimately created a highly industrialized and urbanized nation.

In this book the emphasis is not on isolated facts, but is rather on geographical concepts, patterns, and relationships. We have not provided a complete geographical encyclopedia of Canada. For more detailed information about specific places and industries you will have to turn to other sources. On each major topic we start with a local study, then look at the national pattern, and conclude with some selected sample studies. Although one cannot study geography without learning *where* activities and places are, the emphasis in this book is on *why* they are there. The *why* of geography is the exciting part of the subject. Surprisingly, once you understand the *why* of a geographical pattern or of the location of a certain place or activity, you remember the *where*.

When you understand *why what* is *where*, a natural follow-up is—*so what?* The *so what?* in geography leads you to some of the major problems facing Canada:

(1) How can Canada remain a unified nation with such wide regional differences, two founding peoples with different languages, and a large number of other groups of different ethnic origins?

(2) Can Canada remain an independent nation with its own values and way of life when it is so close to and has so many interactions with the United States?

(3) What is the impact of urbanization and industrialization on the quality of our environment? How adequate are our conservation, resource management, and environmental planning measures? How long can we continue to exploit resources and pollute our environment at the current rate?

(4) Although Canada is a country of wide open spaces, most of our population is concentrated in a small number of large cities. Can urban growth be directed to less populated areas? How can we solve the urban problems of overcrowding, inadequate housing, lack of open space, traffic congestion, and air pollution?

(5) Although Canada has one of the highest standards of living in the world, there are several million Canadians living in poverty in both rural areas and in urban slums. There are some regions where the majority of people are living in or near poverty. What has been done, is being done, and could be done to fight regional disparities and poverty?

(6) What impact will the development of mineral resources in the Far North have on the native people?

We are not suggesting that this book has the solutions to all of these problems. However, the beginning of solutions lies in the knowledge of the origin, nature, and extent of the problems. We hope that this book provides some of that information and points the way to a greater understanding in the future.

Among the many people whose suggestions helped make this revised edition a better book, the authors wish to acknowledge two people in particular. Mr. Harold Flaming, a graduate student at the University of Waterloo, did much of the updating of the statistical data. Dr. J. Lewis Robinson, Department of Geography, University of British Columbia, thoroughly reviewed the first edition and made numerous corrections and suggested improvements. Because of Dr. Robinson's input, this is a much more useful and accurate volume.

Figure 1.1 NORTHERN VIEW OF THE WORLD. Hold this map 45 cm from your eyes. The world would look something like this from 40 000 km out in space.

1 THE GLOBAL VIEW

WORLD LOCATION

To astronauts orbiting the earth at a height of about 40 000 km, our planet would look something like the drawing in Figure 1.1. Although Canada is usually considered to be in the Western Hemisphere, it is clear from this illustration that Canada is also in the Northern Hemisphere. It is also obvious that Canada extends north into the Arctic pack ice, almost to the North Pole, and that some of our nearest neighbours are north of us, across the Arctic Ocean.

The best way to appreciate Canada's global location is to use a globe. The intersecting lines of *latitude* and *longitude* help you to describe the location of any place on the earth. The *parallels of latitude* indicate how far a place is north or south of the *equator*. The equator is 0°, the North Pole is 90° North Latitude, and the South Pole is 90° South Latitude. The *meridians of longitude* indicate how far a place is east or west of the *prime meridian,* which is an imaginary line running through Greenwich, England. Places to the left of the prime meridian have a west longitude, and places to the right have an east longitude.

After examining parallels of latitude and meridians of longitude on the globe, study the two diagrams in Figure 1.2. When you have answered the questions about diagram B, you will be ready to use the globe to describe Canada's location in terms of its latitude and longitude. Your description should include the answers to the following questions.

QUESTIONS

1. What is the latitude and longitude of the most southerly point in Canada? of the most northerly point in Canada?
2. What is the latitude of the southern boundary of the Prairie Provinces?
3. What meridian separates most of Alaska from Canada?
4. How many degrees of longitude is St. John's, Newfoundland, west of Greenwich, England?
5. What is the longitude of the most westerly boundary of Canada?
6. Which states of the United States lie north of the parallel of latitude that runs through Pelee Island, Ontario?
7. Compare the latitude of Montreal with those of London, England; Paris; and Copenhagen.
8. Compare the latitude and population of Whitehorse, Yellowknife, and Churchill in Canada with those of Oslo, Leningrad, and Archangel. Why are large cities located farther north in Western Europe than in Canada?

Not only does a globe help you to describe the location of places in the world, but it also makes it possible to measure distances between places. However, to measure distances on a globe, or even on a map, you need to know something about *scale*.

Since the circumference of the earth at the equator is approximately 40 000 km, you can work out the scale used on any globe. If you are using a globe with a 63.5 cm circumference, then every centimetre on the globe represents (40 000 ÷ 63.5) 630 km on the earth. If your globe has a circumference of 102 cm, then the scale is 1 cm represents 392 km (40 000 ÷ 102).

Because of its curved surface, it is not possible to measure distances on a globe with a straightedged ruler. If your globe does not have a special measuring device attached, you may use a cloth tape measure. After you have calculated the scale of your globe, answer the following questions.

QUESTIONS

1. How many air kilometres is it from Ottawa to Paris, France? to Moscow? to Peking?
2. What is the shortest distance between Vancouver and Tokyo? between Halifax and London, England?
3. Find the shortest water route from Montreal to Adelaide, Australia. How many kilometres less is the shortest route than other routes you may choose?

Although the globe is the best means of showing Canada's world position, a wall map or atlas has some advantages. On a globe you can

DIAGRAM A is a drawing of the globe showing parallels of latitude and meridians of longitude. In diagram A the point **a** has a latitude of 40 degrees north and a longitude of 100 degrees west. This is usually written 40°N. Lat., 100°W. Long.

DIAGRAM B is an enlargement of the area enclosed in the rectangle in diagram A. The location of point **b** is approximately 56°N. Lat., 75°W. Long. What is the latitude and longitude of points **c, d, e,** and **f** ?

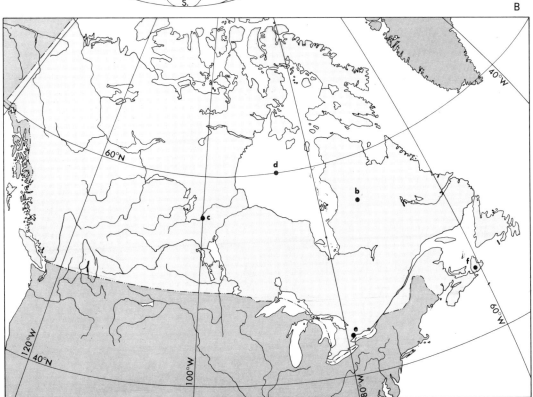

Figure 1.2 CANADA'S LOCATION IN TERMS OF LATITUDE AND LONGITUDE.

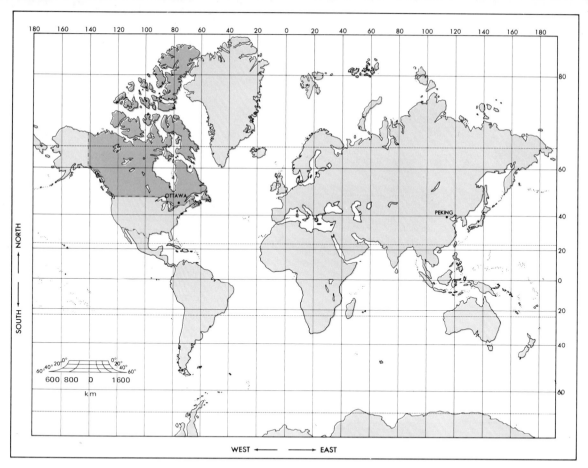
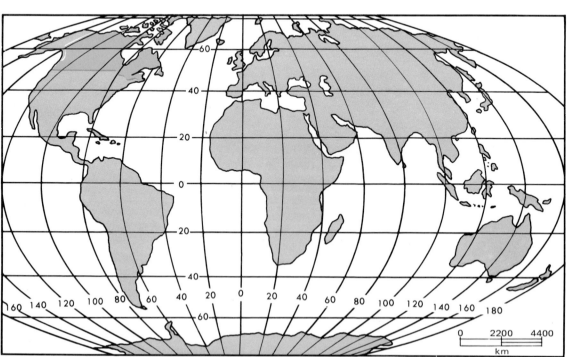

8

Figure 1.3a MERCATOR MAP OF THE WORLD and Figure 1.3b EQUAL AREA MAP OF THE WORLD (Facing page). Note that in the Mercator map the scale changes at different latitudes. Compare the size and shape of Greenland and the United States (i) on the Mercator map, (ii) on the equal area map, and (iii) on a globe. Which map shows the *area* of high latitude countries better? Which map shows the *shape* of high latitude countries better? Neither of these maps shows the shortest distance between Canada and the Soviet Union which lies across the Arctic.

see only one-half of the world at a time; a map can show the whole world at once. The parallels of latitude and meridians of longitude are usually easier to identify on a map than on a globe, and measurements can be made more precisely. On a large-scale map, latitude and longitude can be measured to the nearest minute or even the nearest second (one degree equals 60 minutes; one minute equals 60 seconds). Answer the latitude and longitude questions on the previous page more precisely by using the largest scale maps available instead of a globe.

In measuring distances on a map, you must be sure that you are using the right kind of map. It is difficult to measure distances on some maps because the scale differs from place to place. For example, on the Mercator map (as shown in Figure 1.3a) at the equator, 2.5 cm represents about 7200 km, while at 60°N Latitude 2.5 cm represents about 3520 km.

A Mercator map does not show the north polar region at all. Thus, this kind of map will give you a mistaken idea of the shortest route between Ottawa and Peking, which really lies over the Arctic.

The Importance of Canada's Location

TRANSPORTATION

On a globe, find the shortest distance between Los Angeles and Moscow; between Chicago and Stockholm; and between Boston and London, England. Notice that all of the shortest routes cross part of Canada because Canada lies on most of the great circle routes between the United States and Europe.

A *great circle* is any line continuing completely around the globe which divides it into two equal parts. The shortest distance between any two places on the globe lies along a great circle. Canada happens to lie on some of the most important great circle routes in the world. This fact explains the location of the air routes shown in Figure 1.4, and also explains why a remote place like Gander, Newfoundland, has an important airport and why an air base was maintained by the U.S.A. at Goose Bay for many years. (Canada took over control of the Goose Bay air base in 1973.)

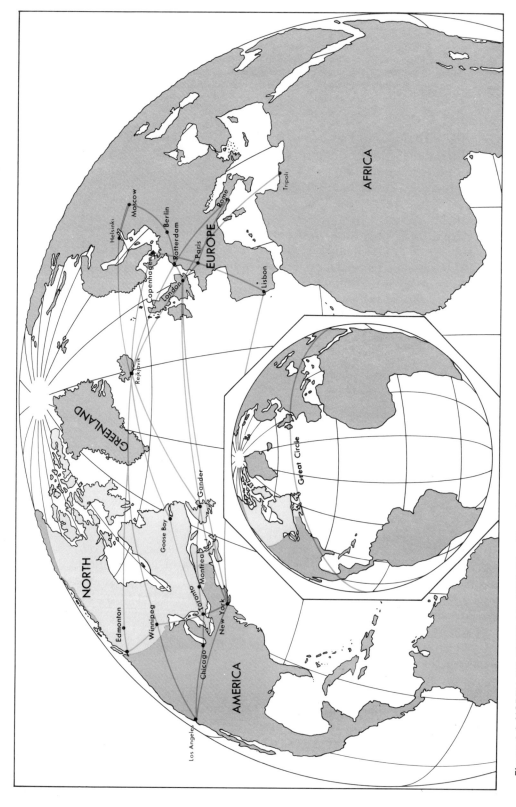

Figure 1.4 NORTH ATLANTIC AIR ROUTES. Most air routes across the North Atlantic Ocean follow the paths of great circle routes. A great circle is the shortest distance between any two air terminals or other locations. A great circle divides the earth into two equal parts. To prove this, join any two places on opposite sides of a globe with a piece of string as shown in the inset. This map shows only a few of the scores of North Atlantic flight paths.

The airport at Gander, which is on a great circle route from many points in North America to Western Europe. In the late 1940's and early 1950's, piston-engined aircraft had to stop to refuel at Gander on transatlantic flights. With the coming of jetliners, Gander declined in importance. However, in the late 1960's, aircraft originating along the west coast of North America started using Gander as a halfway stop to Europe. In 1972, more than 400 000 passengers landed at Gander. In the future, Gander may act as a collecting point for feeder airlines bringing passengers for transfer to huge transatlantic supersonic jetliners.

In the decades following World War I, both the Soviet Union and the United States built up their intercontinental bomber fleets. Because Canada lies on the shortest air route between these two countries, through a joint defence pact Canada and the United States built lines of radar warning stations running in great arcs across the Canadian north. The purpose of these installations was to give warning of the approach of planes or missiles over the Arctic. Modern missiles have already outdated some of these radar installations, and a "thaw" in the "Cold War" between the United States and the Soviet Union has reduced the need for these air defence installations. Nevertheless, their location underlines Canada's strategic global location between the two most powerful nations in the world.

The shortest ocean routes between the great industrial areas in eastern North America and in Western Europe also follow great circle routes across the North Atlantic. Fortunately, Canada's industrial heart is located on the Great Lakes-St. Lawrence waterway, which is almost perfectly in line with a great circle route to Europe. This location of the industrial heart of the country provides a great transportation advantage for Canada. For example, Toronto, which is about 1600 km from the ocean by the St. Lawrence Seaway, is about the same shipping distance from England as is New York, which is right on the Atlantic coast.

CLIMATE AND THE SEASONS

Because of its great range of latitudes, Canada has a diversity of climates. In the extreme south, the summers are hot enough to require air-conditioning for comfort. In the northern Arctic, the length of time without frost is measured in weeks instead of months. Throughout much of Canada, the summers are very warm and the winters are very cold. The climates in the interiors of large land masses in high latitudes always have extremes of temperature and are often called *continental climates*.

Most of Canada experiences four distinct seasons. In a science book or an encyclopedia, you will find explanations for the different seasons found at your latitude. You can demonstrate the causes of the seasons by using your classroom globe. Figure 1.5 shows in which hemisphere and at what latitude the sun is directly overhead at different times of the year, and shows how this affects the angle of the sun's rays over a region.

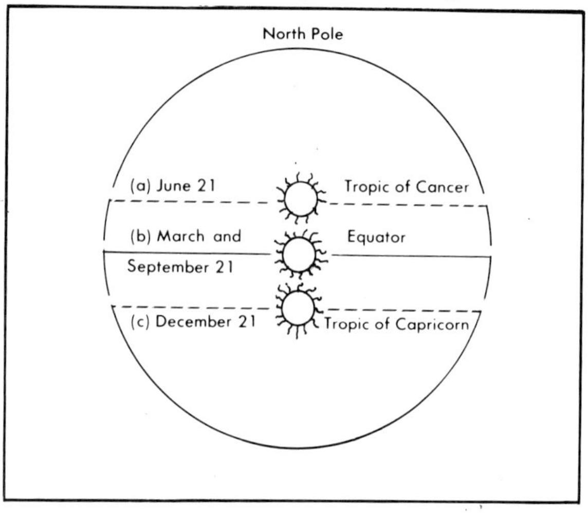

Figure 1.5a THE SUN'S POSITION ABOVE THE EARTH. During the summer season in Canada (a), you can see the sun higher in the sky than at any other season (b) or (c). High sun means warm temperatures and long hours of daylight. A low sun brings colder days with more darkness. About what time does the sun rise and set in your area on June 21st? December 21st?

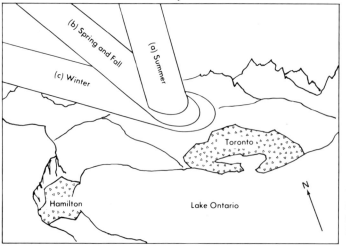

Figure 1.5b THE ANGLE OF THE SUN'S RAYS. In summer in the Northern Hemisphere the sun's rays strike the earth more directly (perpendicularly), travel a shorter distance through the atmosphere, and fall on a smaller area than they do in other seasons. These very intense rays create hot temperatures.

Any area receives more heat from the sun when the sun is more directly overhead. This explains why Canada receives more of the sun's heat in summer and also why the northern parts of Canada receive less heat than the southern parts.

Latitude also affects the length of day. In summer, the northern latitudes have longer days than the southern latitudes. In winter, the reverse is true. This fact can best be explained by using a globe. Can you explain why the area north of the Arctic Circle (66½° N. Latitude) has continuous daylight during a period in the summer and continuous darkness during a period in the winter?

CANADA'S SIZE

Canada is a very large country. This fact is obvious from looking at the globe and identifying this country's extent in latitude and longitude. In area, Canada is second only to the Soviet Union (which is more than twice the size of Canada). What other three countries have areas comparable to that of Canada, that is, between eight and ten million square

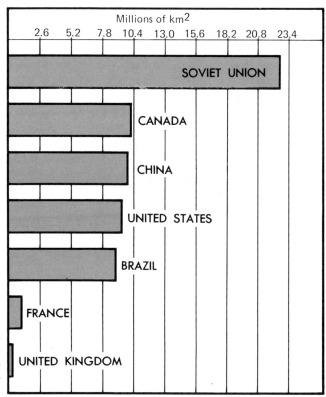

Figure 1.6 CANADA'S SIZE COMPARED TO OTHER COUNTRIES. Canada is second in area only to the Soviet Union.

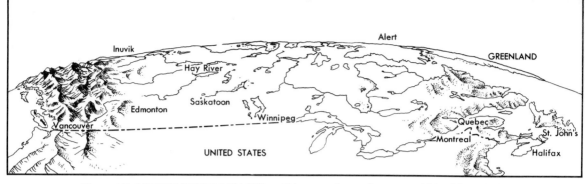

Figure 1.7 ACROSS CANADA BY RAIL. This map shows the great width of Canada. It would take you about five days and nights to travel by train from Newfoundland to British Columbia. Canada's great north-south distances do not show up well on this map.

kilometres? How many times is Canada larger than France? Great Britain? (See Figure 1.6.)

It is very difficult to realize how large Canada really is. Those who have made long trips across the country have some idea of the size of Canada. After travelling all day by car, one seems to have made little progress when one checks the journey on a map of Canada. The distance between St. John's, Newfoundland, and Vancouver, British Columbia, is approximately 7400 km. If it were possible to travel nonstop at 100 km/h for twenty-four hours a day, how many days would it take to travel from St. John's to Vancouver? The fastest time by train between St. John's and Vancouver is about five days and nights (Figure 1.7).

Another trip from Vancouver to the extreme northern tip of Canada would also require a considerable length of time. After several days, you would reach the end of the railroad at Hay River on Great Slave Lake. You still would have twice as far to go to reach Alert on the northern coast of Ellesmere Island. To travel to the extreme northern islands of Canada, you would have to use an airplane. If you preferred to take a shorter journey, you could take a slow barge trip of 1600 km down the Mackenzie River to Inuvik. By this time, you would be convinced that Canada is truly as broad as it is long.

Canada's large size is important because it provides many different kinds of resources as well as a wide variety of scenery, wild plants and animals, and various types of recreational activities. Size, however, can sometimes be a disadvantage. Canada's vastness has caused some difficulties in tying the country together into a unified nation. For example, it is difficult for someone in British Columbia who has never been east of the Rockies to understand the problems of the people living in the Gaspé Peninsula of Quebec.

The long distances, combined with rugged terrain, have added to Canada's cost of living. For example, in the early 1980's it cost about $30.00 to transport one tonne of wheat from Saskatchewan to Montreal, and about $300.00 to ship an automobile from Toronto to Vancouver.

Modern transportation is helping to overcome the great distances in Canada. It is now possible to board a plane at Toronto and arrive in

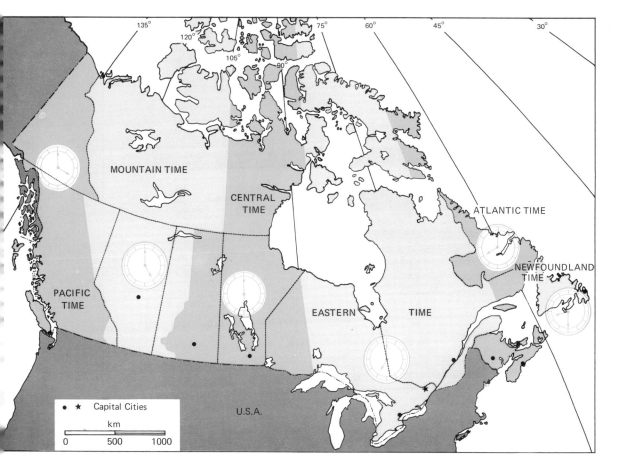

Figure 1.8 TIME ZONES OF CANADA. If it is twelve noon in London, England, the correct A.M. times for different parts of Canada are shown above. Students in eastern Canada are just getting ready for school while in western Canada most people are still asleep.

Vancouver, B.C. or St. John's, Newfoundland, in about four hours. However, if a person plans on arriving at a certain time, he should check the time carefully because, in travelling from Toronto to Vancouver, he would gain three hours. To understand why this is so, look at Figure 1.8. Canada extends over so many degrees of longitude that it has six different time zones.

QUESTIONS

1. Through how many degrees of longitude does the earth rotate in one hour? Canada extends over how many degrees of longitude? On the basis of the answers to these questions, how many time zones should Canada theoretically have? How many time zones does it actually have? How many hours difference is there between Canada's most easterly and most westerly time zones?

2. Why do the boundaries of the time zones in Canada not follow meridians of longitude in all places? Why is a part of Northern Ontario in the same time zone as Manitoba?

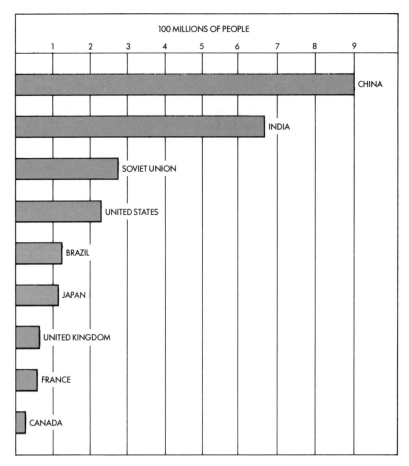

Figure 1.9 CANADA'S POPULATION COMPARED TO OTHER COUNTRIES. Canada is large in area, but has a relatively small population.

There is more than one way of measuring the size of a nation. We have been discussing size in terms of area of land. But, how big is Canada in terms of population? Figure 1.9 shows that Canada has a small population when compared with many other nations. How many times greater is India's population than Canada's? Using an atlas, find ten other countries with a population within one million of that of Canada.

PROBLEMS AND PROJECTS

1. On a globe, locate where you live. What place do you find on the exact opposite side of the globe? Give the latitudes and longitudes of these two places.

2. Compare the network of meridians of longitude and parallels of latitude on a globe with those on a Mercator map of the world (Figure 1.3). What differences do you see in the following:
   (a) the distances between parallels of latitude as you go from the equator to the North Pole?
   (b) the distances between the meridians of longitude as you go from the equator to the North Pole?
   From these observations, explain why the scale of a Mercator map changes from latitude to latitude.

3. The following locations illustrate Canada's vast dimensions. Find the latitude and longitude for each place.
   (a) St. John's, Newfoundland and Mount Logan, Yukon Territory.
   (b) Alert, Ellesmere Island and Pelee Island, Ontario.
   (c) Frobisher Bay, Baffin Island and Victoria, British Columbia.

4. Where do you think the geographic centre of Canada is? The following directions give you one way of checking your estimate.
   (a) Mark each place listed in question 3 on an outline map of Canada. Draw lines joining the pairs of places in parts (a), (b), and (c) of question 3.
   (b) Describe the location of the triangle enclosed by the lines. This area comes close to marking the geographic centre of Canada.

5. Could the British Isles be placed in a body of water the size of the Great Lakes?

6. Calculate the distance from your hometown to the east and west coasts of Canada. Consult railway and airline timetables to find out how long it would take you to travel to each coast. How long would it take you to drive the same distances in an automobile?

7. (a) In which time zone do you live? If it is midnight in London, England, what time is it in your hometown?
   (b) Buckingham Palace announces that the Queen will speak to the Commonwealth countries at 08:00 Greenwich Time, Sunday evening. To hear the broadcast, when would you turn on your radio?
   (c) If the Grey Cup Football Game is televised live from Toronto starting at 13:00 on Sunday, at what time would people in St. John's, Winnipeg, and Victoria turn on their sets to see the beginning of the game?

8. On September 21, the sun is directly over the equator. At this time you can determine your latitude by measuring, at noon, the angle between the sun and your southern horizon and subtracting this angle from 90 degrees. If the noon sun is 30° above your southern horizon, your latitude (90°−30°) is 60° North. At night, you can determine the latitude by measuring the angle between the North Star and the northern horizon. The angle and the latitude are the same. For instance, if the angle of the North Star is 50° above the horizon your latitude is 50° North.

   Estimate the latitude of your hometown by using both the methods just described. Compare your estimate with the latitude given in your atlas.

2 CANADA AND THE WORLD

CANADA'S NEIGHBOURS

A glance at a political map shows that Canada's nearest neighbour is the United States. Moreover, a population distribution map of Canada shows that most Canadians live in a relatively narrow strip along the United States boundary. Because of the geographical location and, in the case of English-speaking Canadians, the similar heritage, culture, and language, the relationships between Canada and the United States have been very close. The number of contacts between Canadians and Americans is suggested by the following questions.

QUESTIONS

1. How many relatives do you have living in the United States?
2. How often do you or your parents visit the United States?
3. What proportion of the movies you have seen in the last year was made in the United States?
4. What proportion of your favourite television shows originates in the United States?
5. What proportion of this year's "hit" songs originated in the United States?
6. How many of the front-page items of a newspaper from the largest city in your province originate in the United States?
7. What Canadian book-publishing companies have their head offices in the United States?
8. What proportion of the books you borrowed from the library last month was printed in the United States?
9. Which of the furnishings in your school were made in the United States?
10. Which furnishings were made by companies with head offices in the United States?
11. How many prominent Americans can you name who were born in Canada?

In most places a sign is the only indication that a boundary separates Canada and the United States. So great is the trust and co-operation between our countries that no guns, troops, or guards have to protect the border. In this photograph, a country road crosses the undefended border between Quebec, Canada and Maine, U.S.A.

After answering a few of these questions you will discover that Canada and the United States are friends as well as neighbours. People, news, ideas, and products flow freely across the international border.

Canada and the United States have signed agreements on all manner of things. They have made agreements on trade, the use of boundary waters, the use of each other's highways and railroads, the conservation of specific resources, and defence. They fought together in two World Wars, and in the Korean War. The 6400 km boundary between Canada and the United States has been undefended since the War of 1812, except for customs and immigration officials at major border crossing points.

Canada and the United States are good friends, but this does not mean that they always agree with one another. Canada has not always agreed with the policies of the United States. You will find a record of these differences of opinion in old newspapers or news magazines. Sometimes, Canada voices these differences publicly, but more often, the Canadian ambassador in Washington talks privately to United States Government representatives.

United States businessmen have invested large amounts of money in Canada, and consequently, they own a great deal of Canadian industry. While this invested money greatly helps Canada to grow and become prosperous, it means that a foreign country controls the policies of many industries. These policies are not always in the best interests of Canada. For example, one automotive company in Canada lost a sale of trucks to Communist China because the parent company in the United States was prevented by law from selling goods to a Communist country. When an international oil shortage occurred, an American-based oil company diverted to the United States an oil tanker originally bound for Canada. When United States parent companies are having economic problems and need to reduce production, they sometimes close branch plants in Canada, laying off Canadian workers.

In some cases the normal growth of industrial plants is stunted by the parent company. For example, some branch plants are not permitted to export their products. Industrial research is usually carried out in the

United States instead of Canada. This research does not always meet Canada's needs and Canadians miss out on research jobs. In the case of resource industries, American ownership may result in exploitation that is not in Canada's long-term interests. The raw materials tend to be refined or processed in the United States, so that jobs are created there instead of in Canada.

Some Canadians feel strongly that United States investment in Canada should be greatly restricted, and that attempts should be made to have American companies bought out by Canadians. The Canadian Government has taken a more moderate view. Believing that American investment is required to keep the Canadian economy healthy, it has not restricted investment, but has established "Guidelines of Good Corporate Citizenship" and a screening procedure to ensure that new American investments are in Canada's best interests. In a few cases, the Canadian Government has refused permission for American investors to buy out Canadian companies. The Canadian Government has also encouraged Canadian ownership of industry by establishing the Canadian Development Corporation (C.D.C.) and by giving tax concessions to Canadian-owned oil companies.

Because Canadians feel free to criticize the United States, Canada is sometimes accused of being anti-American. Nothing could be further from the truth. The record of friendship, co-operation, and agreement far outweighs the disagreements. The future welfare of the two nations is closely linked.

Canada's relations with the Soviet Union, its giant neighbour to the north, have not been so close. Although Canada and the Soviet Union fought on the same side during World War II, in the "Cold War" against the Communist world, Canada was firmly on the side of the West. The Soviet Union has disliked Canada's military co-operation with the United States and the countries of Western Europe.

Nevertheless, there has been some co-operation between Canada and the Soviet Union. Canadian diplomats and business representatives have visited the Soviet Union to see if they could increase Canadian exports; Russian farmers have visited Canada to learn about Canadian farming methods and to take home varieties of Canadian grain. Scientists from both countries have exchanged visits and scientific data.

Because of modern methods of travel and communication, not only the United States and the Soviet Union, but also every other country in the world is a neighbour of Canada. Whether we like it or not, Canadians have to be concerned when something happens in some other part of the world. Whether it is a famine in China, a civil war in Africa, a boundary dispute in Europe, a revolution in Latin America, or a labour strike in the United States, it will have some effect on Canada. By reading the newspapers, by watching television, or by listening to the radio every day, you will learn about world events that either affect Canada as a nation, or affect at least some Canadian citizens.

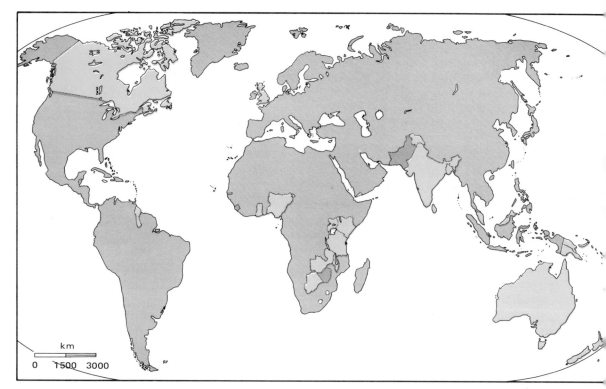

Figure 2.1 CANADA IN THE COMMONWEALTH. Countries belonging to the Commonwealth (in green) cover almost one-fifth of the land surface of the globe. (A number of Commonwealth islands are too small to show on this map.) About one in every four persons in the world lives in a Commonwealth nation.

International Organizations

Canada belongs to a number of important international organizations, and Canada's role in a few of these is discussed on the following pages.

THE COMMONWEALTH OF NATIONS

The Commonwealth of Nations includes countries from many parts of the world (Figure 2.1). The peoples of these countries have different-coloured skins, as well as different customs, religions, and languages. Why then do they belong to the Commonwealth? What do they have in common? Of what good is the Commonwealth to Canada? Why does the United States not belong?

The answers to these questions may be found in the intriguing history of Britain and her colonies. Since this book is concerned with the geography of Canada, we will have to make a long story short.

A few hundred years ago, Britain acquired a large number of colonies all over the world by being the first to either explore, settle, or conquer the land. For many years, Britain governed these colonies directly from

London, England. However, early in the twentieth century, a number of these colonies began to mature politically and wanted to become more independent. Canada played an important part in winning independence for the more mature colonies. After raising an army of over a half million soldiers to fight in World War I, Canada insisted on having a separate vote from Britain at the postwar peace conferences. Over the next few years, Britain realized that some of the colonies wanted more independence, and started transferring certain powers to them.

In 1931, Britain passed the Statute of Westminster which stated that members of the Commonwealth of Nations were free and independent countries bound together by allegiance to the Crown. In other words, Canada is free to do as it wishes, but symbolizes its interest in the welfare of other Commonwealth members by showing its allegiance to the Queen.

The Commonwealth prime ministers meet regularly to discuss common problems. Sometimes, the countries help one another by favouring each other in trade agreements. At other times, they discuss how they may be able to keep the peace or to feed the hungry in various parts of the world.

It may shock you to learn that almost half of the world's people are poorly nourished. At a meeting in 1950 at Colombo, Sri Lanka, the Commonwealth foreign ministers decided to do something about this problem. They established the Colombo Plan to help developing countries of Southeast Asia, and other countries, including the United States, were invited to join the Colombo Plan. Through the Colombo Plan, Canada has given over a half billion dollars to improve the living standards of many millions of people in Southeast Asia. This aid has been extended to both Commonwealth and non-Commonwealth countries. A similar Commonwealth aid program has been set up for Africa. Through this plan, Canada is sending teachers to help African Commonwealth countries with their educational programs. Lack of education is one of the greatest handicaps of many developing nations.

Canada has also participated in the Commonwealth Scholarship and Fellowship Program which helps university students to study in sister Commonwealth countries. This exchange enables students to understand how people in other parts of the world live. Lack of knowledge and understanding about different countries is often at the root of many world problems.

As one of the senior Commonwealth members, Canada is expected to give leadership to the organization. Canada encouraged the Commonwealth to enforce membership rules that would require South Africa to withdraw because of the discrimination against coloured people in that country. (At the time, many Canadians said that people in glass houses should not throw stones. What do you think they meant?) Canada also supported the setting up of a permanent Commonwealth office so that the organization could continue to operate between the meetings of

Commonwealth ministers. The first man to head the Commonwealth office was a Canadian, Arnold Smith.

NATO, NORAD, OAS, UN

The above acronyms stand for the names of other international organizations to which Canada belongs. What do the letters stand for? Who are the members of these organizations? Why were they formed?

After World War II, the Soviet Union gained control of a number of Eastern European countries. Countries in Western Europe were afraid that Soviet-supported Communism might be forced on them also. To protect themselves against this, a number of European nations joined with the United States and Canada to form a defence organization called the North Atlantic Treaty Organization, commonly known as NATO. Figure 2.2 shows you which countries belong to this organization.

Through NATO, Canada has joined Great Britain and the United States in sending troops to Europe. In this military alliance, Canada not only helps to protect Europe, but also gains for itself the military protection of the NATO countries. Some people now believe that NATO is no longer necessary as a military alliance, and that the NATO countries should concentrate on helping each other economically. In 1970, Canada began a phased reduction of its armed forces in Europe. However, political and military situations can change quickly and membership in NATO will continue to be significant.

Canada has also joined the United States in a North American defence pact (NORAD). By the NORAD agreement, the United States and

Figure 2.2 NATO ALLIANCE. In 1949, countries of the North Atlantic community (in green) took defensive action and formed the North Atlantic Treaty Organization.

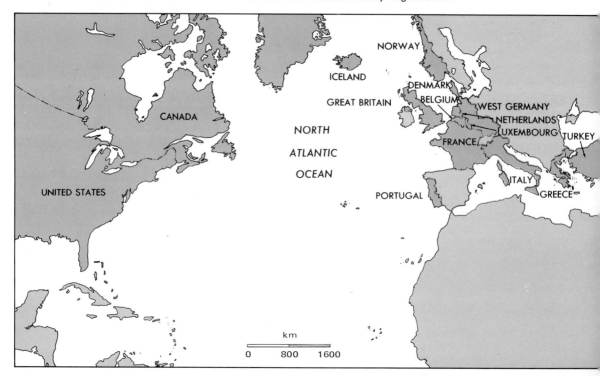

Canada agree to help defend each other if the need should arise. Thus, Canada has permitted the United States to build several defence installations on Canadian soil, including the Distant Early Warning (DEW) line. The NORAD headquarters is located at Colorado Springs, Colorado. The Canadian operational control centre is located at North Bay.

Canada belongs to a number of other world organizations. There is one, however, that Canada has refused to join. The Organization of American States (OAS) includes almost every nation in both North and South America except Canada. There is still a place reserved for Canada

This radar base stands as a northerly sentinel to warn against possible enemy plane and missile attack. It is one of many such bases linked in the 8000 km Distant Early Warning (DEW) System.

If hostile aircraft were to approach Canada from the north, radar bases of the DEW line would signal NORAD headquarters at Colorado Springs. From Clear, Alaska, and Thule, Greenland, high-powered radar antennae can give the alert of an intercontinental ballistic missile attack when it is about 4800 km away.

in the OAS whenever it decides to join. For many years, Canada has given its Commonwealth ties as a reason for not joining the OAS. It has never been stated officially, but students of Canadian foreign affairs believe that Canada does not want to become entangled in the complicated political affairs of Middle and South America. It would be particularly embarrassing for Canada to be forced to oppose the United States policy in many Latin American problems.

The most important international organization that Canada belongs to is the United Nations. Its forerunner, called the League of Nations, was formed after World War I (1914-1918) to help keep world peace. After World War II (1939-1945), the United Nations (UN) was formed to help prevent war and to assist the needy nations of the world (Figure 2.3). From its beginning in 1949, Canada has been a strong supporter of the United Nations. Canada has always paid its United Nations dues, and sometimes, has helped pay for debts that other nations have refused to share.

Since World War II, Canada's policies in foreign affairs have been based upon collective action through the United Nations. In the early 1950's, Canada joined other UN countries in sending troops to Korea. In 1956, Canada suggested the establishment of, and contributed troops to, the UN Emergency Force to keep peace between Egypt and Israel. Canada was again involved in the peace-keeping tasks after the Arab-Israeli wars in 1967 and 1973. Canada was a member of the international truce teams in both Laos and Vietnam, and supported the UN peace forces in both the Congo and Cyprus. A glance at your globe or atlas at this point will show you how far-flung Canada's peace-keeping activities have been.

The United Nations does more than keep countries from fighting one another. It does a great deal to relieve suffering and help the poor in all parts of the world. Canada has had a hand in helping the sick and the hungry of the world through UN agencies such as the World Health Organization (WHO), the Food and Agricultural Organization (FAO), and the United Nations Educational, Social and Cultural Organization (UNESCO).

Foreign Aid

Instead of "trick-or-treating" at Hallowe'en, many children collect money for UNICEF (United Nations International Children's Emergency Fund). The money they collect helps to buy milk, food, medicine, and books for children in more than one hundred countries. For instance, one dollar collected for UNICEF buys enough vaccine to protect many children against the dreaded disease called tuberculosis.

Many Canadian families contribute to organizations such as the Canadian Save the Children Fund or the Foster Parents' Plan. These are only two of the more than seventy-five agencies in Canada that help the underprivileged in many parts of the world. Many churches and welfare

Figure 2.3 UNITED NATIONS MEMBERSHIP (in green). Canada has always been a strong supporter of the United Nations. With the admission of the People's Republic of China in 1971, the United Nations represents a vast majority of the world's population.

clubs have their own foreign relief committees that organize used-clothing drives and collect money to help needy people overseas.

The Canadian Government also has a very active foreign aid program. Through the Canadian International Development Agency (CIDA), in recent years Canada has been contributing hundreds of millions of dollars annually to help other countries. This money has been used to provide scientists, technical equipment, agricultural implements, food, and clothing for the needy nations. Canada has also helped developing countries to build factories, huge dams, hydroelectric plants, and atomic reactors. Canada is a founding member of the World Food Program and is the second largest contributor to this program. Also, Canada has been making substantial contributions to the UN Population Control Fund, because overpopulation is a problem in many developing countries.

Many of these aid programs are designed to help the people to help themselves. It is much better to help the hungry to use better agricultural methods than it is to give them food. We can do this by providing the people of developing countries with better seeds, fertilizer, and dams that store water to irrigate crops. Then, by teaching people how to use these things properly, we help them to grow enough food to feed themselves. Technical assistance and long-term loans help the countries to develop their resources and build up their manufacturing industries so

Lumber, grain, and farm machinery are among the many goods that Canada often donates to needy nations, especially developing countries in South America, Africa, and Southeast Asia. For example, Canadians try to help relieve hardships such as earthquakes, floods, and food shortages.

Facing page: Canada provided money for the construction of this hydroelectric power project in India.

that they can produce more goods for themselves and also can increase exports to pay for food imports.

Although Canada has made a commendable effort in foreign aid, some Canadians feel that Canada is not doing enough. The late Lester B. Pearson, a former Prime Minister of Canada, said in a United Nations report that each of the developed countries should strive to allocate one percent of its gross national product to foreign aid. In order to achieve this goal, Canada would have to more than double its foreign aid budget from what it was in the early 1980's. That does not seem too much when one realizes that in 1981 Canada's foreign aid expenditures came to about $30 for every Canadian. You may wish to compare this amount with what is spent per person in your family for entertainment each year.

In helping others, Canada is often helping itself. Sometimes Canada gives to other countries products that are overproduced. In other cases, a certain product may be shipped as foreign aid primarily to help reduce Canadian unemployment in a particular industry. Moreover, in the long run, Canada knows that if the poor nations become richer, they will want

to buy more Canadian products. Because exports are so important to Canada's economy, foreign aid is really good business.

For many years developing countries have been claiming that they need "trade, not aid". They argue that, whenever one of their industries becomes competitive on world markets, the developed countries freeze out their products by raising tariffs. For example, Canada maintains a high tariff on shoes and clothing because countries in Southeast Asia can produce them at a lower cost than Canadian manufacturers can.

Discussions about economic relations between developed and developing countries have been termed the "North-South Dialogue", in which Canada has played a leading role. Canada's Prime Minister has travelled world-wide discussing North-South issues with other national leaders. In 1981, Canada hosted an international conference at which it was agreed that much more emphasis must be given to changing the international economic order so that developing countries can gain a greater share. Cynics have said that Canada has given more leadership in talk than in action. Do you agree? Give some specific examples to support your answer.

CANADA'S WORLD TRADE

QUESTIONS

1. Make a class list of canned goods used in your homes. Opposite each item, name the country where the product was grown and where it was packed.

2. Make a class list of your personal possessions such as watches, jewellery, bicycles, radios, clothing, and so forth. Again, name the country where each item was produced.

3. Find out which products grown or manufactured in your community are shipped to a foreign market. Where do these products go?

The above surveys will give you some idea of the importance of foreign trade to people in your community. Some communities are much more dependent on trade than others. For instance, Elliott Lake almost became a ghost town when Canada's export sales of uranium suddenly dropped in the early 1960's. If the fruit growers of the Annapolis Valley in Nova Scotia, or the Okanagan Valley in British Columbia, were to lose their exports of apples, large numbers of them could not make a living. The Prairie wheat farmers receive a large share of their income from the export of wheat.

If Canada did not import many products, our way of living would greatly change. We would not be able to buy some manufactured goods at all, and many others would be more expensive. The variety of our food would be greatly restricted. There would be no lemonade, no banana splits, no orange juice or grapefruit for breakfast, no cinnamon for cinnamon rolls, and no coffee for coffee breaks—just to mention a few of the foods Canada imports.

For the country as a whole, foreign trade is extremely important. In the early 1980's, the total annual value of foreign trade was about 150 billion dollars, which amounted to more than $6000 for each Canadian. To see how important Canada is as a world trading nation, study Figure 2.4. Compare this graph with Figure 1.9. Why does Canada rate as a more important trading nation than many other countries with greater populations? Note that the United States comes ahead of Canada in total trade. However, it is fair to say that trade is more important to Canadians, because the value of trade per person is almost three times as much for Canada as it is for the United States. There are several reasons for this. Canada has a relatively small population and cannot make use of all of its vast mineral and forest resources and agricultural products for which there is a high world demand. Thus large amounts of raw materials are exported. As a highly industrialized nation, Canada exports many manufactured goods to different parts of the world, but because of the high level of income of most of its people, Canada also imports a wide variety of manufactured goods. Some imports can be explained by the fact that

Figure 2.4 LEADING COUNTRIES IN WORLD TRADE. This graph shows the total value of all commodities that each country imports and exports to other countries. In what order would these countries be listed according to their per capita trade value (i.e. total trade divided by population)?

United States	$$$$$$$
West Germany	$$$$$$
France	$$$$$
Japan	$$$$
United Kingdom	$$$$
Italy	$$$
Netherlands	$$$
Canada	$$$
U.S.S.R.	$$

Each symbol represents $50-billion worth of trade.

Figure 2.5 CANADA'S LEADING IMPORTS. Most of these imports are manufactured goods. However, crude oil, also a major import item, is a raw material and has been excluded from the graph because of the rapidly rising world prices for this commodity. Why does Canada import so many automobile engines and parts, as well as machinery?

MOTOR VEHICLE ENGINES & PARTS	🚛🚛🚛🚛🚛🚛🚛
AUTOMOTIVE VEHICLES (mainly cars)	🚗🚗🚗🚗🚗🚗
MACHINERY (industrial, business, household)	⚙️⚙️⚙️⚙️⚙️
METAL ALLOYS & PARTS (mainly iron & steel)	▬▬▬▬
FOOD, FEED, BEVERAGES & TOBACCO	🍎🍎🍎
MINERAL ORES, CONCENTRATES, & SCRAP	⛰️⛰️⛰️
CHEMICAL PRODUCTS	🧪🧪🧪
TEXTILES & CLOTHING	👕👕
FARM TRACTORS & IMPLEMENTS	🚜🚜
WOOD PRODUCTS & SPECIAL PAPERS	📄

Each symbol represents $1 billion.

Figure 2.6 CANADA'S LEADING EXPORTS. Most of these exports are raw materials, including crude oil and natural gas, which have been excluded from this graph. Sales of crude oil to the United States are being phased out, while rapidly rising prices for natural gas make it an unfair comparison with other more stable export items. To which countries do most of each of the exports in the graph go?

Note: The values shown in these graphs are averages for the years 1978 to 1980 because the exact amounts of trade for some items (and countries) differ a great deal from one year to another. For the most recent figures, consult the latest edition of the *Canada Year Book* or *Canada Handbook*. (Source: Statistics Canada.)

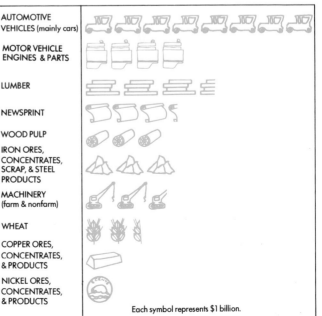

Many manufactured goods that Canada imports from foreign countries arrive at our major seaports in large metal containers. These rectangular boxes are then loaded on to flatbed rail cars or transport trucks before continuing their journey inland. Container trucks and trains can be seen passing through most communities. Besides processed goods, Canadians depend on food and raw material imports from abroad. Items stamped "Made in Japan", "Made in U.K.", or "Made in U.S.A." are found in most Canadian homes.

Canada has too small a domestic market to make the production of certain products economic. Canada has to import special products, such as fruits and vegetables in winter and tropical foods, because of its relatively cold continental climate.

Figures 2.5 and 2.6 show Canada's leading imports and exports. Try to explain why Canada has both large imports and large exports of some of the same commodities, such as automotive vehicles, engines, and parts.

Canada and the United States are each other's best customers. The closeness of the two countries makes transportation of goods quick, cheap, and easy. A common language and a similar standard of living encourage trade. The United States needs for its industrial society many

Winches load large bundles of lumber aboard a cargo ship in Vancouver for shipment overseas. For many years, lumber has been an important Canadian export.

Figure 2.7 CANADA'S TRADING PARTNERS. This table shows that Canada conducts about twice as much import and export trade with the United States as with all other countries combined. Canada appears to have a trade deficit with the United States, but actually has a more even trade balance if the percentage figures are converted to dollar values. With which countries does Canada have a surplus? an even trade balance? For more recent data, see the latest issue of the *Canada Year Book* or *Canada Handbook*.

	Imports	Exports
(Percent of trade in 1978-1980 averages)		
United States	69	66
Japan	5	6
United Kingdom	3	4
EEC (other than UK)	7	8
Central & South America	5	5
Asia (other than Japan)	3	3
Other Countries	8	8
Total	100% (or about $62 billion)	100% (or about $65 billion)

Source: Statistics Canada, 1978-1980 averages.

of the vast amounts of forest and mineral resources available in Canada, and Canadians want the wide variety of manufactured goods produced in the United States. United States advertising reaches the Canadian people freely and increases the demand for American products. Because of the large volume of production there, many United States products cost less than similar items produced in Canada. Also, some products from the United States, such as citrus fruits and fresh winter vegetables, are not available in Canada unless imported.

At border crossing points, inspections by the customs officials help keep a check on the flow of goods and people between Canada and the United States. Each country sets a limit on the value of goods that its residents can bring back from the other country duty-free.

For many years Canada had a large trade deficit with the United States, which it tried to balance by building up trade surpluses with other countries. Canada has used customs duties or tariffs to encourage its own manufacturing industries to produce many of the goods imported from the United States. Some companies have greatly increased their sales to the United States by developing unique, high-quality Canadian products. Others have increased sales in the United States by increasing their advertising in American magazines and on American television, and by establishing sales offices in the United States. Increasingly, Canadian companies are processing raw materials in Canada before exporting them, thus adding to the value of their exports. Despite all of these efforts, Canada continued to import more from the United States than it exported to that country, until the auto trade pact had been in force for several years. This agreement, which opened the United States market duty-free to the Canadian automotive industry, resulted in sales to the United States large enough nearly to balance the trade between the two countries.

Even with a more balanced trade between Canada and the United States, some Canadians think that too much of its trade is with the United States, thus making Canada's economy too dependent on the economy and trade policies of one country. For this reason, Canada has been encouraging trade with many other countries (see Figure 2.7).

In recent years, Canada's trade with Japan has been increasing rapidly. Canada has been importing large quantities of Japanese automobiles, cameras, sports equipment, and many other manufactured products. On the other hand, Japan has been importing vast quantities of raw materials, such as lumber and coal, from British Columbia. Canada's trade has also been constantly increasing with the Soviet Union and the Chinese People's Republic. To date, this has been primarily one-way trade in the form of grain shipments to those countries.

Traditionally, a large proportion of Canada's trade was with the United Kingdom and the Commonwealth of Nations because of preferential tariff agreements. In recent decades the preferential tariffs have been reduced or cancelled so that Canada no longer has a preference in exporting to Commonwealth countries.

In 1971, the United Kingdom joined the trading bloc known as the European Economic Community (EEC), sometimes called the European Common Market. By joining the EEC, the United Kingdom not only abolished all tariff barriers between it and the other member countries, but also had to agree to accept a common tariff on imports from other countries, including Canada. Thus, Canada is finding that tariffs imposed on exports to the United Kingdom are resulting in prices for many Canadian goods that are too high to be competitive. Other EEC countries have a distinct advantage in selling to the United Kingdom because they have free access to the British market while Canada faces a tariff barrier.

Figure 2.8 EUROPEAN TRADING BLOCS. What is the significance of these trading blocs to Canada?

Canada's greatest loss of exports to the United Kingdom has been in agricultural products. (See Figure 2.8.)

Other countries in Europe belong to the European Free Trade Association (EFTA). Membership in this group means that the countries can trade with each other without tariff barriers, but each can set its own tariff rates on imports. Canada has had limited trade with EFTA members.

In view of these developments, Canada has been negotiating favourable terms with these two trading blocs. The fact that the European Economic Community has become an important trading partner to Canada has helped to keep the amount of our trade with the United States from increasing. Therefore, at international trade conferences, Canada will continue to press for reduction of trade barriers while, at the same time, vigorously pursuing new export markets.

Although Canada is in favour of freer trade, if it is achieved, some Canadian industries will suffer if their products are inferior to, or more expensive than, those produced in other countries. To solve this problem, the Canadian industries will have to become more efficient, or specialize in producing unique Canadian products of high quality. For some industries which have natural disadvantages such as low quality resources, too harsh climate or an inferior location, the only solution may be to close down. In such cases it may be necessary for the Canadian Government to provide incentives for a more competitive industry to locate in the community affected.

PROBLEMS AND PROJECTS

1. What do you know about your neighbour?

It has often been said that people in the United States know little about Canada. How much more do Canadians know about the United States?

To compare your knowledge of the United States with your knowledge about Canada, answer the following questionnaire without referring to books.

Ask your teacher to duplicate this questionnaire and send it to a classroom of your grade in some American state bordering Canada. Compare their answers with yours.

The United States
   (a) What is the population to the nearest million?
   (b) Name the present president.
   (c) Who was the first president?
   (d) Name the vice-president.
   (e) Name the governor of any state.
   (f) Name the largest (in area) state in the union.
   (g) How many states are in the union?
   (h) Name the youngest state in the union.
   (i) Name one of the New England states.
   (j) Name the largest city.
   (k) Name an important steel centre.
   (l) Name an important iron-ore range.

Canada
   (a) What is the population to the nearest million?
   (b) Name the present prime minister.
   (c) Who was the first prime minister?
   (d) Name the Governor General.
   (e) Name the premier of any province.
   (f) How many provinces are there in Canada?
   (g) Name the province that joined Confederation most recently.
   (h) Name the province largest in area.
   (i) Name one of the Prairie Provinces.
   (j) Name the largest city.
   (k) Name an important steel centre.
   (l) Name an important nickel-producing area.

2. Divide all the Commonwealth countries into groups according to the continent in which they are located.
   (a) Which continent possesses the most people living in Commonwealth countries?
   (b) In which Commonwealth countries is English an official language?
3. From a daily newspaper, clip articles about Canada's relations with other countries. With what countries does Canada have most contacts?
4. (a) Find the names of organizations in your community that collect money to help needy people in various parts of the world.
   (b) In what way and in what countries do these organizations help the poor?
5. Canadian foreign policy is not static. It changes with changing international conditions and new philosophies developed by our government. You can update your information on Canadian foreign affairs by sending for material from External Affairs Canada, and by reading major newspapers and listening to news broadcasts.
6. Explain what each of the abbreviated organizations means and how Canada is connected with each of them: EEC, OPEC, NATO, OECD, and UN.
7. The Canadian Foundation for Economic Development (Toronto) publishes numerous pamphlets on Canadian and international economic topics such as employment, productivity, and trade. Write to 252 Bloor Street West, #5560, Toronto, M5S 1V5 for more information.
8. Two magazines to look at for information about Canada's current involvement in international affairs are: *Canada and the World* (Toronto: Maclean Hunter) and *Cooperation Canada* (Ottawa: Canadian International Development Agency).
9. The following three references provide working examples of basic map, photo, and statistical skills that show Canada's position in the world:
   W. Derry and M. Horner, *Geolabs* (Toronto: McGraw-Hill Ryerson, 1981), I and II.
   A.R. Grime, *Map Concepts and Skills* (Toronto: Clarke Irwin, 1979).
   W.R. Kemball, *Canada and the World* (Toronto: Oxford University Press, 1981), Books 1 and 2.

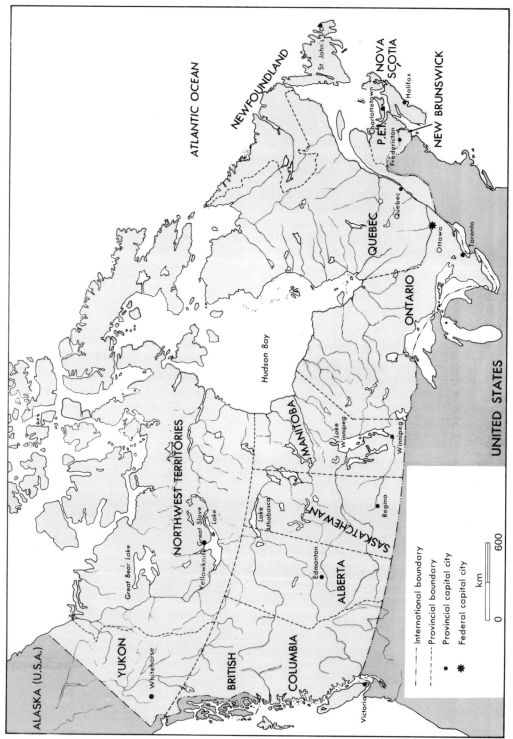

Figure 3.1 CANADA TODAY. Ten provinces stretch across Canada from ocean to ocean. The first provinces did not officially become part of the Canadian nation until 1867. That is why Canada celebrated its Centennial in 1967.

3 CANADA FROM SEA TO SEA

In Chapter 1, we looked at Canada's global position and in Chapter 2 we discussed Canada's relationships with the rest of the world. Now we shall briefly explain how Canada became what it is today. In other words we shall attempt to explain how the settlement patterns came about, how the present political map evolved, and how Canada became an independent nation.

In order to explain the present geography of a country, it is necessary to understand its history. It is equally important to study geography if one wishes to understand the history of a country.

THE FIRST "CANADIANS"

North America was first explored and settled by people who came from the northern and eastern parts of Asia. The details of exactly where they came from, where and how they travelled, and how long it took them to migrate, are still unknown. Piece by piece, archaeologists are fitting the puzzle together. Each discovery of an ancient skull or weapon provides another clue. Sometimes, an original discovery is made by a worker excavating for a new building, by a farmer ploughing a field, or by an observant student while on a hike. If they are notified, archaeologists investigate the area thoroughly to see how many more relics can be

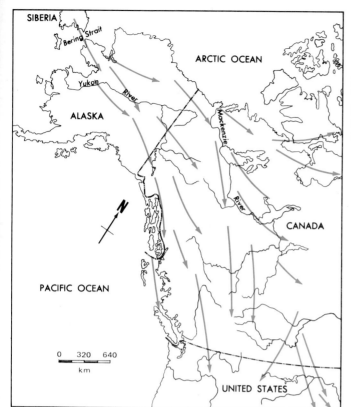

Figure 3.2 EARLY INDIAN AND INUIT MIGRATIONS. From Siberia, Inuit moved eastward and settled the Arctic coast and islands. Indians pushed south and east into the heart of North America.

found. Many times, nothing new is added to the knowledge we already have, but once in a while, there is a discovery that fills in another piece of the puzzle of the past.

At some time between 11 000 and 40 000 years ago people crossed the Bering Strait from Siberia into Alaska. Over a period of many years, some of these people followed the Yukon River and found their way south by travelling through the valleys between the mountain ranges of western North America. Others cut across from the Yukon to the Mackenzie River, which opened a way to the vast plains of the interior, and then, they travelled on to the Great Lakes and St. Lawrence River. Still others followed the northern coast of Alaska to inhabit the Arctic regions as far east as Greenland and the Labrador coast (Figure 3.2).

Different groups of these early North Americans came from different parts of Asia and descended from different races. For these reasons, their appearance and their ways of life differed considerably. The group that settled in the north were the *Inuit*. They were short and broad, with round heads, Mongolian slanted eyes, yellowish skins, and straight, black hair. The rest of the immigrants from Asia are called *Indians*, because when Columbus reached the islands now called the West Indies, he mistakenly thought he had reached the East Indies of Asia. The Indians' proper scientific name is *Amerind*, a contraction of the words *American Indian*.

When Europeans first came there were only about 200 000 Indians living in what is now Canada. These Indians were divided into two

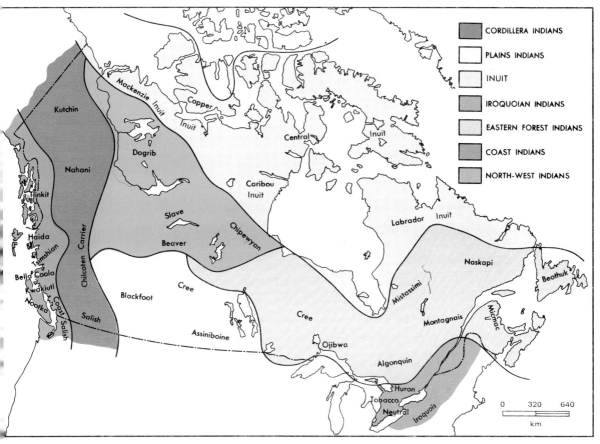

Figure 3.3 INDIAN AND INUIT TRIBES OF CANADA. This map shows the approximate location of Indian and Inuit (formerly, Eskimo) tribes when Europeans first began to explore Canada. Indians and Inuit have differing ways of living from region to region. The physical geography of the regions had a great influence on their civilization and way of life.

main groups. Those found around the Great Lakes and along the St. Lawrence River were tall and slim, with long, narrow heads, and a deep copper-coloured skin. The Indians of the north and west were shorter and stouter, with broad, short heads, slanted eyes, higher cheek bones, and lighter-coloured skin. Within these groups of peoples, the way of life varied from region to region. The physical geography (which includes the climate, the landforms, the vegetation and wildlife) of the regions had a great influence on their way of making a living and on the nature of their dwellings and clothing.

The Inuit and the two major groups of Indians were divided into many tribes. There were no distinct boundaries separating the tribes and they often fought with one another over control of the hunting country that lay between them. Figure 3.3 gives the names of some of the Indian and Inuit tribes and shows the general area where these tribes lived.

The Indians and Inuit did not effectively occupy all of what is now Canada. There was only about one Indian or Inuit for every 50 km² of land. In some areas, such as along the west coast, the land was quite densely populated with Indians. There were, however, thousands of square kilometres where no Indians could be found.

QUESTIONS

1. Which of the Indian tribes have you heard most about in your studies of Canada or in Western stories you have read?
2. Explain why there are so many different tribes along the Pacific coast.
3. From a history book and other library books, find out how the various tribes differed in their methods of hunting, travelling, working, building, worshipping, and playing games.
4. What effect did the physical geography of the different regions of Canada have on the way of life of the Indians and Inuit?
5. Do you think there are more or fewer Indians today than there were in the days of the early explorers? (See Chapter 7.)

EUROPEAN EXPLORERS

It has been said that the Indians and Inuit are the only true native Canadians. The rest of us are new Canadians who have been recently transplanted from other parts of the world, particularly from Europe. There is some truth to this claim, because the Indians and Inuit roamed the land that is now Canada for over 10000 years before any Europeans had even seen the country.

As far as we know, the first Europeans to see the North American Continent were the Norsemen or Vikings, who originally came from Scandinavia, and had settled in Iceland and Greenland between 800 and 1000 A.D. On a trip from Iceland to Greenland, one of these Norsemen, called Bjarni, was driven off course by the wind and sighted a forested shoreline that we believe was the Labrador coast. A few years later, in the year 1000 A.D., Leif the Lucky landed at a number of places along the east coast of North America (Figure 3.4).

The stories of Leif's voyages, which were written into the sagas of the Norsemen, contain the first written geography of parts of Canada. Leif called the land where he first stopped Helluland, which meant the "Land of Stones", as it had rocky, treeless hills and was covered with snow. This may have been the northern part of Labrador. Leif landed later at a place where there were broad beaches of white sand and thick forests along the shore. This place, which he named Markland or the "country of trees", may have been in present-day Nova Scotia. Still farther south, Leif landed

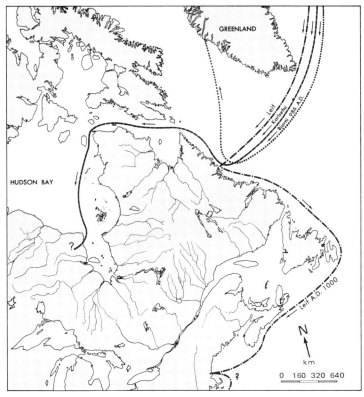

Figure 3.4 NORSEMEN EXPLORE CANADA. The first European explorers were Norsemen who had sailed from Scandinavia by way of Iceland and Greenland. Canada was really discovered from the north.

at a place called Vinland. Some historians believe this was as far south as the New England coast.

It is believed that the Norsemen also reached the shores of Hudson and James Bays. However, who they were, where they landed, and how long they stayed, are still unsolved mysteries. We will have to find more clues before we can fit that story together.

The stories of the Norse explorations make exciting reading. However, the Norsemen were not very important in the story of the development of Canada because they left no permanent settlements. In fact, when explorers from other European countries stumbled upon North America by mistake, they had never heard of the Norsemen's trips.

For many centuries, Europeans had used overland caravans in their trade with Asia. By the middle of the fifteenth century, the Turks had cut off these overland trade routes and European explorers began looking for new routes to the spices and wealth of the East. Christopher Columbus, believing that he could reach the East by travelling west, reached the West Indies in 1492.

In the years that followed, explorers kept running into North America as they attempted to reach the east by crossing the Atlantic Ocean. In 1497, Cabot reached Newfoundland and Cape Breton Island, claiming them in the name of the King of England. Cartier sailed up the St. Lawrence River as far as Montreal, where he heard tales of great wealth lying in the country beyond. Champlain explored Canada as far as Georgian Bay on Lake Huron. A host of other explorers penetrated deep

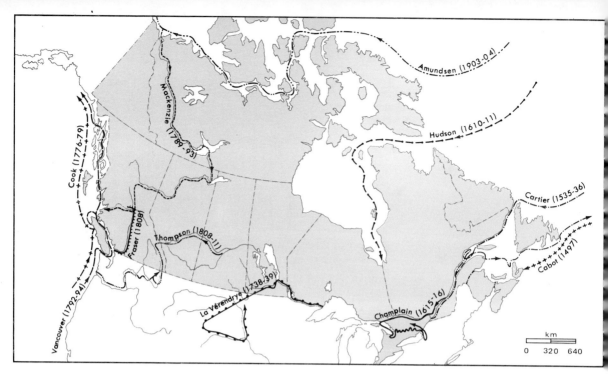

Figure 3.5 EUROPEANS EXPLORE CANADA.

into the interior of the continent. La Vérendrye pushed into the heart of what is now the Prairie Provinces. Mackenzie followed the river that now bears his name to the Arctic Ocean, and then worked his way through the rugged mountains to the Pacific coast. Shortly afterward, Fraser followed the river named after him to its mouth, and Thompson travelled the Columbia River until he reached the sea.

While some explorers were pressing through the interior toward the northern and western seas, others were looking for a northwest passage through the Arctic. Hudson and James came to a dead end in Hudson and James Bays. Others tried to find a sea route through the Arctic, but without success. It was not until early in the twentieth century that Amundsen completed a tortuous voyage through the ice-choked sea passages dividing the islands of the Canadian Arctic.

In the late 1700's, Vancouver and Cook explored Canada from still another direction. They approached North America by way of the Pacific Ocean and explored the western coast of Canada from Vancouver Island to Alaska (Figure 3.5).

The adventures of these explorers are as exciting to read as any fiction. The hardships and tortures they endured are hard to believe. Were their efforts worthwhile? What did they really accomplish? Of what importance are they to a study of the geography of Canada?

The explorers were the first professional geographers. Their rough maps and field notes provide us with the first geography of our country. Samuel de Champlain held the title of "Geographer to the King of France." He was so highly thought of as a geographer that the

This is a photograph of the original astrolabe which Champlain used in exploring Canada. Around the circumference of the astrolabe, a scale was marked off in 360 degrees and a rotating arm was attached to the middle of the disk. When it was suspended from a tree, the explorer sighted the North Star to determine his latitude, then sighted two or more known stars along the movable arm to determine his longitude. Sightings had to be made either at dawn or dusk when the horizon was visible. Champlain drew incredibly accurate maps based upon the sightings from his astrolabe.

Geographical Branch of the Canadian Government used a picture of Champlain's astrolabe as a crest on its publications.

The map shown in Figure 3.6 has been redrawn from Champlain's original map of New France made in 1632. Compared with a modern map of eastern Canada, it does not look very accurate. However, considering the date of the map and the crude instruments available to Champlain, it is an exceptional piece of work.

Figure 3.6 THE FIRST MAP OF NEW FRANCE. Champlain drew this remarkable map with crude instruments in 1632.

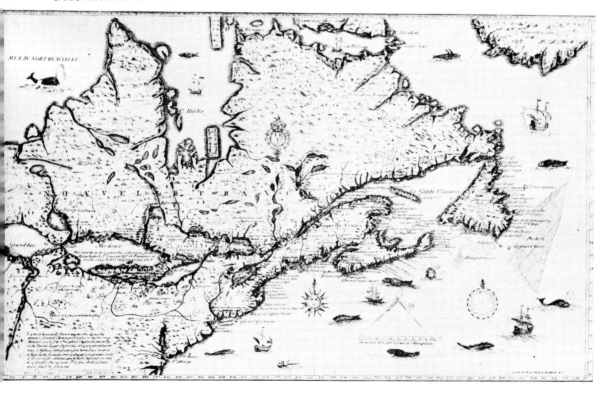

A cairn to David Thompson in southern British Columbia reminds Canadians of the first European to journey down the mighty Columbia River. Memorials such as this have been erected by the Board for Historic Sites and Monuments all across Canada to honour the important explorers and leaders of our country.

David Thompson, who has been called "Thompson, the Map Maker," was another explorer renowned for his ability as a geographer. His maps were even more accurate than those of Champlain.

The explorers did more than make maps. They claimed the new country as colonies for the motherland and opened the way for fur traders and permanent settlers.

EARLY COLONIZATION AND SETTLEMENT

Although Cabot found little of value in Newfoundland, he reported that the fish off the coast were so thick that they almost stopped his ship. News of the good fishing soon spread. By the middle 1500's, fishing boats from many European nations visited the fishing grounds off the coast of Newfoundland. In the late 1500's, there were 300 English fishing vessels sailing regularly to Newfoundland. The English began salting and drying their fish on shore before shipping it home. At first, they built no shelters but landed merely to process their fish and to obtain supplies of wood. Later, some fishermen built cabins, and eventually, some began staying in Newfoundland all year, leading to the establishment of fishing villages in the more sheltered bays.

It was not until 1583 that Sir Humphrey Gilbert claimed Newfoundland and the adjoining mainland coast for the Queen of England. Although Gilbert's attempt at settling Newfoundland was unsuccessful, he established Newfoundland as the first British colony in North America. Control of Newfoundland gave Britain its first toehold in its march toward the control of most of North America. A glance at the map of Canada shows the strategic location of Newfoundland at the entrance of the major waterway leading to the interior of the continent.

It was the French who established the first permanent settlement in North America. In 1605, Champlain and de Monts chose Annapolis Royal

At Port Royal, now called Annapolis Royal, Champlain built his *Habitation*. This is a replica or model of what the first permanent settlement in North America looked like. Champlain designed the small colony for maximum protection against the harsh winter, wild animals, and hostile Indians.

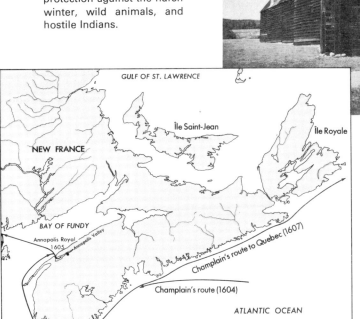

Figure 3.7 CANADA'S FIRST SETTLEMENT. Annapolis Royal, Canada's first permanent settlement, was established by Champlain and de Monts in 1605. Although this was a good site for a settlement, it was cut off from the fur trade of the interior. For this reason Champlain and de Monts moved to establish a trading fort at Quebec, on the St. Lawrence River.

on Annapolis Basin — an excellent location for a colony (Figure 3.7). The winters were not as harsh as in other parts of the country and there was an abundance of fish, game, and fruit. The soil was fertile and the growing season was long.

The new settlement, however, had a major geographical handicap. The Bay of Fundy and the Appalachian Mountains cut the colony of Annapolis Royal off from the great expanse of the fur-trading country in the interior. For this reason, Champlain and de Monts obtained a charter from France to trade up the St. Lawrence River, and in 1608, they established a fortified trading post at Quebec. From Quebec, the French moved towards the vast fur-trading region of the Great Lakes. Figure 3.8 shows how they used the natural waterways and portage routes already discovered by the Indians. Because Champlain had made enemies of the Iroquois, who controlled the Lake Ontario area, the French were forced to use the Ottawa River-Lake Nipissing route to the Upper Great Lakes

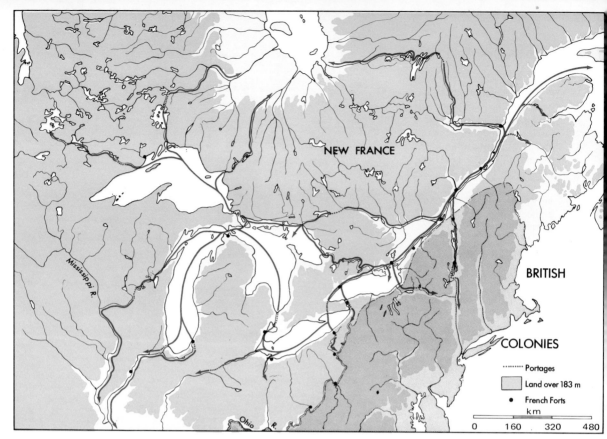

Figure 3.8 FRENCH FUR-TRADING ROUTES. The Great Lakes and major rivers made it possible for French fur traders to penetrate deep into the heart of the continent. Can you name the waterways they used?

of Superior and Michigan. Later in the 1600's, when the Iroquois had been defeated, the French began using the Lake Ontario-Lake Erie route.

By 1700, the French had built trading posts and forts throughout the St. Lawrence and Great Lakes area. They also had forts as far north as James Bay, as far west as Winnipeg, and as far south as the mouth of the Mississippi River on the Gulf of Mexico.

The French had outmaneuvered the English. They had taken much better advantage of the geography of the continent in using the Great Lakes and Mississippi waterways to provide easy access to most of the interior. The English, on the other hand, had settled along the Atlantic seaboard and found the way through the Appalachian Mountains to the West very difficult. When the English did start penetrating through the Appalachians, they found that the French had cut them off from the interior by building fur-trading forts along the Mississippi. Furthermore, the English fur-trading business in the Hudson Bay area found that its best route to Europe through the St. Lawrence waterway was controlled by the French.

For many years, the French did little to settle the vast area over which they had gained control. The fur traders were interested in trading guns, knives, trinkets, and other products for the furs that brought a good price in Europe. They did not want settlers to come in to clear the land. By the middle of the 1600's there were fewer than 2000 Frenchmen in North America, one-third of whom lived in the town of Quebec.

After Canada was made a French Province in 1663, France encouraged settlement. Jean Talon, an ambitious intendant, did many things to encourage population growth. He even taxed bachelors and brought women out from France to encourage men to marry and to settle down in the new land. Most settlement took place along the St. Lawrence River, and the *habitants,* as the French settlers were called, used this river as their main highway.

THE BRITISH-FRENCH STRUGGLE

Early in the 1700's, Britain made gains in its struggle with France for colonies in North America. At the end of a war in Europe, France recognized British control over Nova Scotia (except for Cape Breton Island) and the vast Hudson Bay territory which was known as Rupert's Land. The French were given rights to dry their fish along the north shore of Newfoundland. The map in Figure 3.9 shows that there were still some areas disputed by both countries even after the Treaty of Utrecht in 1713.

Figure 3.9 FRENCH AND BRITISH TERRITORIES, 1713. The Treaty of Utrecht in 1713 gave Britain control over Rupert's Land and Nova Scotia. There were still large areas claimed by both Britain and France.

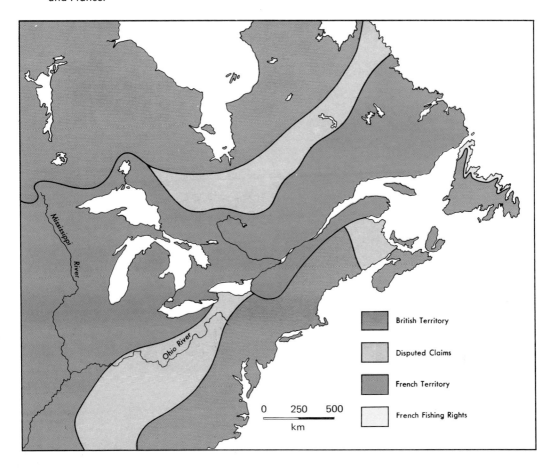

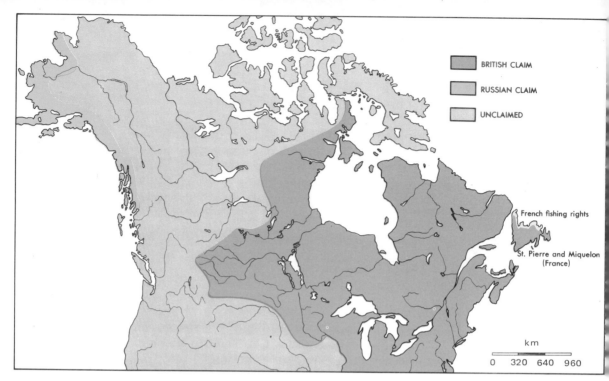

Figure 3.10 BRITISH POSSESSIONS IN NORTH AMERICA, 1763. By the Treaty of Paris, 1763, France lost almost all of her North American possessions to Britain. British influence in the western part of the continent at this time was so weak that the Russians were able to claim part of the west coast.

The French and British fought whenever they came in contact with one another. The French helped the Hurons and Algonquins fight the Iroquois, so the British helped the Iroquois. When the British pushed their settlements through the Appalachian Mountains, they met and fought with the French who had built forts along the Ohio and Mississippi Rivers. In the north, there were fights between the French fur traders and those of the British Hudson's Bay Company.

In the area that is now Nova Scotia, the French Acadians, accused by the British of being unco-operative and of stirring up the Indians, were expelled from their land and moved to different places in other British colonies. British settlers were encouraged to settle in Acadia to replace the expelled Acadians.

All of these skirmishes between the British and French happened without any declaration of war. Finally, fighting in other parts of the world led to the Seven Years' War (1756-1763). At the end of that war, New France was turned over to Britain. The only places left to France were the islands of St. Pierre and Miquelon. French fishermen also retained the right to land on the coast of Newfoundland. In 1904, France gave up its rights on the coast of Newfoundland, but has retained St. Pierre and Miquelon to this day (Figure 3.10).

By the early 1770's, Britain was having difficulties with the Thirteen Colonies. To prevent similar trouble in Quebec, Britain passed the

Quebec Act, in 1774. This Act extended the territory of Quebec to include the Ohio and Upper Mississippi valleys. It also permitted the French to use their own civil law in the courts and gave the Roman Catholic Church some special privileges. The Quebec Act is very important because it made it possible for the French Canadians in Quebec to keep their traditions and customs. This Act helps explain why Quebec is so different from other provinces in Canada today.

CANADIAN-AMERICAN RELATIONS

Twenty years after the fall of New France, Britain lost its Thirteen Colonies. After a bitter war between the colonies and Britain, the colonies gained their independence and became the United States of America. At that time, a new boundary between the United States and the British territory was drawn along the Great Lakes, as shown in Figure 3.11.

Figure 3.11 UNITED STATES AND BRITISH NORTH AMERICA, 1783. The Thirteen Colonies had gained their independence from Britain. A boundary was drawn roughly along the Great Lakes to separate the United States from British North America.

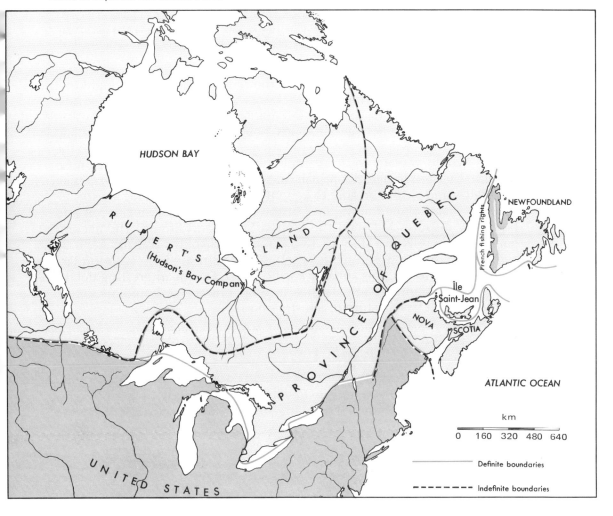

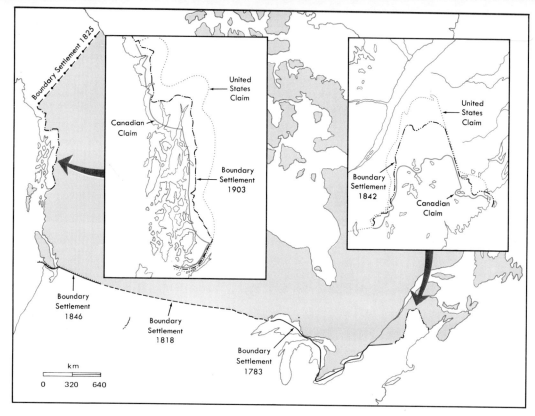

Figure 3.12 CANADA'S INTERNATIONAL BOUNDARY SETTLEMENTS. A number of boundary disputes developed between the United States and Canada. Since Canada was not yet an independent nation, Britain did most of the negotiating. Note that in each case the final boundary was chosen between the two claims.

During the American Revolution, the Thirteen Colonies thought that the British colonies to the north (Canada) should join them in fighting the mother country. However, the Canadian colonies remained loyal to Britain, and the French and Indians helped British troops to defend Canada against several attacks by the Americans. It seems strange that the French would remain loyal to their old enemy, Britain. They did so because they felt that under British rule they would have greater freedom of religion and would be able to keep their own language and laws.

The Treaty signed at the end of the Revolutionary War did not end boundary disputes between the United States and Britain. The accompanying map tells the story of two boundary disputes (Figure 3.12). By agreement, the forty-ninth parallel was established as a boundary between the Great Lakes and the Pacific Coast. For the interesting details of the quarrels and arguments that accompanied these boundary agreements you will have to turn to a history book.

LATER SETTLEMENT

During and after the American Revolution, thousands of people who remained loyal to Britain fled from the United States to Nova Scotia, New

Brunswick, and Quebec (which also included Southern Ontario at that time). These refugees, who were called United Empire Loyalists, became excellent settlers. Many of them cleared the land for farming, some set up factories, and others ran stores or worked in trades. In this way, agriculture, manufacturing, and commerce became more important than fur trading and fishing. So many Loyalists came to the British colonies that two new provinces, New Brunswick and Prince Edward Island, were established. The old province of Quebec was divided into Lower Canada (now Quebec) and Upper Canada (now Ontario).

In the early 1800's, plenty of cheap land in Upper Canada attracted people from Europe and the United States. From Pennsylvania came the "Pennsylvania Dutch." (They were originally of German descent, but were called Dutch because they were *Deutsch*, the German word meaning "German.") Other settlers came over directly from Germany. However, the largest proportion of the Upper Canada pioneers came from the British Isles.

In the later 1800's, settlers began moving into the vast, open spaces known as the prairies. They settled first along the Red River and the Saskatchewan River. The building of railroads encouraged settlers to move farther west, and many small towns and villages sprang up along the railroads. Settlers were attracted to the prairies because they could obtain huge farms cheaply from which no trees had to be cleared. The land was divided into sections of 640 acres each. For only ten dollars, a settler could claim a 160-acre farm known as a quarter-section. Many farmers were able to buy an entire section.

The Indians and the Métis opposed the coming of the settlers. They tried to prevent newcomers from building the railroads that would open up the country to settlement and ruin their hunting. There was some fighting, and occasionally, railroad construction and settlement were delayed. In the long run, the non-Indians won. The buffalo herds were depleted and the Indians were put on reserves, while the settlers started growing wheat and grazing large numbers of cattle on the land.

British Columbia owed its early settlement boom to the discovery of gold. This precious metal was found in the sandbars of the Fraser River in 1858. Within a few months, 25 000 people, mostly from the United States, poured into British Columbia to "strike it rich." Following the prospectors came merchants, auctioneers, real estate agents, bartenders, and entertainers. Everyone spent freely, and while some became rich, many more went home penniless. A new wave of people came when more gold was discovered a few years later in the Cariboo Mountains near the source of the Fraser River.

When the gold petered out, most of the people returned home. However, about 10 000 people remained as permanent settlers in the colony. This settlement resulted in the establishment of the Province of British Columbia, made up of Vancouver Island and the mainland west of the Rockies.

When the CPR, the Canadian Pacific Railway, was extended westward in the late 1800's, railway builders encountered many special construction problems. Because they lacked large, modern pieces of equipment, they found the rugged land surface north of Lake Superior and in the mountains of British Columbia a real obstacle. Trains are unable to run up and down steep hills or around curves as automobiles can; railway tracks must be built on smooth, gently sloping ground. Therefore, the construction engineers laid out routes that had to cut through solid rock barriers and had to bridge deep river valleys with trestles.

Gold miners dig for nuggets in the bed of a small tributary in the upper Fraser River Valley. Digging or tunnelling down to the gold was much harder work than panning or washing gold dust from sandbars along the lower Fraser. The Cariboo gold strike in 1861 helped open the interior of the British Columbia territory for permanent development and settlement.

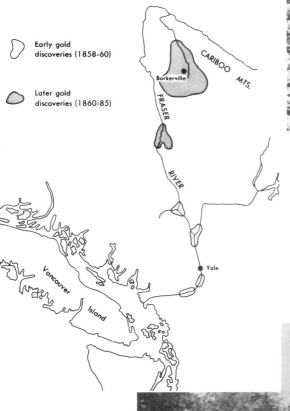

After the initial strike, a twisting, primitive road was built from Yale to Barkerville, once a booming city of 40 000 people. Miners and later settlers carried their supplies into Cariboo country by wagon train pulled by teams of mules. Stage coaches, which were used in most of the settled parts of Canada before railways were built, also transported passengers, mail, and gold along the road.

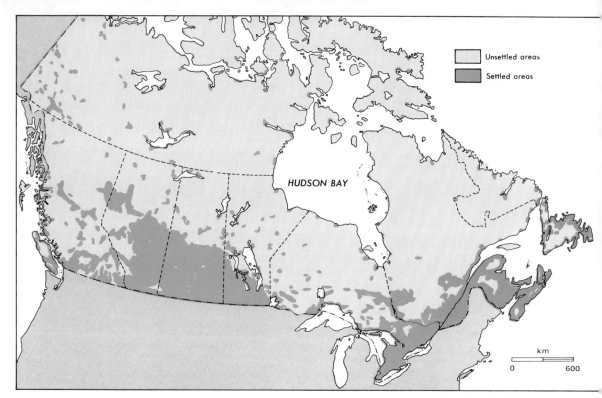

Figure 3.13 SETTLED AREAS IN CANADA. Why do many people live in the southern parts of Canada and so few live in the north?

New land continued to be settled in western Canada in the early years of the 1900's. People from eastern Canada, the United States, and from all over Europe poured into the Prairie Provinces (Alberta, Saskatchewan and Manitoba). A number of Chinese and Japanese settled on the west coast. Most of the settlers who took up land in the interior valleys of British Columbia were British. Large numbers of immigrants have continued to come to Canada right up to the present time. Since World War II, most of the immigrants have settled in the large cities where it is easier to find jobs.

There are large areas of Canada still not densely settled today. In the north, there are a few isolated farming areas, towns and villages based on the primary resource industries of mining, lumbering and fishing, as well as scattered Indian and Inuit settlements. In between these settlements are vast areas where a person could travel for hundreds of kilometres without coming across even a small village. However, much of this land that appears to be unoccupied is actually used by Indians and Inuit for hunting, fishing and trapping. (See Figure 3.13.)

FROM SEA TO SEA

Canada has not always extended from the Atlantic to the Pacific. The name "Canada" was first officially used when Lower and Upper Canada

Figure 3.14 THE GROWING MAP OF CANADA. To see how Canada's political boundaries changed, look at the above four maps.

were formed. Later, these two were joined to form the Province of Canada. However, Canada's real birthdate was 1867 when the British North America Act (B.N.A. Act) created the "Dominion of Canada" that included the provinces Nova Scotia, New Brunswick, Quebec, and Ontario. Fear of being taken over by the United States was one of the reasons for Canadian Confederation. The Fathers of Confederation were afraid that other parts of British North America would be captured by the United States. For this reason, they included a clause in the B.N.A. Act that welcomed additional provinces.

The story of the territorial growth of Canada can best be told by maps. Look at the "Growing Map of Canada" (Figure 3.14). From this series of maps, a Canadian history book, and an atlas, find the following information.

QUESTIONS

1. In 1870, Manitoba looked like a postage stamp in the broad expanse of the Northwest Territories. Consult a history textbook to discover why Manitoba became a province of Canada and why it was so small.

2. To attract British Columbia into Confederation, Canada promised to build a railroad to the west coast. When was the promise fulfilled? What was the railroad called? What were the major construction problems in Northern Ontario? across the Prairies? in British Columbia?

3. Newfoundland, the oldest colony, is the youngest province. Why did Newfoundland join Confederation in 1949?

4. The boundary between the Labrador Coast of Newfoundland and Canada was disputed for many years. The dispute was apparently settled in 1927, but recently has again gone under judicial review. By studying a large-scale map, find the clue to explain the irregular nature of the western boundary of Labrador. What mining development in the years after World War II made the precise location of the Labrador boundary more important? (See Chapter 10.)

5. Name the two provinces that do not border on sea water.

6. In what year did Canada claim all the Arctic islands as far as the North Pole? What is the most northerly group of islands called? To whom do the islands in Hudson and James bays belong?

7. By consulting later chapters in this book and other references, explain (a) the role of the following natural resources in Canada's growth: cod, furs, white pine, wheat, gold, iron ore, pulpwood, hydroelectricity, and petroleum; and (b) the role of these methods of transportation in Canada's settlement and development: ships (sailing, canal, freighters), railways (steam *versus* diesel), roads (wagons to motor vehicles), and aircraft (bush planes to jetliners).

CANADA COMES OF AGE

Confederation, in 1867, marked the birth of Canada, but the infant still had a lot of growing up to do. To be a strong and independent nation, a country must have its own government, must be free to have its own foreign policy, and must have enough wealth to be able to look after itself. The people living in the different parts of the land must have a feeling of unity, and a loyalty to their country.

In the early 1900's, Canada had reached the adolescent stage in its growth to nationhood. Canada had its own government elected by the Canadian people, it collected and spent its own taxes, and it made its own laws. But, in its dealing with other countries, Canada was not an adult. When Britain was at war, Canada was at war. If there was a boundary dispute between Canada and the United States, Britain did the negotiating. Canada could not sign treaties with other countries, had no representatives at international meetings, and had no foreign embassies.

All of this changed shortly after World War I (1914-1918). After the war, Canada insisted on an independent seat at the peace conferences and Canada was given membership in the League of Nations. In 1920, Canada established its first treaty with a foreign country. After the Statute of Westminster in 1931, Canada became an independent nation.

This independence did not mean that Canada cut off all relationships with Britain. As a member of the Commonwealth of Nations, Canada regards the monarch of England as its sovereign; the monarch's representative in Canada is the Governor General who is appointed by the monarch on the advice of the Canadian Government. The Governor General is a symbol of unity who has no real powers, but opens Parliament, signs bills that have been passed by Parliament, welcomes important guests to the country, and attends special functions on behalf of the monarch.

For many years, the Canadian Government could not decide upon a national flag. Both the Union Jack and the Red Ensign were flown. In 1964, Canada chose a new flag with a red maple leaf on a white background. The Union Jack, however, is still used to show Canada's connection with the Commonwealth, and a number of the provinces have their own flags.

Over the years, the British North America Act provided the constitution for the way Canada was governed. Because the B.N.A. Act was passed by the Parliament of Britain, any requests for change had to be made to it. In 1981, after much debate among the federal and provincial governments, the Canadian Parliament voted to alter parts of the constitution and to bring it home, so that the last trace of Canada's colonial status will disappear. The revised constitution contains a charter of rights for all Canadian citizens and spells out how the constitution can be changed in the future.

NATIONAL UNITY

An independent nation is not necessarily a unified nation. Many nations have been seriously weakened because of quarrels between different parts of the country.

Even after Confederation, not all of the provinces were convinced that they were better off being a part of Canada. Particularly in Nova

The completion of the Canadian Pacific Railway in the late 1800's did much to unify Canada. It was Canada's first transportation system. After World War I, the Canadian Government purchased a number of small railway companies to form the Canadian National Railways. It has grown to become the largest railway company in North America, operating about 48 000 km of track. Today, streamlined trains, similar to the ones shown in these photographs, help link the various parts of Canada together.

Until the 1960's a motorist could not travel completely across Canada on a hard-surfaced road at all times during the year. The Trans-Canada Highway, officially opened in 1962 but not entirely finished until 1967, now makes this trip possible. Just as the railway engineers encountered difficult construction problems on the CPR almost 100 years ago, so did the highway builders. They had to blast their way through the rock of the Canadian Shield north of Lake Superior as well as carve a route through the Cordillera as shown in the photograph.

Air Canada and Canadian Pacific Airlines have established regular trans-Canada flights that help to overcome the great distances separating parts of the country. Canadians have come to depend upon aircraft for rapid transport, passenger and mail service. Besides linking distant parts of Canada in a matter of hours, both companies operate long-range flights to countries in other parts of North America and the world.

The St. Lawrence Seaway is a complex system of canals, rivers and lakes that allows large freighters to penetrate halfway across Canada.

Scotia and New Brunswick, some people felt that they should be part of the United States. This feeling was partly a result of geography.

The Maritime Provinces were closer to the United States than to the rest of Canada. The easiest transportation routes and the natural directions for trade were north and south rather than east and west. Canadian tariffs seemed to be aimed at protecting the industries of Ontario and Quebec. The Maritimers found that they were forced to buy Canadian-made products instead of lower-priced products from the United States.

Likewise, British Columbia felt isolated from the rest of the country. Even after the completion of the Canadian Pacific Railway in 1885, British Columbia had more frequent contacts with the United States than with eastern Canada. The rugged Western Cordillera, the vast expanse of prairies, and the rocky Canadian Shield were formidable barriers to unity that took many years to overcome.

Rugged terrain and vast distances can be overcome by modern transportation and communication. East and west are now connected by two major transcontinental railroad systems, the Trans-Canada Highway, and two trans-Canada airlines. The St. Lawrence Seaway provides cheap water transportation from the Atlantic Ocean to almost halfway across the country. Pipelines transport oil and gas from Alberta to industrial centres in Ontario and Quebec. An excellent telephone and telegraph system allows a person to contact any place in Canada within minutes.

The Canadian Broadcasting Corporation's radio and television systems, along with private networks, provide Canadian news, public affairs programs, and entertainment in every province. Some major magazines and newspapers are sold from coast to coast. All of these transportation and communication links help to tie the country together into a unified nation. Nevertheless, there are still problems of disunity in Canada.

Serious problems of disunity can result from differences in language and culture. Some people consider themselves to be Polish Canadian, English Canadian, or French Canadian instead of simply Canadian. There is considerable lack of understanding between English-speaking Canadians and French-speaking Canadians.

The B.N.A. Act made both French and English the official languages of Canada. Look at a Canadian dollar bill and you will see that it has both English and French on it. Members of Parliament in Ottawa may speak in either language and all federal government reports are published in both languages. The federal government provides its services in both official languages right across the country.

However, many French-speaking Canadians (in Quebec called Québecois) believe that they have not always been treated fairly. Even in Quebec they have found it difficult to obtain high-level jobs in industry and business. This situation results partly from the fact that many of the firms in Quebec are offshoots of companies from English-speaking Canada or the United States.

If Québecois families wish to move to another province, they will probably have to become fluent in English. They may not be able to find French-language schools for their children, French-language radio, television, or local newspapers. If any of the family have to appear in court, the proceedings will be conducted in English. For these reasons, Québecois do not feel that they can move comfortably to other parts of Canada.

Québecois are also afraid that they may lose their cultural identity. Traditionally, their culture had three foundations: attachment to the land, obedience to the church, and their language. All three of these foundations have eroded since World War II. Quebec has become sufficiently industrialized and urbanized that attachment to the land has been reduced. The influence of the Roman Catholic Church over society has lessened considerably. The 1971 Census of Canada showed that the Quebec birthrate was the lowest in Canada, sometimes called the "shock of the empty cradle": a direct reflection of deep changes of attitude. Moreover, many French Canadians were learning English and moving to other parts of Canada, and immigrants to Quebec did not want to use French as a first language. Quebec appeared to be losing its language, a strong foundation of its culture.

In addition, some Québecois feel that Quebec is a conquered nation and that it should be "liberated" to become a sovereign (independent) national state. In the 1976 provincial election, Quebeckers elected a majority government of the Parti Québecois, whose major policy platform was "sovereignty association". By this it meant political separation of Quebec from Canada, making it an independent country. However, it proposed that economic ties like free trade and a money system common to the rest of Canada be kept.

In 1980, the Parti Québecois government held a referendum in which Quebeckers were asked if they would give the government a mandate to negotiate "sovereignty association" with Canada. The vote was 60 percent "No", 40 percent "Yes"; the citizens of Quebec turned down any form of separation. However, it must be remembered that slightly more than half of the French-speaking Quebeckers voted in favour. The issue may arise again at some time in the future. Perhaps a long-term solution will be a modified constitution that provides opportunities for the Québecois to achieve their aspirations.

Canadian unity has also suffered from differences in job opportunities, levels of income, and standards of living in different parts of the country. Economists call this problem *regional economic disparities* (see Chapter 13). Traditionally, the Atlantic Provinces have lagged behind the rest of the country in per capita income. Some people think this is true because the Canadian confederation has not permitted the Atlantic Provinces to share in the economic growth of the country; some even suggest they would be better off as a separate country or as part of the United States.

Separatist feelings have also been expressed in the West. For many years the Western Provinces felt disadvantaged by federal trade and transportation policies. In the 1980's, since their wealth has risen dramatically, these provincial governments are resisting federal government price regulations and resource taxes. This is particularly true of Alberta which has benefited most from rapidly rising oil prices. The Alberta Government believes that its own citizens should benefit most from oil profits; the federal government insists that increased wealth should be shared by all Canadians.

Regional debates like these have been occurring for over a century. So far the problems have been solved through negotiation and compromise. In the final analysis, the forces of unity have been greater than those of separatism.

PROBLEMS AND PROJECTS

1. Transportation routes are necessary before a new country can be settled. To discover how important railroads have been in the settlement of Canada: on an outline map of Canada, draw all the main lines and important branch lines of Canadian railroads. Colour all areas that lie within 40 km of a railway. Compare your map with the map of the settled areas of Canada in Figure 3.13.

2. Why was Ottawa chosen as Canada's capital city? What are the present advantages and disadvantages of Ottawa's location as a capital city?

3. There has been discussion about the Atlantic Provinces uniting to form one province. What advantages might come from such a union?

4. At one time, the western part of Northern Ontario was claimed by Manitoba. What justification would Manitoba have for this claim? Consider the following: (a) What is the time difference between that part of Ontario west of the 90° meridian of longitude and Winnipeg? (b) Compare the distances between Kenora and the capital cities of Ontario and Manitoba. (c) What is the Canadian city with over 500 000 people nearest Thunder Bay?

5. List factors that unify Canada and then list factors that divide Canada. Which of these lists do you think provides the most powerful factors?

6. Have a class debate on the statement, "Canada is a nation despite its geography." Before beginning, make sure that there is general agreement on the meaning of the words *nation* and *geography*. A particularly useful source for this debate is the first chapter in J.W. Watson's *Canada: Problems and Prospects* (Don Mills: Academic Press, 1966).

4 THE FORMATION OF THE LAND

In Chapter 3, we discussed how the political map of Canada evolved and how the country became an independent nation. In the chapters to follow we shall discuss the natural landscape—what it is like and how it came to be. This chapter describes and explains the origin of the landforms. Succeeding chapters describe the what, where, and why of climate, natural vegetation, wildlife, and soils. After that, we will discuss how well people have made use of the natural environment.

HOW WAS THE LAND FORMED?

The story of the formation of the land, or terrain, of Canada is difficult to believe.

QUESTIONS

1. How could what is now the top of the Rocky Mountains once have been at the bottom of a sea?
2. In the age of the dinosaurs was the western part of Canada really a tropical marshland with no winters?
3. What proof have we that the great oil deposits of Alberta were formed from millions of little sea creatures?
4. How do we know that the coal beds of Nova Scotia were created from ancient forests buried in rock?
5. Can you possibly imagine glacial ice up to 1600 m thick in the place where you now live?

The secrets of how the country was formed are buried in the land and in the rocks that compose it. Geographers and geologists have been working to unlock these secrets for many years. Some of the answers have been found; others may have to await another generation of scientists.

You will find it interesting to try your own hand at explaining the origin of the terrain in your area. Your first job as a scientist is to make observations and record them.

LOCAL STUDY

The next time you go for a ride or hike in the country, keep your eyes open and record your impressions of the terrain and its many different kinds of landforms. You will find that describing terrain is very difficult. Sketches and diagrams will help. If you have a camera, you can take pictures to illustrate your description. You can best describe where special landforms are by outlining their location on a map.

You may be able to describe only a small area of land yourself. If your class works together on this description, it may be possible for you to describe an area as large as 80 km². The following questions will provide a checklist of some of the kinds of things you should look for. Classroom discussion may help you pose questions more suitable for your particular area.

1. How much of the area is flat? gently sloping? steeply sloping? (You may have some disagreement with your classmates about "how steep is steep". After some discussion, set your own rules.)

2. How many metres is the highest point above the lowest point in the area?

3. From the highest point, describe the horizon in all directions. Ignoring trees and buildings, sketch the general shape of the horizon.

4. If there are hills or mountains, in what direction do they run? Is there any bedrock showing on the hills?

5. How much below the surface is the bedrock in most of the area? What kind of rock is it? (If the bedrock is not exposed anywhere, you may be able to find out how deeply wells have to be drilled to get into the bedrock.)

6. What is the composition of the earth material over the bedrock?

7. What different soil textures are found in different parts of your area?

8. Sketch the location of the streams and rivers. How wide are the stream and river valleys? How much of the valley is used by the stream channel? In what direction do the streams flow? Do they flow slowly or rapidly? Are there waterfalls? Do the streams run in a fairly straight course or do they wind back and forth a great deal? Is the water muddy or clear? Does the amount of water in the stream differ greatly in different seasons? Are there any valleys with no water in them?

9. Measure the depth and width of a small gully. Drive in stakes to mark where you made the measurements. After a heavy rain, make your measurements again. If possible trace where the eroded material went.

10. How many lakes and swamps are there in your area? How big are they? How large is their drainage area? Are they on the higher land or in the lower parts? How deep are the lakes? Are the slopes of the banks steep or gentle?

11. Describe any spectacular landforms in your area.

Using Air Photos and Maps in Landform and Terrain Description

Your class project has likely resulted in some good descriptions of individual landforms. However, you probably found it difficult to describe the whole terrain in such a way that a complete stranger would know what it looked like without ever seeing it. It is very difficult to give a good description of the terrain of an area by travelling across it by car or on foot, since you see only a small part of it at any one time. It is much easier to describe terrain, or even an individual landform, if you can see all of it from a high hill or from an airplane, but this kind of view is not always possible.

Geographers use air photos to see what the land looks like from a height. By using *stereoscopes* they are able to see the landforms in three

The geographers in this photograph are using stereoscopic pairs of air photos for terrain analysis, as a first stage in a resource inventory.

You may obtain stereoscopic pairs of air photos of your area from the National Photo Library, Ottawa, as well as from certain provincial government departments and private air survey companies. Stereoscopes are available from a number of instrument companies. The mirror stereoscope in the photograph is expensive. However, small 'pocket' stereoscopes are quite inexpensive.

It takes practice to see depth on air photos. Once you have learned the technique, the landforms seem to "jump out" at you. Mountains seem higher and slopes steeper in the photos than they really are on the landscape.

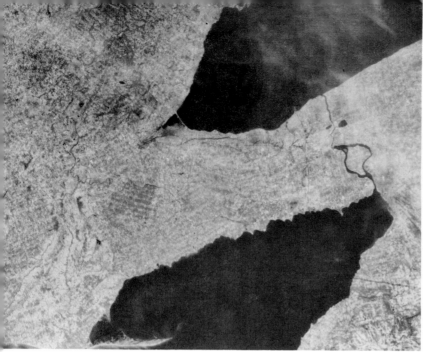

Only broad land-use patterns in the Niagara Peninsula between lakes Ontario and Erie (in black) can be seen on this small scale photograph taken from an overhead satellite. Numerous resource satellites orbit over Canada daily to monitor changes in crop, forest, and other land areas over time. Early black and white photography has now been replaced by colour imagery. For more information about this kind of resource analysis, write to the Canada Centre for Remote Sensing (Energy, Mines and Resources Canada).

dimensions. They can also determine the steepness of slopes, the nature of earth materials, the degree of soil drainage, and the nature of natural vegetation and agricultural land uses. Because landforms, soils and vegetation are interdependent, knowledge about one often helps in interpreting the other.

Different types of maps are also very useful in describing and interpreting terrain. Topographic maps show elevation, local relief, steepness of slopes, and general patterns of landforms, water bodies and streams. Because topographic maps usually cover a larger area than air photographs, they give a better appreciation of the terrain patterns and the relationships between different landforms. Geological maps provide details about the nature of the rock underlying the surface. Soils maps provide information about the texture, natural drainage, stoniness and depth of the soil. Geographers use all of these kinds of maps, as well as air photos and field observations, when doing a terrain description and analysis.

You may obtain air photos of your area from the National Photo Library, Ottawa, and large-scale topographic maps from the Map Distribution Office of Energy, Mines and Resources Canada in Ottawa. The best way to learn how to use air photos and topographic maps is to take them out-of-doors, locate your position on them, and then try to identify on the map or photograph what you see around you.

Earth Materials

The earth is made up of rocks in many different forms. The terrain with its various landforms is composed of either solid bedrock, or rock that has been ground up into pieces that range in size from particles of dust to

Top left: Granite — an igneous rock with large brightly-coloured mineral crystals.

Top right: Lava — an igneous rock with small, dark mineral crystals that formed during volcanic eruptions.

Middle: Sandstone — a sedimentary rock formed by the cementing together of layers of sand deposits.

Bottom left: Fossil — a sedimentary rock containing the imprint of a leaf. The leaf became trapped in layers of silt and was pressed into rock before it could decay.

Bottom right: Marble and Gneiss — metamorphic rocks. The light-coloured rock is marble formed from limestone. The colour-banded rock is gneiss that originates from granite.

house-sized boulders. In the following pages, we describe and explain the origin of three different kinds of rocks.

IGNEOUS ROCKS

Far below the surface of the earth are pockets of rock that are so hot that they are in a melted form (*molten* rock). When molten rock cools and hardens, it is called *igneous* rock, which means "fire rock".

When molten rock stays far below the earth's surface, it cools slowly. When it cools slowly, the resulting igneous rocks have large mineral crystals that you can easily identify without a magnifying glass. The most common igneous rock with large crystals is *granite*. See if you can identify the following minerals in a piece of granite: quartz (white), feldspar (pink), hornblende (black), mica (black or transparent).

When molten rock comes through to the surface, it cools very quickly. The igneous rock that forms then is composed of very small mineral crystals. Lava that flows from a volcano and then hardens is a good example of an igneous rock with small crystals. *Basalt* is one kind of hardened lava. Most fine-grained igneous rocks are darker in colour than granite.

SEDIMENTARY ROCKS

Solid rock is broken up into many little pieces by the process called *weathering*. There are four basic ways in which weathering occurs:

(1) Minerals expand when heated, and contract when cooled. Frequent expansion and contraction resulting from changes in the weather help to crack the rocks.
(2) When moisture accumulates in cracks in the rocks and freezes, it expands and breaks off pieces of rock.
(3) Carbon dioxide in the air, when mixed with water, forms an acid solution that dissolves certain minerals, thus weakening the rock.
(4) Pieces break off rocks when they slide down the sides of mountains or hills.

As a result of these weathering processes, rock is finally broken down into many particles of different sizes. The smallest rock pieces are carried by streams, into the oceans, where they are laid down or *sedimented,* in the bottom of the sea. After millions of years, many metres of these sediments are pressed together into *sedimentary rocks*. If the rock particles are very fine, then *shale* is formed. Sand deposits cement together to form *sandstone*. If a large number of skeletons of ocean life are deposited, *limestone* is formed. This fact explains why limestone often contains many fossils, which are the remnants of animal or plant life that have turned to stone.

METAMORPHIC ROCKS

When igneous or sedimentary rocks are exposed to tremendous pressure and heat, they change form. These changed rocks are called *metamorphic* rocks, because the word metamorphic means "changed." Most metamorphic rocks were formed millions of years ago during mountain-building periods. When the earth's crust was squeezed, bent, and broken, the tremendous heat and pressure changed many sedimentary and igneous rocks into metamorphic rocks.

The very common limestone is turned into *marble* by metamorphic processes. *Slate* is the metamorphic form of shale. Sometimes you may find a rock that looks like granite, but has bands of layers of different coloured minerals. That rock is likely a granite *gneiss* (pronounced "nice"), which is the metamorphic form of granite.

LOCAL STUDY

The following directions will help you make useful observations about the rock material found in your area.

1. Make a rock collection. Divide the rocks into the three main groups: igneous, sedimentary, metamorphic. Are the rocks rough with sharp edges or are they smooth with rounded edges? The smooth and rounded rocks have been carried long distances by water or have been washed for many years by waves on a beach. Keep a record of where you found the rocks.

2. Examine any exposed bedrock. Note its location on a map. To what family of rocks does it belong? Consult a science book or encyclopedia to see if you can name the specific rock type. If the rock is igneous, does it have large, or small mineral crystals? If the rock is sedimentary, are the layers of rock horizontal? If not, describe the way in which the layers slope. Does the rock contain fossils?

3. Gather samples of different ground-up rock materials such as gravel, sand, silt, and clay. These materials will be easy to find in gravel pits, road cuts, excavations, or along gully or stream banks. Take your samples from a depth of several metres so that the rock does not contain any topsoil. Record the location of your samples on a map. Examine the ground-up rock material carefully. Note its colour, the way it feels when rubbed between your thumb and fingers, and the way it breaks into pieces when you squeeze a chunk of it. If the material contains different-sized particles, mix some of it thoroughly in a litre jar of water and let it settle. The larger pieces (gravel) will quickly sink to the very bottom of the jar. Then in turn a layer of sand, silt, and clay will be deposited. The latter is so fine that it will stay suspended in the water for some time before settling. When all the material has settled, note the proportion that is gravel, sand, silt and clay.

4. What relationships can you find between the kind of rock material in a given area and (a) the stream pattern? (b) the kind of natural vegetation? (c) the land uses?

5. Compare the kinds of rock materials with the landforms you previously described. Are the hills all made of the same material? Where did you find the gravels and sands, as opposed to the silts and clays?

Landform Processes

By this time, you know something about how the terrain in your area looks and what materials are in the landforms, but you still have not explained how the landforms were made. To understand how the land was formed, one must know something about landform processes. Scientists have been doing research in this field for many years. The major ideas are easy to understand, but it takes much study and research to understand all of the details.

It is difficult to imagine the hard, rocky crust of the earth warping, bending, and breaking; but this is exactly what happens in mountain-building periods. Great earth forces raise or lower whole continents. They bend the rock layers into *folds* like those found in corrugated cardboard. Sometimes, the pressures are so great that the rock breaks. These breaks are called *faults*. In zones where there is much folding and faulting, molten rock usually comes to the surface to form *volcanoes*. Folding, faulting, and volcanic activity are all mountain-building processes, which have taken place over periods of millions of years.

As soon as mountains have been built and continents have been raised above the sea, weathering begins to break up and to weaken the rock. The weathered rock is then carried away, or *eroded* by running water (streams), moving ice (glaciers), and wind. After millions of years, these forces of weathering and erosion destroy the mountains and wear down the continents to low plains near sea level. Vast amounts of rock material are carried to the sea where they form new layers of sedimentary rock.

The great pressure of thousands of metres of sedimentary rock, deposited in the bottom of the sea, presses the bedrock downward and forces the continents up again. The new mountains thus created are attacked and worn down by the forces of weathering and erosion, and so the cycle continues.

How Old Is Canada?

The rocks on which the land is built are so old that we measure their age not in decades or centuries but in hundreds of millions of years. How long is a hundred million years? In an earlier chapter we mentioned that people crossed the Bering Strait to North America 11 000 to 40 000 years ago. That was 9000 to 38 000 years before Christ. With simple arithmetic

Figure 4.1 FORMS OF MOUNTAIN BUILDING.

Folding: The bedrock is often folded into a series of parallel mountain ridges and valleys. (Note: Upfolds are called *anticlines*, while downfolds are known as *synclines*.)

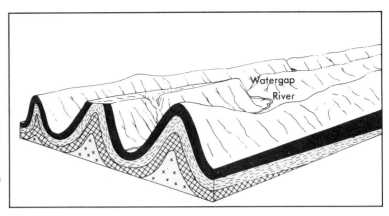

Faulting: The bedrock breaks to form a steep-sided escarpment. (Note: Not all escarpments are caused by faulting. For example, the Niagara Escarpment has been caused by erosion. See p. 114.)

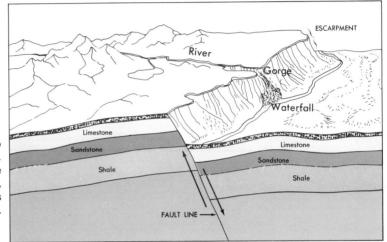

Volcanic Activity: Lava and cinders build volcanic mountains. (Note: In all these diagrams, the original bedrock is some form of sedimentary rock.)

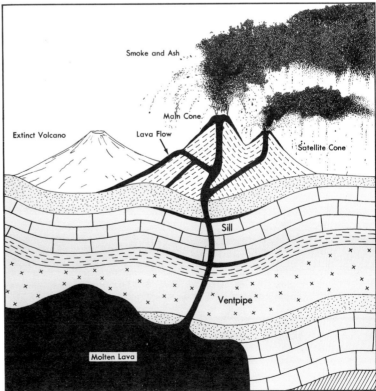

AGE IN MILLIONS OF YEARS	ERA	PERIODS OF MOUNTAIN BUILDING AND SEAS	BEGINNING OF GROUPS OF LIVING THINGS
1 — 60 —	CENOZOIC (Modern Ages) (Ice Age)	Pacific Coast Mountains Second Rocky Mountains	Man Mammals
200 —	MESOZOIC (Middle Ages)	First Rocky Mountains Cretaceous Sea Jurassic Sea Triassic Sea Innuitian Mountains	Dinosaurs Ancient Birds Coniferous Trees
500 —	PALEOZOIC (Ancient Ages)	Appalachian Mountains Permian Sea Carboniferous Sea Devonian Sea Silurian Sea Ordovician Sea Cambrian Sea	Insects Ferns Amphibians Fishes Animals with shells
Over 4500	PRE CAMBRIAN (Prehistoric Ages)	Laurentian Mountains (Canadian Shield)	Single-celled plants and animals

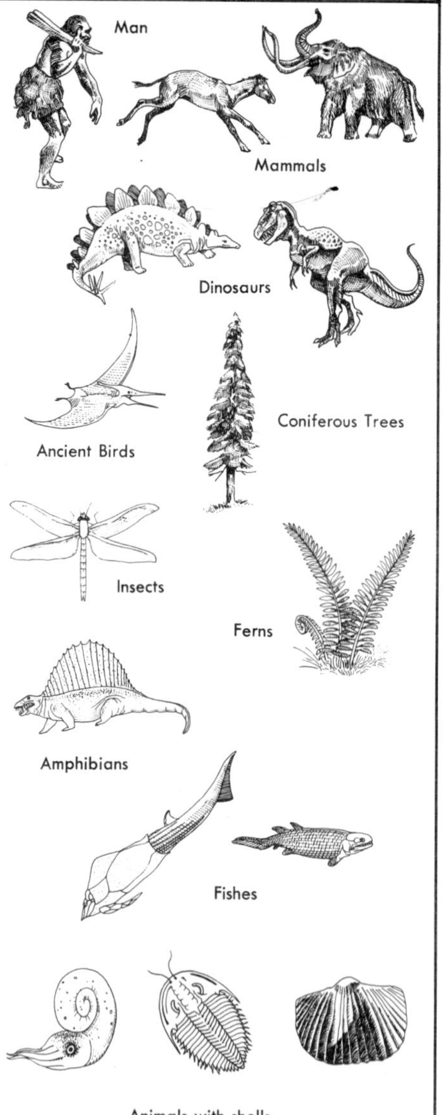

Figure 4.2 GEOLOGICAL TIME SCALE. For several billions of years before the Precambrian era, the earth's surface was cooling to form the rocks that make up the crust.

calculation, you can figure out how many times longer a hundred million years is than the period since the coming of the Indians.

Because the history of the earth's landforms involves such a vast span of time, scientists have grouped the ages of rocks into periods. These time periods make up what is known as the Geological Time Scale (Figure 4.2). The oldest rocks at the bottom of the column are called *Precambrian*. They may be considered as prehistoric because there is no record of what happened in that era. The records in the rocks have been erased over the hundreds of millions of years that have passed. The *Paleozoic* era may be considered as ancient earth history, the *Mesozoic* as the Middle Ages, and the *Cenozoic* as modern history. The *Pleistocene* period, or Ice Age (about a million years ago), may be considered as "current events."

Now, let us return to the question, "How old is Canada?". The earliest rock foundations of Canada were laid in the prehistoric times called the Precambrian era. By looking at the Geological Time Scale, you will now be able to give the approximate date of the laying of the cornerstone of our country.

THE HISTORY OF CANADA'S ROCK FOUNDATIONS

The Canadian Shield

The platform on which Canada is built is a large mass of rock formed in the Precambrian era. If you dig deeply enough in any place in Canada you will eventually come to Precambrian rock. Outside of the Canadian Shield (also known as the Laurentian Shield or the Precambrian Shield), the Precambrian rock is covered with thousands of metres of younger sedimentary rock. In the Canadian Shield, however, the Precambrian rock is right at or near the surface.

Much of the Canadian Shield is composed of igneous rocks that once were the roots of mountains towering higher than the Rockies. During the mountain-building periods, there was much folding, faulting, and volcanic activity. Numerous bands and domes of molten rock were intruded into existing rock. Geologists believe that high mountain ranges were built three times, only to be levelled by weathering and erosion each time. In between the mountain-building periods, the low-lying land was covered by the sea and sedimentary rocks were formed. In this complex process lasting hundreds of millions of years, much of the igneous and sedimentary rock was changed into metamorphic rock, and large deposits of metallic minerals, such as gold, nickel, silver, and lead, were formed.

Each time the Canadian Shield became a mountainous island, the process of weathering and erosion began to break up the rock and carry it to the surrounding shallow sea. There, the sands, silts, and clay were sedimented into rock at the bottom of the sea (Figure 4.3). During the Paleozoic era, following the Precambrian, thousands of metres of sedimentary rock, such as sandstone, limestone, and shale, were formed in the surrounding sea. The first sedimentary rock formed close to the Shield. Then, as the Shield kept rising the older sedimentary rocks rose above the sea and the younger sedimentary rocks were formed farther out (Figure 4.4). Thus, today, the older Paleozoic rocks are found close to the Canadian Shield and the younger rocks farther away.

Figure 4.3 EROSION OF SHIELD MOUNTAINS. Hundreds of millions of years ago, rivers began to erode the mountains of the Canadian Shield. The sediments carried by the rivers were deposited in the surrounding shallow seas.

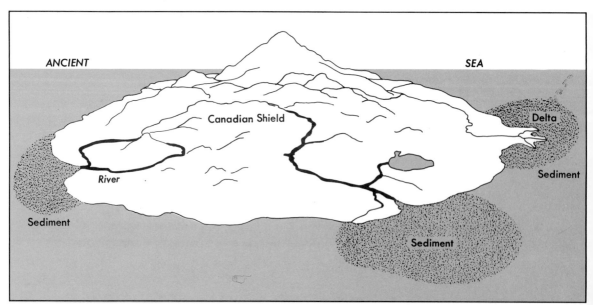

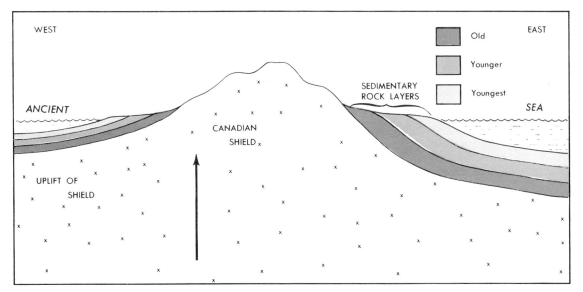

Figure 4.4 SEDIMENTARY ROCKS FORMED IN THE SHALLOW SEAS. The sediments carried to the sea by the rivers were cemented into layers of sedimentary bedrock. The older rock formed next to the Shield. As the Shield rose, younger rock was formed farther away.

The great thicknesses of rock, extending for hundreds of kilometres on all sides of the Canadian Shield, caused the sea's floor to sink, forming great depressions that were hundreds of kilometres wide. Rock continued to be formed in these depressions for hundreds of millions of years during the Paleozoic era.

The tremendous mass of sedimentary rock pushed down the earth's crust and set up great horizontal forces that squeezed the sedimentary layers into mountain ridges. This activity happened on all sides of the Canadian Shield. Its rocks were bent and broken, and molten rock came near the surface to form new igneous rock and new mineral deposits. The eastern edge was tilted up to form the high Torngat Mountains that can be seen today in Labrador.

Along the south and west, great breaks occurred in the rock and large sections were pressed downward. These lowlands are now occupied by the Gulf of St. Lawrence, the Great Lakes, and the series of western lakes that ring the Shield from Lake Winnipeg to Great Bear Lake. Major faults along the northern edge of the Shield created lowlands that were flooded to form the straits and channels of the Canadian Arctic Islands.

The Appalachian Mountain System and the Innuitian Mountains

In the latter part of the Paleozoic era, there was a great push on the layers of sedimentary rock to the southeast of the Canadian Shield. The relatively soft and flexible sedimentary rocks were pushed up against the Canadian Shield. Throughout most of the area, the layers of rock were steeply folded, or severely faulted, leaving row upon row of high

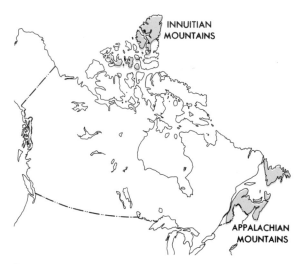

mountain ridges and deep valleys (Figure 4.5). This event was the beginning of the Appalachian Mountain system that presently extends from Alabama in the United States to the Atlantic Provinces of Canada. The original Appalachian Mountains have been eroded several times over the ages to form a low, gently sloping surface; but each time, they have been uplifted again into a series of ridges and valleys.

Because the sedimentary rocks of the Appalachians were deposited during the time of the great fern forests, they contain large deposits of coal. Other metallic minerals, such as iron, lead, and zinc, have been formed where igneous intrusions have come in contact with sedimentary and metamorphic rocks. Nonmetallic minerals, such as asbestos, salt, and gypsum, are also found embedded among the sedimentary rock layers of the Appalachians.

After the great Appalachian "squeeze", new pressures were exerted on the sedimentary rocks that had been laid over the northern part of the Canadian Shield. By the middle of the Mesozoic era (Middle Ages), a range of fold mountains was formed (Figure 4.5). These mountains, that extend for about 1280 km across the most northerly islands of the Arctic

Figure 4.5 THE FOLDING OF ANCIENT MOUNTAINS.

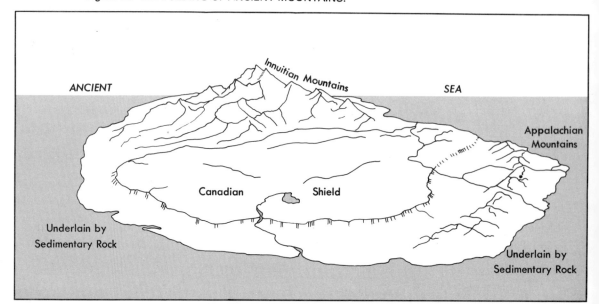

Archipelago, are known as the Innuitians. They are very similar to some parts of the Appalachian Mountains. It is probable that they contain similar minerals, but to date, little mineral investigation has been carried out.

The Western Cordillera

With the uplift of the Appalachian Mountains to heights of over 6000 m, the continental drainage divide moved nearer to the east coast (Figure 4.6). Now, the major rivers were carrying their loads of sands, silts, and clays to the west. Large volumes of sediment were deposited in the Western Sea. During the latter part of the Mesozoic era (Middle Ages) and the beginning of the Cenozoic era (Modern Age), great earth forces thrust up

Figure 4.6 RIVERS CARRY SEDIMENT WESTWARD. After the Appalachian Mountains formed, rivers began carrying sediments into the Western Sea where they formed sedimentary rock layers.

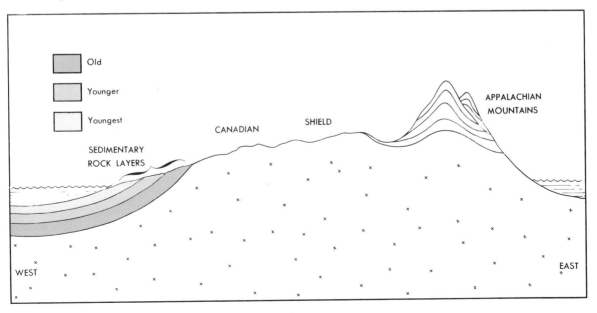

giant mountains near the western margins of the continent. Some of the rock layers were folded as if they were paper. There was much more faulting and volcanic activity than in the Appalachian mountain-building period. Rock layers that were thousands of metres thick broke and were thrust upward as tilted mountain blocks reaching 3000 to 6000 m in height. Some blocks of rock were pushed down, forming great trenches. Volcanoes spewed out huge quantities of lava, and great masses of molten rock were pushed upward to cool into igneous rock. The resulting series of mountains, valleys, and plateaus along the western side of North America is known as the Western Cordillera. The most easterly mountain system of the Western Cordillera is known as the Rocky Mountains.

The Rocky Mountains rose to great heights in the early part of the Cenozoic era, and then, over a period of tens of millions of years, they were eroded by glaciers and streams to a series of low mountains and hills no higher than 600 to 1000 m. The modern Rocky Mountains were again built up during a later period of the Cenozoic era. They rose more slowly this time. As their height increased, streams and glaciers eroded the softer rocks, leaving the rugged Rocky Mountain terrain that we know today.

The mountains along the coast of British Columbia, as well as the plateaus and mountains between the Coast Mountains and the Rockies, were formed at about the same time as the Rockies. The coastal mountains have more igneous intrusions than the Rockies. The interior plateaus are uplifted blocks into which the rivers, such as the Fraser, have eroded deep canyons.

Because of its many rock types, the Western Cordillera contains a wide variety of both metallic and nonmetallic minerals.

The Interior Plains and the Great Lakes-St. Lawrence Lowlands

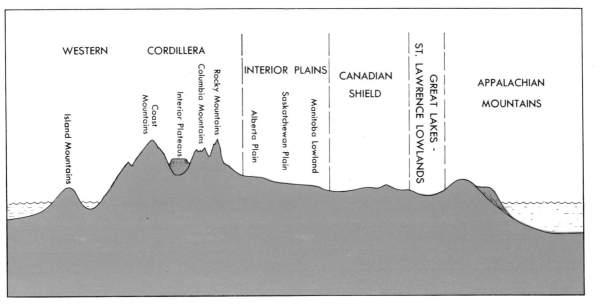

Figure 4.7 THE ROCK FOUNDATIONS OF CANADA. After studying this diagram, turn to Figure 4.10 showing the landform regions of Canada.

With the formation of the Western Cordillera, the major rock structure of Canada had been built (Figure 4.7). However, a large part of the story of the rock foundations of Canada has not been told because we have made no mention of the plains of western Canada or the lowlands of the Great Lakes and the St. Lawrence River.

These interior lowlands and plains were not greatly affected by the great mountain-building periods. Their bedrock layers are relatively horizontal, with only gently sloping arches and basins. One example is the gently sloping, saucerlike basin that underlies both Lake Huron and Lake Michigan. The outside edge of this basin can be seen in the Niagara Escarpment in Ontario.

On a number of occasions during the Paleozoic era, the Interior Plains and Great Lakes Lowlands were flooded by seas for vast periods of time (Figure 4.2). During these periods of flooding, sedimentary rock was laid in the bottom of the sea to depths of up to tens of thousands of metres. Not all of the interior was flooded at the same time, and as a result, not all of the areas have rocks of the same age.

During the time of the Devonian Sea (the middle of the Paleozoic era), the climate of North America was warm and humid (Figure 4.8). In this near-tropical climate, billions upon billions of tiny marine plants and animals lived in the shallow seas. They were so small that one would not have been able to see them without a microscope. As these tiny plants and animals died and fell to the bottom of the sea, they were often covered with silt before they could decay. This silt turned to shale, and more rock was formed on top. Cut off from air permanently, these minute plants and animals could never decay. After millions and millions of years

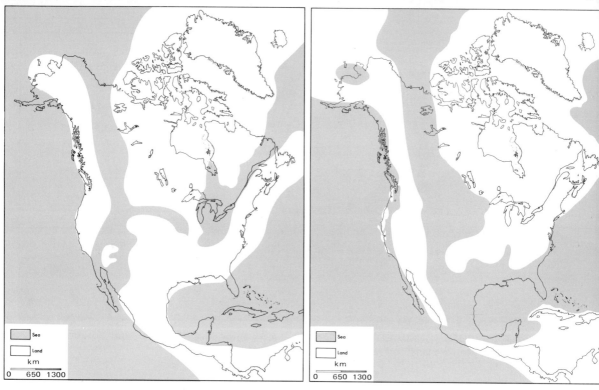

Figure 4.8 EXTENT OF DEVONIAN SEAS. At this time, much of the Great Lakes region, Interior Plains, and Western Cordillera was beneath the sea. The core area of the Canadian Shield was too high to be invaded.

Figure 4.9 EXTENT OF CRETACEOUS SEAS. During the Cretaceous period, shallow seas advanced from the Arctic Ocean and the Gulf of Mexico to cover the Interior Plains. Eastern Canada was the only large area above sea level.

the particles of these plants and animals, known as *hydrocarbons*, were chemically changed to form oil and natural gas. These pools of oil and gas collected in parts of the bedrock where they could not escape. This fact explains the oil found in small quantities in Southern Ontario and in vast quantities in Alberta and Saskatchewan.

The huge volumes of salt found in Southern Ontario were also formed by the salty seas during this period of time.

The Cretaceous Sea (late Mesozoic era) covered only the Interior Plains. This period marked the last time that a large portion of the continent was under seawater. At its greatest extent, this sea reached from the Gulf of Mexico to the Arctic (Figure 4.9). It lasted thirty to forty million years.

As the Cretaceous Sea began to withdraw, it left vast, shallow swamps along its shores. With a hot, humid climate, ferns, trees, and shrubs grew up, similar to those found in the tropical swamps of today. From time to time, the sea returned, and the trees and shrubs died and fell over, covering the swamp bottom with tangled mats of vegetation. The wood could not decay because first water, and then, sands and silts cut it off from the air. Later, deposits of rock pressed the vegetable matter into peat, and finally, coal. This process explains the large amounts of soft coal found in Alberta.

Right: Life in the Paleozoic Seas that once covered much of North America. A variety of odd-looking plants and animals thrived in the murky water of the shallow seas. In the water lived tiny plants and animals that eventually formed oil and gas deposits.

Left: A steaming Mesozoic swamp. The trees, ferns and shrubs that grew thickly in the shallow water provided the material for the formation of extensive coal deposits in the Interior Plains. Huge dinosaurs lived in and around the swamps.

During the time of the Cretaceous Sea, several more deposits of oil were made in the Interior Plains. The greatest of these deposits is found in the famous tar sands along the Athabasca River.

The time of the Cretaceous Sea also was the time of the giant dinosaurs. They lived in and around the swamps. Some of them reached lengths up to 12 m and masses of as much as 18 t. The skeletons of the dinosaurs that became mired in the swamp have been preserved in the rock that later covered them. These fossil skeletons are still being found in parts of Alberta. Dinosaur skeletons are on display in a number of museums.

A scientist unearths the skeleton of a dinosaur buried in soft sedimentary rock. Fossil discoveries such as this add more pieces to the puzzle of the past.

Late in the Cenozoic era, many millions of years after the Cretaceous Sea, great rivers brought sediments from the Rockies and deposited them on the Interior Plains. These sediments formed a layer of Cenozoic rock that lay over the older rock which had been formed during the time of the Cretaceous Sea. Since that time, most of the Cenozoic rock has been carried away by streams.

The Hudson Bay and Arctic Lowlands

The Hudson Bay and Arctic Lowlands are composed of slightly folded Paleozoic sedimentary rocks that have been laid on top of the Canadian Shield foundation. In this way, these lowlands are much like those in the Great Lakes area. It is therefore not surprising that both salt and oil have been discovered in the Arctic Islands.

Summary

The Precambrian or Canadian Shield is the platform of all of Canada. Deep layers of sedimentary rock were laid in shallow seas surrounding it. Pressures in the earth's crust caused mountains to be built to the south-east (Appalachians), to the north (Innuitians), and to the west (Cordillera). The plains in the interior have been little affected by the mountain building. Thus, the rock foundations were laid for the major landform, or *physiographic,* regions of Canada (Figure 4.10).

Figure 4.10(facing page) LANDFORM REGIONS OF CANADA. Some geographers call these physiographic regions. Which region covers the largest area? Which region contains the oldest mountains? the highest mountains? In which regions are mineral fuels (coal, oil, gas) and metals (gold, silver, nickel, lead) found? Can you explain why?

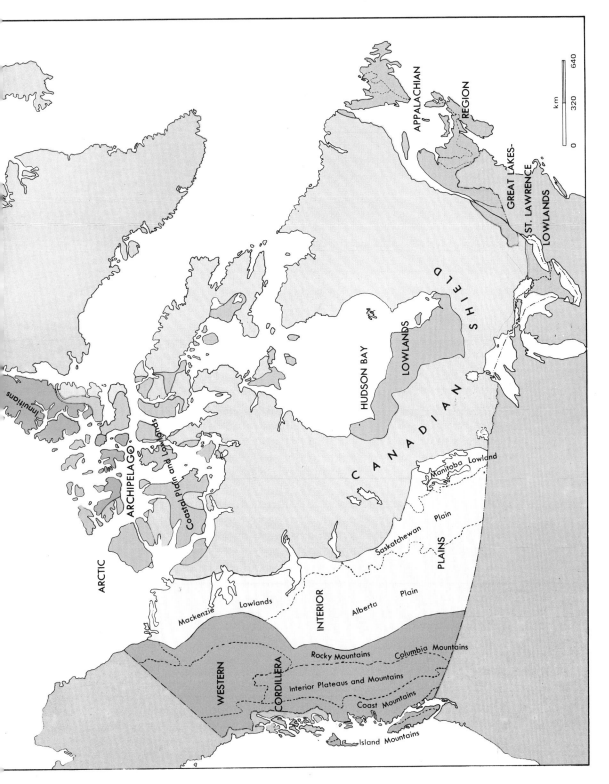

SCULPTURING THE LANDSCAPE

So far, we have described only how the rock foundations of Canada were built. This explanation is like describing a roughly shaped slab of rock from which a sculptor intends to carve a statue. One can tell that the statue is going to be a man, but the details of the face, body, and limbs have not yet been chiselled out. The landform details of the face of Canada have been sculptured by the agents of erosion. The most important of these have been running water and glaciers.

The Work of Running Water

Of all the processes of erosion, running water has been the most important in carving the details of the Canadian landscape.

As soon as the rock land arose from the sea, the elements of nature began to break the rock up into pieces of various sizes. This weathered rock was easily carried away by the running water resulting from rainfall. Streams, from tiny rivulets to giant rivers, began carrying the weathered rock to the sea. Valleys were carved and the slopes were worn down. After millions of years, the original Precambrian mountains of the Canadian Shield were eroded down to low, gently sloping plains.

The Appalachian and Innuitian Mountains have been so eroded by streams that only the low stumps of the original mountains can be seen. The softer rocks have been eroded to create valleys, while the harder rocks remain as ridges. The Annapolis Valley of Nova Scotia, for instance, is a depression caused by stream erosion of the soft sedimentary rocks lying between the hard rocks of North Mountain and South Mountain. The ridges and valleys of the Notre Dame Mountains in the Gaspé Peninsula have been formed by streams wearing away the softer rocks of the steeply-folded sedimentary layers.

The mountains of the Western Cordillera are higher than the Appalachians, not because of greater mountain-building forces, but because they are much younger, and therefore, less eroded. The many V-shaped valleys show that the streams have not given up in their struggle to carry the mountains to the sea. The many muddy rivers flowing swiftly in rocky canyons provide further evidence of the eroding power of streams.

What is picked up must be laid down again. Streams not only carry away earth materials, thus lowering the hills and cutting valleys; they also create landforms by dropping their materials. The gravels, sands, silts, and clays carried by the major rivers have been building up at the river mouths for thousands of years. Today, these materials form *deltas* extending over hundreds of kilometres. The Fraser and Mackenzie rivers provide excellent examples of deltas in the process of being formed (Figure 4.11a).

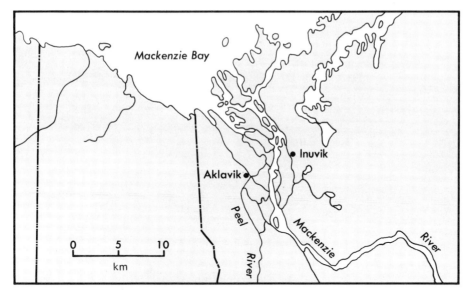

Figure 4.11a THE MACKENZIE RIVER DELTA. Delta deposits are formed in such a haphazard fashion that they force the rivers that created them to follow many twisting channels to the sea. In this way numerous islands, both large and small, have been formed.

From time to time, the streams that flow more slowly overflow their channels. The water then floods across the valley floor, dropping silty materials as it does. Over the years, this silt builds up level plains called *floodplains* (Figure 4.11b). You will likely be able to find examples of small and large floodplains in your local area.

Figure 4.11b FLOODPLAIN. When streams overflow their channels, they drop the silt they have been carrying. After many floodings, a flat floodplain is built. The stream meanders back and forth across the floodplain. Sometimes a meander is cut off to form an ox-bow lake.

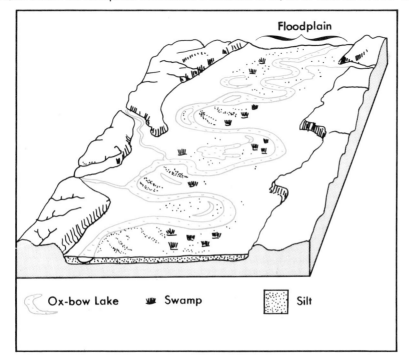

Numerous rivers, such as the Ottawa River you see here, have been at work thousands of years rounding the rocky upland of the Canadian Shield into smooth, low hills.

Long before the Ice Age, streams had been at work wearing down the land. In the plains of the interior, deep valleys had been carved out of the rock. Many tributary streams reached into all parts of the land. But the great ice sheets changed all this surface. The old stream patterns of the Canadian Shield and surrounding plains were completely disrupted, and the streams had to begin their work all over again.

The glaciers left Canada so recently that the streams have just had time to get started on their new job. There are vast areas where the streams have not had time to drain the lakes and swamps, nor to wear down the barriers that create waterfalls and rapids. In millions of years, the streams will have drained the lakes, including the Great Lakes, and will even have eroded away Niagara Falls!

—The Great Ice Age

About one million years ago, a strange thing began to happen in Canada. The climate became a little colder, and in the north, more snow fell each winter than melted during the following summer. This process continued for several thousands of years, and the snow reached thousands of metres in height.

The mass of the snow compressed the lower layers into ice. After a while, the heat created by this great compression made the ice pliable so that it started to flow out in all directions. The centres of these ice sheets were in Greenland, around Hudson Bay, and in the Western Cordillera. The ice slowly moved out from these centres until most of North America was covered as far south as the Ohio and Missouri Rivers (Figure 4.12).

Figure 4.12 NORTH AMERICAN GLACIATION. During the Ice Age continental glaciers covered Canada four different times. The last continental glacier melted from Canada less than 10 000 years ago. An icecap still covers all except the edges of Greenland.

It seems strange that the only parts of Canada not glaciated were certain parts of the Arctic, including the Klondike area of the Yukon. This resulted not from a milder climate, but rather from the lack of enough snowfall to form glacial ice.

It took thousands of years for the ice to cover most of Canada because, in some years, the ice sheet moved only a few metres; in others, a number of kilometres. For years at a time, the ice probably did not move at all, and sometimes there was more melting back than there was advance. The ice did not move along a smooth front; instead, great fingers, or lobes, moved into the lowlands and valleys ahead of the main body of ice.

Near the centre of the glacier, the ice probably reached a depth of about 3200 m. Along the southern edge of Canada, the ice was probably less than 1600 m in depth. At the outside edges, it tapered off to only a few hundred metres.

During the last million years, known as the Pleistocene epoch, or Ice Age, a continental glacier covered Canada, not just once, but four times. In between the glacial periods, the ice disappeared completely except in the far north and in some of the mountain areas. The climate between the glacial periods was warmer than it is today. Some of the periods between the major ice advances were so long that soils developed and trees grew.

The last continental glacier reached its greatest extent only about 25 000 years ago. It melted from Southern Ontario about 10 000 years ago. The fact that the Greenland Ice Sheet is still melting back indicates that our climate is still becoming warmer.

Will the great ice sheets cover Canada again some day? No one really knows. But, if they do, we will have plenty of time to prepare. It would take thousands of years for the ice to reach the southern parts of Canada.

GLACIAL EROSION

The continental glaciers did much to change the face of Canada, for a sheet of ice 1600 m thick exerts a great deal of pressure. As the ice moved over the land, the great pressure caused the clays, silts, sands, gravels, and boulders to become frozen into the bottom of the ice. These rocks of all sizes, in the bottom of the glacier, became giant ploughing, scraping, and sanding tools. They helped the ice to plough into loose rock and soil, to scrape out the softer rocks, and to scratch and polish everything that it moved over.

In many places in Canada, one can still see grooves, chips, hollows, and scratches that were made on the bedrock. The results of the glacial erosion can best be seen in the Canadian Shield where the bedrock is at or near the surface. Many of the hundreds of thousands of lakes in the Canadian Shield are the results of glaciation. The ice scoured out hollows, many of which are now filled with water. By looking at the direction of the ridges, the valleys, and the long narrow lakes, one can easily tell which way the ice moved.

The glaciers eroded away even more material in the areas with softer sedimentary rocks. Huge tongues of ice followed the valleys, deepening and widening them. When these depressions were filled with water, large lakes were formed. This process explains how the Great Lakes were created.

It is sometimes said that glaciers removed entire mountain ranges. This statement is not true. The major mountain systems in the Canadian Shield were worn away by the work of streams hundreds of millions of years before the Ice Age.

The continental glaciers did change the landscape in mountainous areas. As ice was pushed up and over the Appalachian region in Canada, this ice rounded off the tops of the mountains and hills, and scoured and polished the rock. The ice also deepened and smoothed out the valleys.

In other words, it created a "softer" outline for this mountainous and hilly region.

In the Cordillera region, the ice filled all the valleys and trenches and covered all the mountains up to a height of 1800 to 2100 m. Only the highest mountain ranges and individual peaks stood above the sea of ice.

GLACIAL DRIFT

The continental glaciers carried tremendous amounts of rock and rocky material. As more and more rock was picked up and became frozen into the bottom of the ice, the material it was previously carrying was pushed farther up into the ice. In time, there was as much rocky material in the glacier as there was ice.

All the earth picked up by the moving ice was, of course, deposited somewhere else. The rocky debris left by the glaciers is called *glacial drift*. Almost all of the Interior Plains and the Great Lakes-St. Lawrence Lowlands are covered with glacial drift. In some parts, the drift is only a few metres deep; in others, its depth runs to hundreds of metres. In an area just north of Toronto, drift as deep as 240 m has been measured.

Most of the rock material in the drift was carried by the glacier for relatively short distances. It is not true that the glaciers scooped up all the soil and huge amounts of Precambrian rock in the Canadian Shield and deposited it all in the surrounding lowland plains. Most of the rock material picked up in the Canadian Shield was deposited again in another part of the Canadian Shield. There is not much glacial drift in the Canadian Shield because there was little soft rock and loose weathered rock for the ice to pick up. In the surrounding plains of softer sedimentary rocks, the ice was able to pick up and therefore drop large volumes of material. If you examine glacial drift in your local area, you will likely find that as much as sixty to eighty percent of the rocks have been carried less than 160 km.

Some rocks, however, have been carried long distances by the ice. Huge granite boulders have been found hundreds of kilometres from their place of origin in the Canadian Shield. Because the igneous rocks are out of place in an area of sedimentary bedrock, they are called *erratics*.

GLACIAL LANDFORMS

The continental glaciers deposited their rocky debris in many different ways, creating numerous kinds of landforms and a varied, interesting terrain.

When the ice had picked up a bigger load than it could carry, it "smeared" or "plastered" the clay and rock mixture off the bottom. This process created gently rolling plains of clay, studded with rocks of various sizes. The material is called *boulder clay* or *till*. Plains formed in this way are called *till plains*.

Above: Icecap on a rocky island in the Arctic. These icecaps are small remnants of the continental glaciers that once covered much of Canada. A tent in the foreground gives some indication of the depth of the ice tongue flowing down a valley to the sea.

Below: Ice scoured Canadian Shield in the Northwest Territories. Hundreds of small lakes (black) fill depressions that have been hollowed out of the rock surface (grey) by the ice. White patches are larger lakes from which winter ice and snow have not yet completely melted. Can you tell in which direction the glacial ice moved across this surface?

Boulder clay forms the soil material in this farm field. Rocks are so numerous in the till that the land can be used only for rough pasture.

A moraine landscape in Southern Ontario.

An esker appears as a rock, sand, and gravel ridge in the Manitoba countryside. Eskers like this weave across the landscape in snakelike fashion. They are the beds of streams that were formed within the glacial ice.

Sometimes, for some unknown reason, the ice "plastered" boulder clay into oval-shaped hills called *drumlins*. These drumlins vary greatly in size, but usually range in height from 30 to 45 m, and in length from 800 to 1600 m. They are often found in groups, and if you use your imagination, they look like a herd of whales in the till plain sea.

When the margin of the ice was neither advancing nor retreating, the glacier acted like a giant conveyor belt. The ice continued to move out towards the edge, but it melted as fast as it came. As it did so, the ice brought with it large amounts of boulder clay, sands, and gravel. This material was dumped at the margin of the glacier as the ice melted. If the glacier remained stationary for a long time, long ridges called *moraines* were formed. Some of these moraines have gentle slopes that have been cleared and farmed; other moraines present a rugged terrain with slopes too steep to cultivate.

At the margins of the glacier, there was a great deal of water coming from the melting ice. This *meltwater* created huge rivers carrying heavy loads of clay, silt, sand, and gravel. The large valleys created by these meltwater rivers are called *spillways*. Today, much smaller rivers and streams are found in some of the spillways while others have no stream at all. When dumped into water, the boulders, gravel, and sand sink first, while the silts and clays are suspended in the water for some time. In this way, the large volumes of glacial meltwater sorted out materials of different sizes. This process explains why some of the glacial drift is composed mainly of clay and silt, while other drift is mostly sand and gravel.

In some places huge amounts of sand and gravel were washed out of the glacier and deposited in flat *outwash* plains. Where the glacial stream action was more turbulent, gravelly hills called *kames* were formed. A series of these gravel hills is called a *kame moraine*. In some areas, huge blocks of ice were buried with the sand and gravel. When these blocks of ice melted, hollows were formed. In many areas, these hollows have filled with water to create *kettle lakes*.

Streams inside, or at the bottom of the glacial ice, formed very strange-looking landforms. The sands, gravels, and rocks carried by the water were dropped in the bed of the stream. When the ice banks melted, this stream bed was left above the surrounding land, looking something like a winding railway embankment. In some areas, these ridges, called *eskers*, run for many kilometres.

As the glacier gradually became smaller, large lakes often formed along the retreating ice edge. Silts and clays formed in the bottoms of these lakes to form large, almost perfectly flat areas. The flat plains in the Red River Valley area of Manitoba, as well as the Great Clay Belt of **northern** Ontario and Quebec, are the bottoms of these kinds of glacial lakes (Figure 4.13).

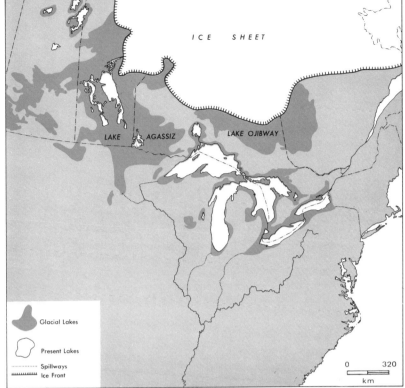

Figure 4.13 GLACIAL LAKES. Because the ice blocked the natural drainage to the north, large lakes formed at the edge of the retreating glacier. Note that the Great Lakes were much larger than they are today and that they drained through the Mississippi and Hudson Rivers. This southward drainage of the Great Lakes began when the St. Lawrence River was blocked by the glacier.

ALPINE GLACIERS

In the Western Cordillera, glaciers remained in the mountain valleys long after the major continental glaciers had disappeared. In fact, there are still glaciers active in some valleys of the Cordillera. The best known of these glaciers are in the Rocky Mountains near Banff and Jasper.

An alpine glacier starts with an accumulation of snow in the upper parts of the valley. The snow turns into ice and moves slowly down the valley like a lazy river. The movement is so slow that you usually cannot notice it at all.

Glaciers change V-shaped valleys created by streams into broad valleys with a cross-section shaped like a U. Along the Pacific Ocean, the

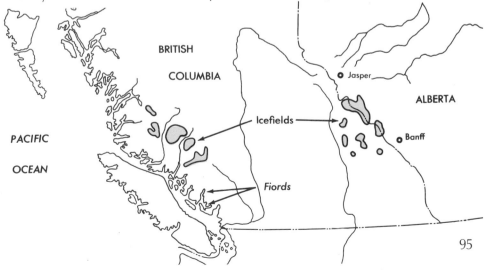

The Columbia Icefields near Jasper, Alberta. Snow that accumulates each year is compacted into ice that slowly creeps down the valley slopes. For a number of years the ice has been melting more rapidly than new snow has been accumulating, and so the glacier has been shrinking or "retreating", revealing both lateral and terminal moraines.

A fiord along the British Columbia coast. In the past, a valley glacier moved down from the mountains, in the background, towards the sea. It followed an existing river valley but gave the valley a smooth outline and made it much deeper. When the glacier melted back, the sea flooded the inlet and a fiord was formed.

deep U-shaped valleys have been flooded to form *fiords*. Thus, the rugged coastline of British Columbia is called a fiorded coastline.

At the upper end of a glaciated mountain valley, there are usually deep amphitheatre-shaped basins, called *cirques*. These cirques, where the snow collects and turns into ice, were caused by a complex combination of weathering and ice erosion. Where two cirques on opposite sides of a mountain meet, a jagged, knifelike ridge results that is called an *arête*. Glaciers may erode cirques on several sides of an individual mountain, creating rugged pyramid-shaped peaks called *horns*. The saddle-shaped gaps between horns are called *cols*. Lakes that form in the bottom of cirques, after the ice has melted, are called *tarns* (Figure 4.14).

Like continental glaciers, alpine glaciers not only erode, but also deposit rocky material. Deposits along the side of a valley glacier are called *lateral moraines*; those at the end of the glacier are called *terminal moraines*.

Figure 4.14 ALPINE GLACIAL LANDFORMS.

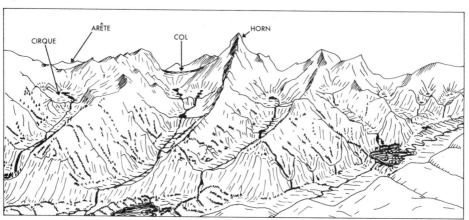

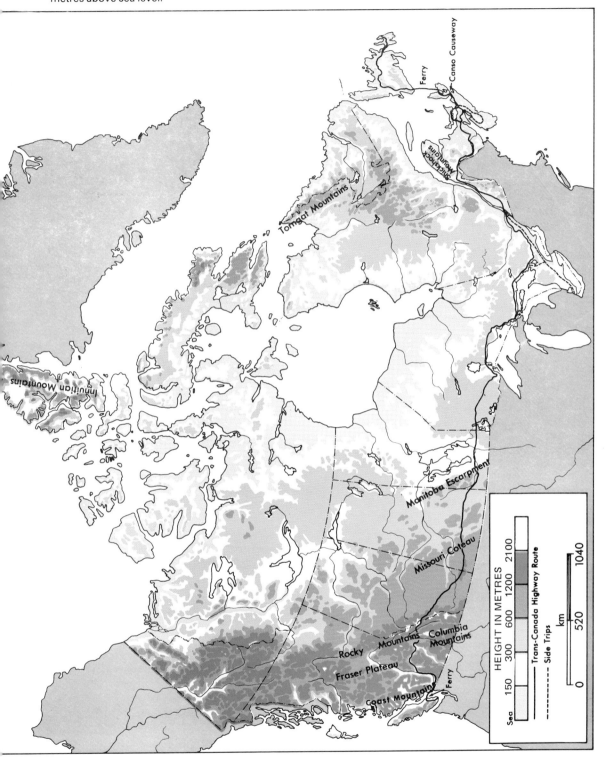

Figure 4.15 A SCENIC TRIP BY TRANS-CANADA HIGHWAY. The map shows the route followed on the imaginary scenic trip described in this chapter. The map also shows general elevation in metres above sea level.

A SCENIC TRIP

In the preceding pages, we have explained the origins of the various landforms found in Canada, but we have not described the landform or physiographic regions in detail. In fact, it is very difficult to paint word pictures that adequately describe a landscape a reader has not seen. The best way to see what the different Canadian landform regions are really like, and to appreciate the diversity and beauty of the Canadian landscape, is to take a trip across the country.

The balance of this chapter attempts to describe, by word, map and photo, the physiographic landscape you would see if you took a scenic trip across Canada, following the Trans-Canada Highway (Figure 4.15). You will find that topographic maps and air photos will also help you greatly in visualizing and interpreting the landscapes of the various regions (see books of maps and photos listed in question 7 of *Problems and Projects*).

Appalachian Region

The coastline of the Avalon Peninsula is indented with bays, coves, and inlets, formed by the flooding which was caused by the rising seas after the Ice Age. Because the headlands between the bays are too rugged and steep to give access to the sea, the fishing villages, called outports, are nestled close to the water in the bays. This irregular and rugged shoreline, which is typical of much of the coastline of Newfoundland, provides many excellent sheltered harbours for the fishing fleets of the island.

As you leave the Avalon Peninsula the highway curves in a great arc towards the north, leading to the Central Plateau which is composed of gently rolling hills and valleys. Next, you travel through a district studded

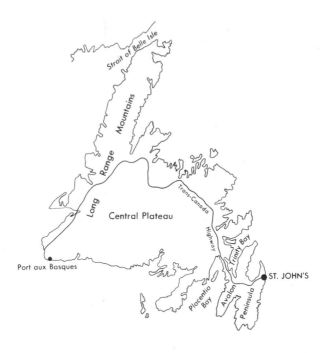

with lakes, streams, and bogs. As you approach Newfoundland's west coast, you sight very old, worn-down ranges of flat-topped hills known as the Long Range. The Long Range is a plateau-like surface that has been eroded into hills by stream erosion. In general, the elevation of this plateau is about 300 m higher than the Central Plateau.

The Long Range Mountains trend in a north-east and south-west direction, as do all the major ridges and valleys of the Appalachian region, because in the Paleozoic era, the region was squeezed up against the Canadian Shield to the north-west. This squeezing action formed many ridges and valleys parallel to the edge of the Canadian Shield.

From Port-aux-Basques, Newfoundland, you travel by ferry to Cape Breton Island. You land at North Sydney, the port for the surrounding coal mines and steel mills. All about you are forested, rounded, and flat-topped hills that rise abruptly from the sea. Around one portion of these Cape Breton Highlands, a special scenic route known as the Cabot Trail has been built. Some of the views from the lookouts on this route are breathtaking.

From the southern tip of Cape Breton Island, you cross to the Nova Scotia mainland by way of the Canso Causeway. At this point, you decide to leave the Trans-Canada Highway to go to Halifax, and then, to cut across the peninsula to see the Annapolis Valley.

Most of Nova Scotia is a hilly upland composed of resistant igneous and metamorphic rocks. The level horizon indicates that this surface is

St. John's Newfoundland occupies the head of an inlet that faces the Atlantic. It is Canada's most easterly port and the beginning of the Trans-Canada Highway. The low hills of the Avalon Peninsula in the background rise above this old city.

Petty Harbour is typical of the small fishing villages scattered along the rocky Newfoundland shoreline. Such communities have usually been built at the head of inlets of the sea, called coves, for protection from the stormy Atlantic. Note how the houses are perched on the steep hillside near the water's edge.

The Cabot Trail winds along the edge of the Cape Breton Highlands that rise to heights of 450 m in northern Cape Breton Island. Motorists can stop at the many camping, picnicking, and lookout areas for a longer look at the spectacular scenery.

The Canso Causeway is a combined road and railway transportation link. It was constructed to join the northern (Cape Breton Island) and southern parts of Nova Scotia. This transportation link has led to large-scale industrial development at Port Hawkesbury on Cape Breton Island.

In the interior of Nova Scotia, forested hills seem to be linked together in an endless chain. To cross the peninsula, roads and railways were built in depressions or breaks in the hilly upland. Only in the larger river valleys and along the coastal plain is the land flat enough for farming and settlement.

So low and gently rolling is the Prince Edward Island landscape, there are few locations where elevations exceed 90 m above sea level. Much of the land is well suited for farming, the most important crop being potatoes.

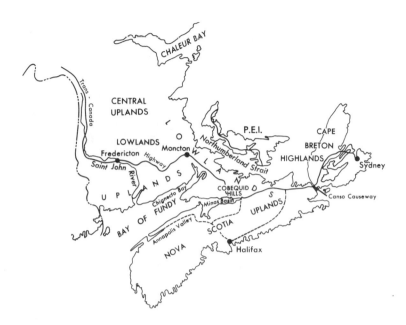

really a plateau in which streams have cut many valleys. The hills are rounded and are covered with forests. The best farmland in Nova Scotia, as in much of the Atlantic Provinces, is in the broad valleys and the coastal plains.

The most important valley in Nova Scotia is the Annapolis Valley. The valley floor lies some 180 m below North Mountain on the north-west and

Here you see some of the flat land along the coast of Nova Scotia where farming and settlement are possible.

In this view of Prince Edward Island, you can see how low and gently rolling the landscape is adjacent to Charlottetown, the capital city of the province.

South Mountain on the south-east. Both of these mountains are composed of hard igneous and metamorphic rocks. Over millions of years, streams have carved the valley between these two ridges.

The Annapolis Valley extends for some 110 km between Annapolis Royal and Wolfville. Its width ranges from 3 to 12 km. It is a striking contrast to the rocky, forested uplands. As you descend into the valley, you see that most of it has been cleared for farming and that orchards dot the landscape.

In going north from the Annapolis Valley, you have to go around the Minas Basin and pass over the Cobequid Hills before you reach the level, low plain bordering the Northumberland Strait. This plain extends into New Brunswick as far as Chaleur Bay.

Prince Edward Island is also a low, rolling plain. The underlying bedrock of Prince Edward Island is a reddish-brown sandstone, which gives a reddish tone to the whole landscape. The beaches are red, the cultivated fields are red, and even the pavement of the roads is red because of the reddish-coloured crushed sandstone. Most of the rolling land is cleared of forest and is intensively farmed. Prince Edward Island is famous for its potato crops.

Near Moncton, New Brunswick, you are able to view the tidal flats of Chignecto Bay, an arm of the Bay of Fundy. Even the stream channels draining into the bay have high and low water periods each day. During the times when the tide is out, the low water level exposes bare mud banks along these streams. The tide of the Bay of Fundy is so high that some streams flow backwards at high tide. The tidal water rushing upstream in the Petitcodiac River is called the *tidal bore*.

At low tide, the water withdraws so far from the Bay of Fundy shoreline that wide tidal flats are exposed along the coast. Places at which boats moor or dock are often left high and dry. At high tide water will cover the flat land you see in these photographs. Can you locate the high tide mark in each photo? Why does the Bay of Fundy have some of the highest tides in Canada?

At some places along the coast, small rock islands, called seastacks, have been carved out of the shoreline. Can you explain how these were formed?

During high tide, water rushes up the channels of rivers that drain into the Bay of Fundy so rapidly that the rivers actually reverse their flow. In this photograph the leading edge of the tidal bore can be seen moving upstream in the Petitcodiac River near Moncton, New Brunswick.

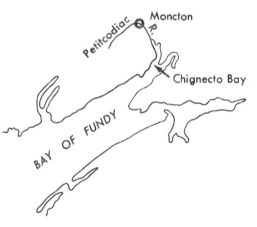

The gently sloping hillsides of the middle Saint John River Valley, north of Fredericton, have encouraged the development of many rich farmlands. The major crops in the Saint John Valley are potatoes and apples.

When the Trans-Canada Highway meets the Saint John River, it turns north-west and follows the valley. Near its mouth, the Saint John valley is very broad and the river flows slowly. Upstream a few kilometres from

Along the upper Saint John River in northwest New Brunswick, the valley becomes quite narrow and the steep slopes are completely wooded.

Fredericton, the river has been dammed to form a large lake called Mactaquac, which serves as a reservoir for a major hydroelectric plant. Around Fredericton, apple orchards can be seen on the terraces part way up the valley slopes, while farther north, the upper slopes are used extensively for potato growing. Still farther upstream, the valley narrows and steep banks rise abruptly from the fast-flowing stream. The Central Uplands of New Brunswick are to the north. Here, rounded and forest-covered hills rise to an elevation of over 600 m above sea level.

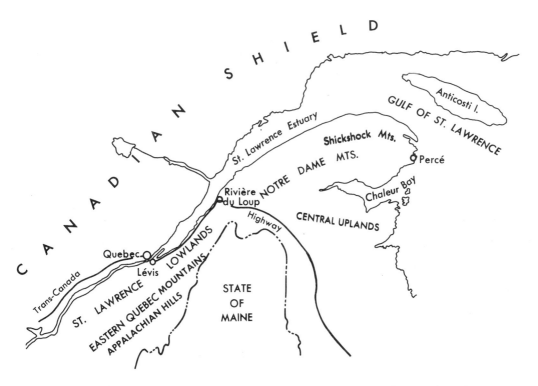

By cutting straight through to Rivière du Loup, you have missed a very scenic drive around the Gaspé Peninsula. The Gaspé Peninsula is made up of a number of parallel, forest-covered ridges and valleys. The central ranges are called the Notre Dame Mountains, and Shickshock Mountains. In the north-east corner of the peninsula are found the highest mountains of the whole Canadian Appalachian region. These mountains reach a height of over 1200 m in some places. The wave-eroded cliffs at Percé are a major tourist attraction.

When you reach Rivière du Loup, you marvel at the St. Lawrence Estuary which is about 24 km wide at this point. You follow the narrow coastal lowland until you reach Lévis, opposite Quebec City. You have now left the hills and valleys of the Appalachian region behind. Ahead of you is a lowland region which is quite a contrast from what you have just seen.

The Shickshock Mountains of the Gaspé Peninsula are much higher and more rugged than any of the upland areas of the Maritimes. On these eroded, steep slopes facing the mouth of the St. Lawrence River, little natural vegetation can grow.

Percé Rock is a famous tourist attraction which lies at the tip of the Gaspé Peninsula. Wave erosion for many thousands of years has separated Percé Rock from the mainland and has carved the steep-sided cliffs you see in the photograph.

Great Lakes-St. Lawrence Lowlands

From Quebec City, you can see the Appalachian Mountains to the south and the hills of the Canadian Shield to the north. The Canadian Shield forms a distinct northern boundary for the whole region. To the south and west, the lakes themselves become the limits of the region.

From Quebec City to Lake Ontario, the lowland is not very wide. You are almost always in sight of the bordering highlands. The plain itself is very flat and makes excellent farmland. South-east of Montreal, the monotony of the landscape is broken by the Monteregian Hills that rise

Here is part of the flat to gently rolling St. Lawrence lowlands at the base of the Canadian Shield on the west side of the St. Lawrence River. These lowlands contain rich soils that are used for mixed farming specializing in dairy production.

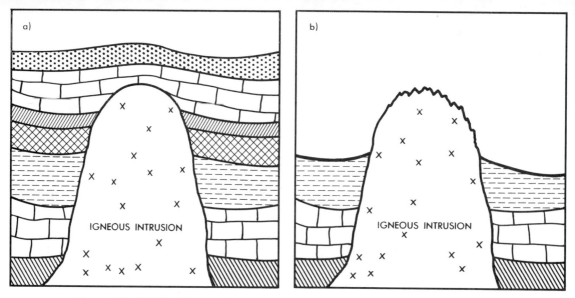

Figure 4.16 ORIGIN OF THE MONTEREGIAN HILLS. Millions of years ago igneous rock was intruded into the sedimentary rock layers in the St. Lawrence Valley. After many years of erosion, the softer sedimentary rocks have been worn away, leaving the igneous intrusions that rise abruptly from the flat plain.

abruptly from the flat plain. These hills are composed of igneous rocks that are harder than the surrounding sedimentary rocks. Thus, they have withstood millions of years of erosion while the softer rocks have worn away (Figure 4.16).

By taking the four-lane Macdonald-Cartier Freeway into Southern Ontario, you have deviated from the Trans-Canada route that takes you up the Ottawa Valley, a valley that is similar in appearance to the St. Lawrence.

When you near the place where the St. Lawrence leaves Lake Ontario, you notice igneous rock exposed at the surface. The many islands in the river, called the Thousand Islands, are also composed of igneous rocks. The occurrence of this rock can be explained by the narrow band of the Canadian Shield that crosses the St. Lawrence River at this point. The Adirondack Mountains on the south side of the river are really a part of the Canadian Shield. The Thousand Islands exist because the granite rock is very resistant to erosion.

As you continue toward Toronto, there is a narrow, flat plain along Lake Ontario to your left. This plain, which extends all the way around Lake Ontario to Niagara Falls, is famous for its orchards. This lake plain is the bottom of an old glacial lake that was larger than the present Lake Ontario. The old lake has been called Lake Iroquois. The shoreline of Lake Iroquois can still be seen clearly in many places. It forms the slope found just south of St. Clair Avenue in Toronto. Highway 8 follows the Lake Iroquois shoreline most of the way from Hamilton to the U.S. border, parallel to the Niagara Escarpment (see sketch map and Figure 4.17).

Immediately to the north of the Macdonald-Cartier Freeway, between Kingston and Toronto, is hilly land rising several hundred metres above the lake plain. These hills are moraines and drumlins deposited by the glacier of the last Ice Age.

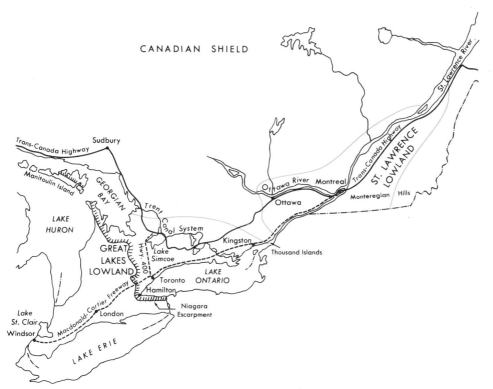

Figure 4.17 LANDFORMS OF SOUTHERN ONTARIO. Almost all of Southern Ontario is covered with glacial drift. There is a great variety of landforms as this map below shows. The only prominent bedrock feature is the Niagara Escarpment.

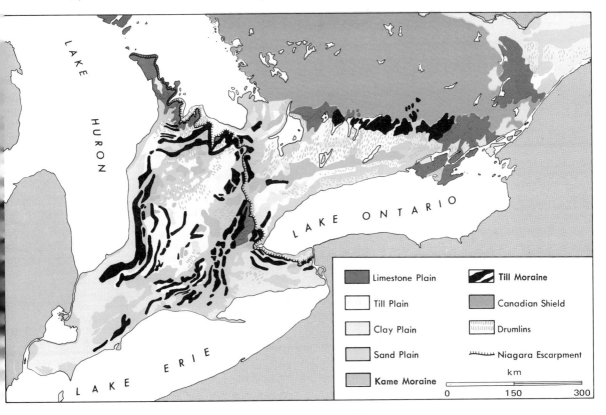

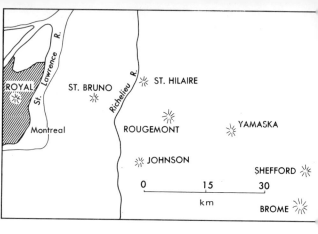

Above, left: Mont. St. Hilaire is one of the Monteregian Hills located in the Eastern Townships of Quebec, a part of the Great Lakes-St. Lawrence Lowlands. The steeply sloping sides of this igneous intrusion are heavily forested in comparison to the more open farmlands of the surrounding flat plain.

Near Kingston, hundreds of small, rocky and forest-covered islands have been eroded from a low portion of the Canadian Shield that extends across the St. Lawrence River. The photograph shows a part of the Thousand Islands. Homes and tourist facilities have been built on some of the islands and along the banks of the St. Lawrence River to take advantage of the scenery.

The low ridge of land you see marks the upper limit of the former Lake Iroquois shoreline. Highway No. 8 follows the shoreline for much of its length from Hamilton to St. Catharines. To the left of the highway, the land drops away toward the present Lake Ontario shoreline.

The Niagara Escarpment is the only major bedrock landform feature in Southern Ontario. In this photograph, the Niagara Escarpment is so precipitous near the top that no trees can gain a foothold. In other places, steep slopes are used for skiing. A hiking trail called The Bruce Trail follows the Niagara Escarpment from Niagara Falls to Tobermory on the Bruce Peninsula.

In some places where rivers have carved broad valleys through the escarpment, or where the bedrock has been completely masked with glacial drift, the Niagara Escarpment is evidenced by undulating slopes instead of cliffs. The photograph below shows the rolling landscape of the picturesque Beaver Valley near Georgian Bay.

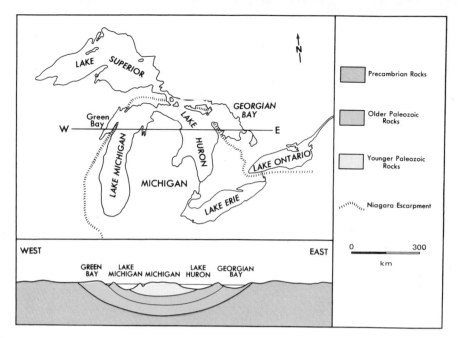

Figure 4.18 BEDROCK OF THE MICHIGAN BASIN. The Paleozoic bedrock underlying much of the Great Lakes area forms a saucer-shaped basin that is centred over southern Michigan. The ridge at the outer edge of this basin is known as the Niagara Escarpment. In the Georgian Bay area the steep slope of the escarpment faces east, while in the Green Bay area it faces west.

Just west of Toronto, you suddenly come upon a steep cliff several hundred metres high that faces east. This formation is known as the Niagara Escarpment and it runs from Niagara Falls to Hamilton and then snakes its way north to the Bruce Peninsula. It also forms the backbone of Manitoulin Island. The Niagara Escarpment is the only prominent landform feature of Southern Ontario in which the Paleozoic sedimentary bedrock of the area is exposed (Figure 4.18).

In the Niagara Escarpment, a layer of very hard limestone, called *dolomite*, lies over softer shales. The soft shales erode more quickly than the hard dolomite. Thus, the dolomite is undermined so that it breaks off, leaving a steep cliff (Figure 4.19).

Figure 4.19 THE RETREAT OF NIAGARA FALLS. As the soft, less resistant shale rock is eroded by the turbulent water, the hard dolomite (a very hard limestone) cap rock is undermined. In time, a piece of the dolomite breaks off and the lip of the falls retreats a few more metres. By cutting back into the escarpment in this manner, the Niagara River has carved out the Niagara gorge.

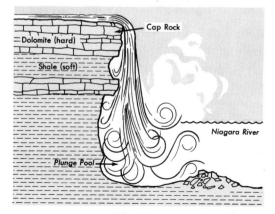

Whenever streams cut across the Niagara Escarpment, waterfalls and rapids are formed. Niagara Falls is the most spectacular example. It has cut back into the Niagara Escarpment several kilometres, creating the Niagara gorge.

West of the Niagara Escarpment is a till plain consisting of boulder clay that was smeared off the bottom of the glacier. Scattered on the till plain are moraines, drumlins, and eskers. Great valleys, called spillways, suggest that at one time torrents of glacial meltwater poured across the area (Figure 4.20).

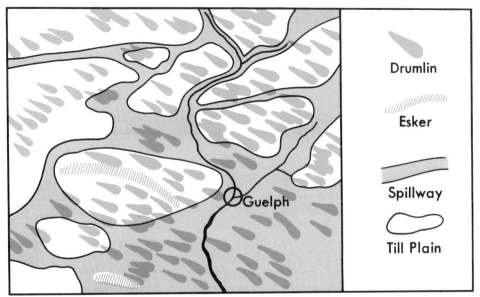

Figure 4.20 PART OF THE GUELPH DRUMLIN FIELD IN SOUTHERN ONTARIO. Drumlins are smooth, elongated hills composed largely of boulder clay. They range from 8 to 60 m in height, with the average somewhat less than 30 m. Most drumlins are between 400 and 800 m in length and they are normally several times longer than they are wide. They are usually found in groups. The spillways are valleys created by glacial meltwater. The eskers are the beds of streams that flowed within the ice of the glacier.

West of London, the rolling landscape gives way to a flat plain which was formed in the bottom of an old glacial lake. The flat land, which extends all the way to Windsor, is important for crops such as soybeans, corn, tobacco, and fruits and vegetables.

The shortest and quickest way to the Canadian West, from Windsor, is through the United States. However, in order to see more of Canada you retrace your route to north of Toronto where you pick up Highway 400 heading north.

A rock island splits the Niagara River into two channels and creates the Canadian Horseshoe Falls on the right and the American Falls on the left. Below Niagara Falls, the river continues flowing towards Lake Ontario through the Niagara gorge.

The Niagara gorge is a steep-sided valley, carved over many years by the Niagara River as it cut back into the Niagara Escarpment. On the left some of the exposed sedimentary rock layers are shown.

The Ottawa Valley with the hills of the Canadian Shield in the background.

The Grand River meandering across a till plain in Southern Ontario.

The Canadian Shield

At the town of Orillia, you return to the Trans-Canada Highway and resume your westward journey. The rocky, forest-covered hills and the road-cuts through granite knobs make you realize that you are in a different landform region.

The Canadian Shield is the largest landform region in Canada. It will take you two full days to reach its western boundary, and you will have covered only a small portion of it. Like a giant horseshoe, the Shield curves around Hudson Bay from the Arctic Ocean to Labrador. Most of the area slopes towards Hudson Bay, and so, most of the rivers drain in that direction. The Canadian Shield surrounds the Hudson Bay Lowlands region that is underlain by the same Paleozoic sedimentary bedrock as the Great Lakes-St. Lawrence Lowlands. The Labrador Trough, despite its name, is not a lowland, but is a highly mineralized area of Precambrian sedimentary, volcanic and metamorphic rocks. It extends for about 960 km south of Ungava Bay.

You are crossing the lowest and least rugged part of the Canadian Shield. Your altimeter will never register over 600 m above sea level. The elevations increase and the landscape becomes more rugged towards the east. The Torngat Mountains in Labrador rise to a height of over 1500 m.

The Canadian Shield contains beautiful scenery. Along your route, north of Georgian Bay and Lake Superior, there seem to be hills everywhere. The multicoloured Precambrian granites show up along the roadsides, on the tops of bare hills, and in the rocky river beds. Lakes of all sizes and shapes, surrounded by rocky, partly-wooded hills, dot the landscape. Thousands of rapid streams seem to connect the lakes in a helter-skelter way. Swampy areas, called *muskegs*, add variety to the landscape. The green spires of the evergreen trees, mixed with the white-barked birches, add to the beauty of the scenery.

The Canadian Shield is a land of rock, forest, rivers, muskeg swamps, and lakes.

The Canadian Shield is more than scenic. It provides a large part of the tree resources for Canada's pulp and paper industry, and it is a treasure-house of minerals. In a few areas, such as the Great Clay Belt in Northern Ontario, limited farming is carried on. The Great Clay Belt is the bottom of a glacial lake (Lake Ojibway) that existed during the latter part of the Ice Age (Figure 4.13, p. 95).

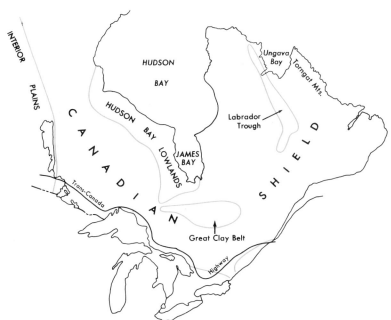

The Interior Plains

Shortly after reaching the Ontario-Manitoba boundary, the highway descends to a broad and extremely flat plain, known as the Manitoba Lowland. It is the beginning of the Interior Plains that stretch all the way to the Rocky Mountains and extend north along the Mackenzie River system to the Arctic.

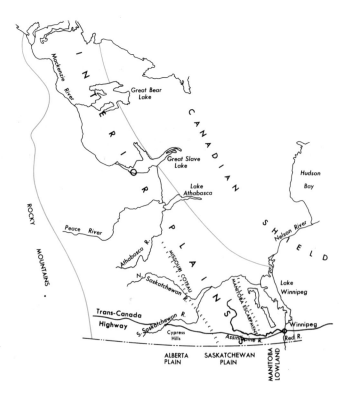

The flat Manitoba Lowland, with its rich, black soil provides a sharp contrast from the landscape of the Canadian Shield.

The Manitoba Lowland is the bottom of glacial Lake Agassiz (Figure 4.13, p. 95). Because it is perfectly flat, you can see for many kilometres. The soil is very black and rich, and there are only a few trees, many of which were planted as windbreaks. In every direction you look, you can see grain fields and towering elevators that are used to store the grain.

The Red River flows north through this area into Lake Winnipeg. The Assiniboine River joins the Red River at Winnipeg. These two rivers used to combine to cause serious floods in Winnipeg before adequate flood-control measures were taken.

Several hours west of Winnipeg, your car climbs a slope that rises about 90 m above the Manitoba Lowland. This ridge, known as the Manitoba Escarpment, marks the eastern edge of the Saskatchewan Plain. The Saskatchewan Plain extends westward to the slopes of the Missouri Coteau, a much eroded escarpment which marks the eastern edge of the Alberta Plain, sometimes called the Alberta Plateau because of its high elevation. These plains form part of the Great Plains of North America.

On the broad, flat to gently rolling Saskatchewan Plain a panorama of farmlands, with few trees and no rock outcrops, stretches before you.

This photo shows the rolling landscape of the Alberta Plateau in sight of the foothills of the Rocky Mountains.

Cypress Hills. The Cypress Hills, located on the southern border of Alberta and Saskatchewan, rise high above the surrounding plain. They stand as remnants of a former surface level, most of which has been eroded away.

Neither the Alberta nor the Saskatchewan Plain is as flat as the Manitoba Lowland. Although some parts of the plains are flat, most of the area is a rolling landscape with long, gentle slopes. Some of the rivers, such as the North and South Saskatchewan, have cut deep valleys into the surface. A dam across the South Saskatchewan River has created Lake Diefenbaker, which provides water for irrigation purposes.

At one time, the plains were higher than they are now. For millions of years, streams have been at work cutting the land down, until only the Cypress Hills remain of this higher level.

During the Ice Age, the Alberta and Saskatchewan Plains were covered with glacial ice. The hummocky, glacial drift has left many depressions that have filled with water. These ponds, called *sloughs*, are scattered among the vast grain fields, and provide excellent habitats for many species of waterfowl.

The Interior Plains are not monotonous as some people claim. To repeat what was said earlier, they are not perfectly flat. Also, there are constant changes in colour of the soil, in the nature of vegetation, and in the kinds of farming. From the black soils of the Manitoba Lowland, you come into the region of the dark brown soils of the Saskatchewan

Coulee. Some of the prairie streams dry up during the hot summer. These dry valleys are called coulees.

Sloughs. Depressions in the rolling prairie surface often become water-filled and are known as sloughs. They are important for the collecting and storage of water from rainfall as well as for valuable nesting and feeding grounds for waterfowl.

Plain. Here, huge fields of wheat, and cattle grazing on the natural grasslands are common sights. West of Regina, the grass becomes shorter, the soils are a lighter brown, and more of the farmland is devoted to cattle grazing.

Along the Red Deer River, west of the Alberta-Saskatchewan border, the badlands of Alberta are found. These badlands have been caused by extreme erosion. The gullies have been cut by streams, but the strange shapes of the landforms have been caused by wind erosion. The wind is a more effective erosion agent where there is little vegetation.

Badlands. In the badlands of south-central Alberta, the hoodoos are an interesting and spectacular sight. These landforms have been carved from sedimentary rock by water and wind. Each mushroom-shaped feature has a hard limestone rock layer on top with softer, more easily eroded shale rock below.

The Western Cordillera

From Calgary, you head northwest towards the mountain land of Canada. After going through hilly country known as the Foothills, you soon begin your climb through a band of mountains of breathtaking beauty. These mountains are the Canadian Rockies that extend from the United States border to the Yukon. The great heights and angularity of the peaks indicate that these are much younger mountains than those of the Appalachian region. Alpine glaciation is evidenced by the many horns, cols, arêtes and cirques. In some places, valley glaciers are still active. Often, the snow-capped peaks are hidden in the clouds. Some of the mountain sides are so steep that they are bare of trees. The whole landscape seems to be an irregular, rugged sea of mountain peaks.

The road twists and turns as it winds through the lowest gaps between the peaks. Your altimeter indicates that you have never reached an elevation much higher than 1500 m, but some of the peaks you have seen are at least twice that high. When you come to Kicking Horse Pass

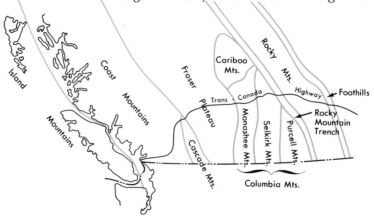

A majestic band of mountains, the Canadian Rockies, marks the western edge of the Interior Plains. The rugged beauty of these mountains has created major tourist attractions such as the one at Banff. The southern part of the Alberta-British Columbia border follows the crestline of the Rockies. The Rocky Mountains form the *Continental Divide* between rivers flowing eastward and rivers flowing toward the Pacific.

Kootenay Valley, just south of Kootenay Lake, occupies the trench between the Purcell and Selkirk Mountains. Notice how steeply the mountains rise from the flat valley floor. In an atlas, trace the course of the Kootenay River from its source to where it joins the Columbia River. Through what major mountain ranges does it cut?

The "flats" (floodplain) of the Kootenay Valley south of Kootenay Lake have been drained and diked to prevent flooding. The "flats" are used for grain growing. Some orchards are found on the terrace slopes.

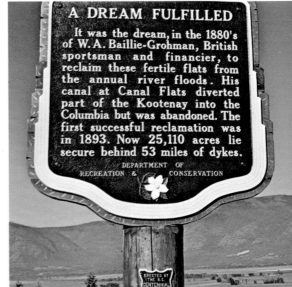

A DREAM FULFILLED

It was the dream, in the 1880's of W. A. Baillie-Grohman, British sportsman and financier, to reclaim these fertile flats from the annual river floods. His canal at Canal Flats diverted part of the Kootenay into the Columbia but was abandoned. The first successful reclamation was in 1893. Now 25,110 acres lie secure behind 53 miles of dykes.

DEPARTMENT OF
RECREATION & CONSERVATION

you have reached the Continental Divide. From here on, all streams flow towards the Pacific.

After descending about 750 m, you break out of the mountains into a broad valley known as the Rocky Mountain Trench. In this trench are found the headwaters of the Fraser and Columbia Rivers. The Rocky Mountain Trench is a remarkable valley. It runs about 1600 km north from the forty-ninth parallel, and it ranges from 3 to 16 km wide. On both sides, mountains rise steeply to dizzy heights. The Rocky Mountain Trench has tremendous potential for hydroelectric power. If dams were built, vast lakes could be created to store water, but in the process, many forest, wildlife and scenic resources would be lost.

You are not through the mountains yet. To the west is the Columbia Mountain System which comprises the Purcell, Selkirk, Monashee, and Cariboo Mountains. On the western side of both the Purcell and Selkirk Mountains, major valleys or trenches are found, similar to, but not as extensive as the Rocky Mountain Trench.

After the Columbia Mountain System come the Interior Plateaus. The southern part of this plateau area is called the Fraser Plateau. Although some of the surface is fairly level, most of this plateau has been eroded by streams into mountains and hills. A major valley to the south, the Okanagan, has become famous for its orchards. The Trans-Canada Highway crosses the Fraser Plateau by way of the Thompson Valley to where the Thompson joins the Fraser River.

The Cascade Mountains border the western edge of the Fraser Plateau. The Fraser River has cut a deep canyon between the Cascade

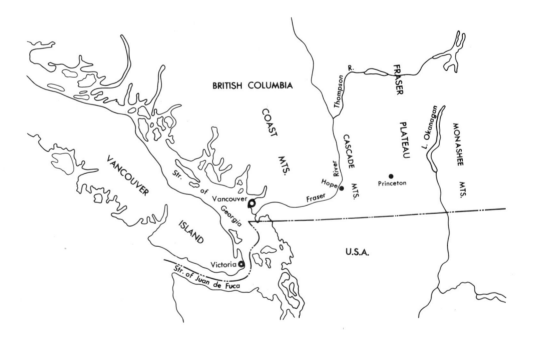

The Okanagan Valley is a major depression in the rolling upland of the Fraser Plateau (see background). Note the steep slopes of the valley and the orchard-covered terraces beside Lake Okanagan.

Near its mouth the Fraser River flows through a broad valley, and the channel is flanked by flat delta deposits. The Fraser Delta is one of the richest farming areas in Canada. Because of the mild climate and fertile soil a greater variety of crops is grown in this single region than in any other in Canada.

Mountains and the Coast Mountains. The canyon is so steep that it is difficult to understand how the highway was ever built along its edge. Sometimes, the road is perched on a narrow ledge; at other times, it passes through tunnels blasted through the rock. The fast-flowing muddy water indicates that the Fraser is still deepening its valley and carrying silt to the sea.

About 40 km north of the United States border, the highway follows the Fraser Valley as it swings westward and broadens out into a flat plain. You have now reached the Fraser Delta which has built up over many thousands of years. In contrast to the forested mountains, the Fraser Delta has been cleared of trees and is intensively farmed. It provides farm products for the cities of Vancouver and Victoria.

All along the west coast, rivers come tumbling out of the Coast Mountains and empty into arms of the sea that often penetrate many kilometres inland. These deep, steep-sided embayments, called fiords, often offer protection to boats in stormy weather.

To see the most westerly landform subregion of Canada, you must board a ferry and cross the Strait of Georgia to Vancouver Island. This island is part of the Island Mountain subregion of the Western Cordillera. A whole chain of islands forms protection for boats using the coastal waters. These mountainous islands seldom reach an elevation of more than 1500 to 1800 m. Because of the very high precipitation, particularly on the western slopes, they are covered with dense forests that provide an excellent source of lumber.

You have now completed your long journey. You have not seen all of Canada, but you have seen enough to know that your country has a wide variety of contrasting and beautiful natural landscapes.

A major rock slide on the Hope-Princeton Highway through the Cascade Mountains. This is not on the scenic route marked on your map (Figure 4.15), but is the southern route between the Okanagan Valley and Vancouver. The whole face of the mountainside slid down into the valley, burying two automobiles that were on the highway at the time. The new highway has been built on top of the rock rubble.

The Fraser and Thompson Rivers have cut deep canyons through the Western Cordillera. The upper photo shows the Fraser cutting through the Fraser Plateau on the eastern boundry of the Coast Mountains. Farther south, the Fraser River separates the Coast and Cascade Mountains. The lower photo shows the Thompson River just a few kilometres east of where it joins the right bank of the Fraser River. In some places sheds have been built to protect the rail line from falling debris and tunnels are used where slopes are too steep for the railroad.

Snow-capped and cloud-shrouded peaks of the Coast Mountains stand guard along Canada's west coast. In many places, this mountain range rises to elevations that are much higher than the awesome Rockies. Before draining into fiords along the Pacific coast, rivers have cut broad valleys through the Coast Mountains.

Here are the rugged mountains that form the backbone of Vancouver Island. Because of heavy precipitation on the west coast, a thick layer of snow blankets the mountain peaks. No deeper snow cover can be found even on the higher Coast and Rocky Mountain systems.

PROBLEMS AND PROJECTS

1. The Geological Time Scale given in this book is not drawn to scale. On a piece of Bristol board draw the Geological Time Scale using the scale of 2.5 cm representing 100 000 000 years. The more exact age of each period can be obtained from a geology book or any good encyclopedia. The 1965 Yearbook of the *World Book Encyclopedia* has an interesting series of maps showing the extent of the various seas and the nature of life in each of the periods.

2. On an outline map, divide the landform region in which you live into subregions. Before you start, consider carefully the criteria you are going to use to delimit the boundaries.

3. What impact has the physiography of Canada had on the political evolution of Canada, its settlement patterns, and economic geography?

4. Problems have resulted from people settling the country with insufficient knowledge of landform processes. What problems have resulted from the following:
   (a) clearing most of the forests in the Great Lakes-St. Lawrence Lowlands?
   (b) building on Great Lakes shorelines and on river floodplains?
   (c) cultivating the dry grasslands and planting them to wheat?
   (d) building roads and pipelines in the Arctic?

5. For a beautifully illustrated booklet on Canada's natural landscape and major land-use problems, see: D.M. Welch, *for land's sake* (Ottawa: Environment Canada, 1980).

6. If you are particularly interested in rocks, landforms, and geological history, you should look at these books available in libraries or Canadian government bookstores:
   B. Bird, *The Natural Landscapes of Canada* (Toronto: Wiley, 1980).
   R.G. Blackadar and L.E. Vincent, *Focus on Canadian Landscapes* (Ottawa: Supply and Services Canada, 1976).
   L.J. Chapman and D.F. Putnam, *The Physiography of Southern Ontario* (Toronto: University of Toronto Press, 1966).
   B. Curtis and J.A. Kraulis, *Canada from the Air* (Edmonton: Hurtig, 1981). This fascinating book contains over 14 000 photographs of Canada taken from the air.
   The Illustrated Natural History of Canada (Toronto: Natural Science of Canada, 1970). Seven separate landform regions are described.

7. If you wish to make a detailed study of sample areas within different physiographic regions, consult one of the following books. They contain reproductions of topographic maps and air photographs, as well as suggestions on map and air photo interpretation and comments on the human use of natural landscapes.
   C.L. Blair and R.I. Simpson, *The Canadian Landscape*, 2nd. ed. (Toronto: Copp Clark, 1978).
   A.R. Grime, *Topographical Maps* (Toronto: Clarke Irwin, 1976 and 1979), Sets 1 and 2.
   J. Koegler, *Canada's Changing Landscape: Air Photos Past and Present* (Toronto: Douglas Fisher, 1977).
   J. Koegler, *Canada's Modern Landscape: An Air Photo Study* (Toronto: Douglas Fisher, 1978).

If you were to glance at the sky each day for a month or more, you would soon get to know what weather to expect from the various cloud formations that you see. High clouds, such as those in the top photo, float 6000 m or more above the ground and are called cirrus clouds. They appear as thin, white, featherlike streaks across the sky. Sometimes they are in thin sheets or layers. High cirrus clouds indicate an approaching storm.

White, fleecy, cumulus clouds, as shown in the middle photo, are indicative of good weather. When cumulus clouds mushroom into giant, billowy thunderclouds, they are called cumulonimbus (*nimbus* means rain).

Nimbostratus clouds, as shown in the bottom photo, are thick, dark, shapeless masses that cover the entire sky and often come down to within a few hundred metres of the ground. They usually bring steady rain or snow.

5 WEATHER AND CLIMATE

WEATHER

The layer of air and moisture surrounding the earth is called the *atmosphere*. What happens in the atmosphere at any one time is called *weather*. *Climate* is a generalization about weather conditions over a long period of time. Scientists who study weather are called meteorologists, while those who study climate are called climatologists. Some geographers specialize in studying meteorology and climatology, and their relationships to other geographical patterns. Such geographers attempt to answer questions such as the following:

(1) How does the climate differ from one region to another, and why?
(2) What accounts for weather and climate differences over short distances?
(3) In what ways do temperature and precipitation affect natural vegetation, soils, and agricultural land uses?
(4) In what ways has human activity, such as clearing forests and building cities, affected climate?
(5) What are the economic consequences of extreme cold? of vast amounts of snow?
(6) How have people adjusted to climates of different regions?
(7) What has been done to overcome climatic discomforts and hazards?
(8) How can knowledge of weather and climate have social, economic and environmental benefits to people?

Answers to these questions show how important weather is to our daily living and how different climate is across Canada. By further study we may find ways to adjust better to the constant changes in weather and climate.

LOCAL STUDY

Meteorologists and climatologists use much complicated and expensive equipment to help them to observe accurately and record weather information. However, with very little equipment, you can record weather observations that will help greatly in describing the weather of your area.

The following checklist will help you in making and recording daily observations. The accompanying drawing shows how you can record some of your observations in diagram form (Figure 5.1). If you do not have the proper instruments, you may have to obtain some of your information from current radio weather reports.

SKY COVER

Is the sky clear or cloudy? If cloudy, what fraction of the sky is cloud covered? Are the clouds high or low? Describe the appearance of the clouds.

TEMPERATURE

Read the temperature from a thermometer located in the shade. If possible, keep a record of the high and low temperature for each day. To do this, you will need a special thermometer that records minimum and maximum temperatures.

PRECIPITATION

Is there any rain or snow falling? Is it light, moderate, or heavy? How much has fallen in a 24-hour period? If the precipitation is in the form of snow (in centimetres), divide the amount that has fallen by 10 to find the equivalent amount of rain (in millimetres).

HUMIDITY

Use a wet-and-dry bulb thermometer or an instrument called a hygrometer to measure the moisture content of the air. Is the moisture content high, medium, or low? Sometimes, the moisture content is measured in percentages called relative humidity. If the relative humidity is 100%, the air can hold no more moisture. A relative humidity of 75% to 100% is considered very humid; 50% to 75%, medium; below 50%, dry.

WIND

From what direction is the wind coming? Estimate its speed. Meteorologists have instruments that measure the exact speed of wind in kilometres per hour. You may wish to use the following categories to estimate the wind speed:

(1) Calm—no noticeable wind,
(2) Breeze—leaves of trees fluttering,
(3) Light Wind—branches of trees bending,
(4) Strong Wind—trunks of young trees swaying,
(5) Gale—violent windstorm.

ATMOSPHERIC PRESSURE

The mass of the air, or atmospheric pressure, is measured with an instrument called a barometer. Make a barometer reading several times a day and record how much pressure has changed between readings.

To see what your weather is like over a period of time, it will be necessary for you to keep records for at least a month. You should take readings at the same time every day, including weekends.

Figure 5.1 A SIMPLE RECORD OF WEATHER OBSERVATIONS.

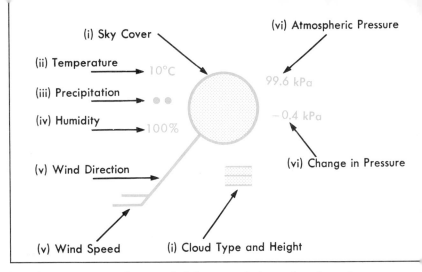

Sky The complete shading of the circle indicates that the total sky is cloudy. If the sky were only three-quarters clouded, then three-quarters of the circle would be shaded. If the cloud is low, the cloud symbol is placed below the circle; if high it is placed above the circle. You may use the following symbols to describe cloud types:

    cirrus (high, white, wispy)
    cumulus (bumpy or fleecy)
    stratus (in layers)
    fog (cloud on the ground)

Temperature Temperature is measured in the shade in degrees Celsius.

Precipitation The following symbols are useful in describing precipitation:
    drizzle   thunderstorms
    light rain   light snow
    heavy rain   heavy snow

Humidity The percentage indicates relative humidity. This is a measurement of the actual moisture in the air calculated as a percentage of how much moisture the air could hold when saturated.

Wind The wind is blowing in the direction that the arrow is pointed. The sky circle represents the head of the arrow. The wind in this example is a southwest wind because it is coming from the southwest. The speed of the wind is shown by the number of feathers on the arrow tail:

    calm (no feather or arrow)
    breeze
    light wind
    strong wind
    gale

(vi) *Atmospheric Pressure* On many barometers the atmospheric pressure is measured in inches. (In more up-to-date models, it is measured in kilopascals.) The minus sign in the above figure shows that the pressure has fallen in the last three hours. This information is very important for weather forecasting purposes. Falling pressure indicates that cloudy, stormy weather is coming: rising pressure indicates that clear weather is coming.

Note: The above symbols are much simpler than those used by official weather forecasters. If you are especially interested in weather maps and weather forecasting, write to your nearest regional office of Atmospheric Environment Services (Canada): see materials in no. 7, p. 161.

If you follow the above directions, you will be recording the weather in one place only. To describe and understand the weather patterns in your area, or across the country, it would be necessary to record observations at many different places all at the same time!—and this is, in fact, what is done.

There are about 150 official weather stations spread across Canada. At most of these stations, weather observations are made every three hours. This weather information is sent by radio or telegraph to a regional office, which then sends it on to a central office where weather maps are drawn.

A weather station has been built to take readings at Eureka, N.W.T. Some of the towers you can see support instruments that measure wind speed and direction. Other meteorological instruments are located in small, white, boxlike structures known as Stevenson Screens. The weather technicians live and work in some of the red buildings. The weather readings they take are transmitted to a central place farther south where they are compiled and put on maps.

More recently, orbitting weather satellites send data every three hours to North American offices for printing weather photographs. These photo images showing current weather patterns help meteorologists compile very detailed weather maps.

To draw a weather map, meteorologists note the weather conditions for every station, using symbols like those shown in Figure 5.1. Then, they draw lines called *isobars* to join areas that have the same atmospheric pressure, and other lines to show the division between warm air and cold air called *fronts*; then, they shade in areas where there is precipitation.

Weather *systems* like those shown in Figure 5.2 continuously move across Canada from west to east. The Low pressure systems, or storm centres, follow general tracks that tend to concentrate in the Great Lakes-St. Lawrence Lowlands and Atlantic Regions.

136

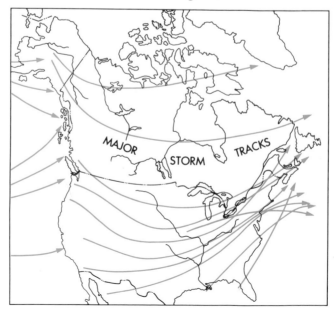

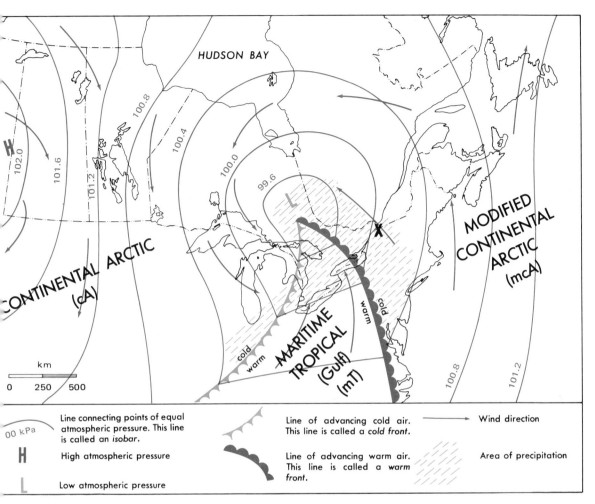

Figure 5.2 WEATHER MAP. This map is similar to ones used in newspapers and on television, which are simplifications of the complicated maps drawn by meteorologists. This map shows a typical weather system moving across eastern Canada. The Low pressure system shown here will move in a northeasterly direction roughly parallel with the St. Lawrence River. The weather conditions now shown occurring over the Prairie Provinces would probably affect Southern Ontario within a day or two. Looking at the weather patterns above, answer the following questions.

1. Describe the precipitation conditions and wind direction at Point X.

2. What will happen to the temperature at Point X when the warm front has passed over? What is the source of air behind the warm front?

3. What will happen to the temperature at Point X when the cold front passes over? Describe how the wind will change. What is the source of the air behind the cold front?

4. Describe the weather (i.e. temperature, wind, and precipitation conditions) at Point X if the High pressure area moves that far.

If you answer the questions accompanying Figure 5.2, you will understand why weather is so changeable. You will be able to do some amateur weather forecasting, and compare your weather forecasts with the official ones for your area.

Weather systems like those shown in Figure 5.2 pass continuously from west to east and are responsible for our weather from day to day and from season to season. The Low pressure systems with their accompanying fronts bring cloud, storms, precipitation, and warm, humid maritime Tropical air from the Gulf of Mexico. The High pressure systems bring cold, clear, dry air—called continental Arctic air—from the north. As it moves southward, the cold air is warmed slightly; then, it is called modified continental Arctic air. The cold air is also warmed as it passes over large bodies of water.

In winter, most of Canada is under the influence of cold continental Arctic air much of the time. In summer maritime Tropical air has a much greater effect since the warm, moist air penetrates far into the north. In the fall, cold surges of Arctic air invade southern Canada. However, the continental Arctic air does not dominate our weather until after an "Indian Summer" that results from a lingering, tropical air mass. In the spring, warm air masses from the Gulf of Mexico and the Pacific Ocean begin to push into Canada regularly, but the Arctic air does not give up its hold on the country without a fight. It is not uncommon to have frost and a snowstorm in southern Canada after several weeks of warm, balmy, summerlike weather.

In much of Canada, precipitation results from the mixing of cold Arctic air with warm tropical air along the fronts in a Low pressure system. Precipitation results from the warm, moist air being forced up over the cold air. As the air rises, it cools and moisture condenses into cloud droplets. When these droplets become too heavy to float, they drop in the form of rain or snow. This is called *frontal* precipitation (Figure 5.3).

The large number of Low pressure systems that pass through the southern part of eastern Canada partly explain why that part of the country has a humid climate. The northern parts of Canada and the Prairie Provinces are affected by fewer Lows and thus receive less of this kind of precipitation.

British Columbia receives precipitation from Lows that move inland from the Pacific Ocean. However, the high precipitation along the west slopes of mountains results from moist Pacific air being forced up over the mountain ranges. This is known as *orographic* precipitation (Figure 5.4).

Many parts of the country receive considerable summer precipitation from thundershowers resulting from air that rises because of heating by the sun (*convectional* precipitation). After a period of hot weather, the air rises to great heights; it cools, forms clouds, and then the moisture falls as rain. The rain from thunderstorms is very unpredictable. One locality

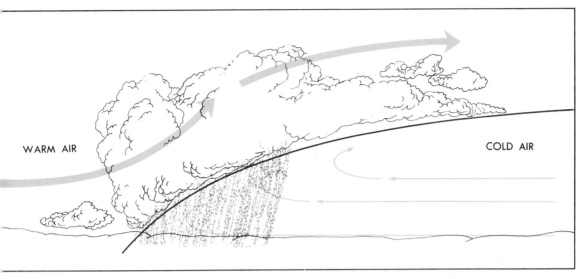

Figure 5.3 FRONTAL PRECIPITATION. The dividing line between a warm and a cold air mass is called a front. When these air masses meet, the warm air is forced up over the cold air. As it rises, the warm air cools, moisture in the air condenses, and clouds and precipitation result. Precipitation usually occurs along both cold and warm fronts (see Figure 5.2).

Figure 5.4 OROGRAPHIC PRECIPITATION. As the moist air from the Pacific rises over the Coast Mountains, it cools, causing clouds and precipitation on the western slopes. As the air descends the eastern slopes it warms up again and becomes a drying wind. The dry eastern slopes and the plateaus in the interior of British Columbia are said to be in the rainshadow. As the air is forced up over the Columbia and Rocky Mountains rain again falls on the western slopes while the eastern slopes receive very little rain. Explain why the western slopes of the Coast Mountains receive more precipitation than the west slopes of the Columbia and Rocky Mountains.

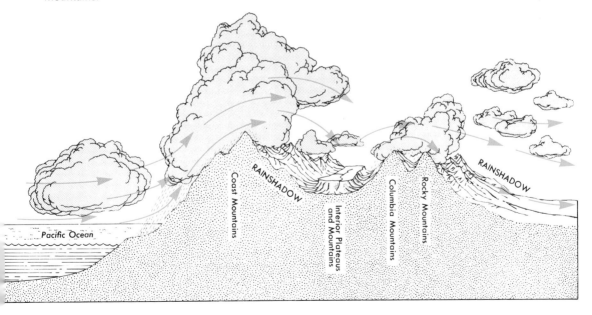

Figure 5.5 CONVECTIONAL PRECIPITATION. Violent thunderstorm activity can result from the heating of the earth's surface that causes air to rise rapidly. As the air rises, it cools and moisture condenses into clouds and precipitation. Thunderstorms are usually accompanied by heavy rain, gusts of wind, lightning and thunder.

may receive as much as 75 mm of rain while a place 16 km away may not get any rain (Figure 5.5).

CLIMATE

Earlier in this chapter, we said that climate is a generalization about weather over a long period of time. If you wanted to describe the climate of your area, you would have to keep weather records for each day over a number of years. One year would not be long enough because it might not be a typical year.

To help you to appreciate how much work goes into the keeping of climate records and the working out of climate averages, try the following project. Since you probably do not have the proper instruments for recording temperatures and for measuring precipitation, you will have to obtain your statistics from the daily newspaper, or a radio or television weather broadcast.

PROJECT

For one month, keep a record of the high and low temperatures for each day and the amounts of precipitation. Compute the daily average temperature (a) by adding together the high and low temperatures for each day, and (b) by dividing this sum by two. Calculate the average temperature for the month (a) by finding the sum of all the daily averages, and (b) by dividing this figure by the number of days in the month.

Calculate the total precipitation for the month by adding the precipitation records for all the days of the month. (Do not forget to convert snowfall readings to rainfall: 1 cm of snow is equivalent to 1 mm of rain.)

Temperature

The following table gives sample high and low temperatures for each day in January and July at Toronto.

Sample January Temperatures at Toronto, Ontario

	2	3	4	5	6	7	8	9	10	11	12	13	14	15	16	17	18	19	20	21	22	23	24	25	26	27	28	29	30	31
Temperature (°C)	4	3	7	0	-4	-5	-6	-1	1	0	-9	-2	1	-2	-10	-7	-16	-12	-3	1	2	1	1	3	3	-12	-10	-6	-6	-13
Temperature (°C)	-9	0	0	-4	-9	-12	-14	-9	-6	-1	-16	-15	-10	-10	-18	-16	-19	-20	-15	-6	-3	-4	-3	-2	-11	-18	-17	-12	-12	-18

Sample July Temperatures at Toronto, Ontario

	2	3	4	5	6	7	8	9	10	11	12	13	14	15	16	17	18	19	20	21	22	23	24	25	26	27	28	29	30	31
Temperature (°C)	24	24	28	28	30	32	30	30	28	25	25	30	25	28	29	24	25	20	23	26	24	30	22	25	25	22	25	24	21	26
Temperature (°C)	16	14	14	20	19	20	21	20	16	18	16	18	16	14	16	16	12	16	12	14	17	18	20	18	20	12	16	16	16	18

QUESTIONS

From the above table, answer the following problems for each month:

1. Calculate the average temperature for each day and the average temperature for the month.

2. What was the greatest range in temperature (that is, the difference between the high and low temperature) for any one day? What was the average for that day?

3. How much higher than the average for the month did the temperature go? How much lower than the average for the month did the temperature go?

4. How many days had the same average temperature as the average for the whole month?

By answering these questions you will have noticed that averages hide details. A monthly average temperature of 20°C would seem to indicate a very comfortable average temperature. In fact, though, the temperature may be at 20°C for only a small part of the month. In one day, the temperature may rise from a chilly 10°C to a hot and humid 30°C and may remain there during most of the day.

The average temperatures for a month do help to compare the weather of one month with that of another. At Toronto, for instance, it is obvious that January is considerably colder than July. Monthly averages also help to compare the weather of one place with that of another.

Figure 5.6 AVERAGE JANUARY TEMPERATURES. The lines on this map are called isotherms and join places with the same average January temperature. What other places in Canada have the same average January temperature as the place where you live?

142

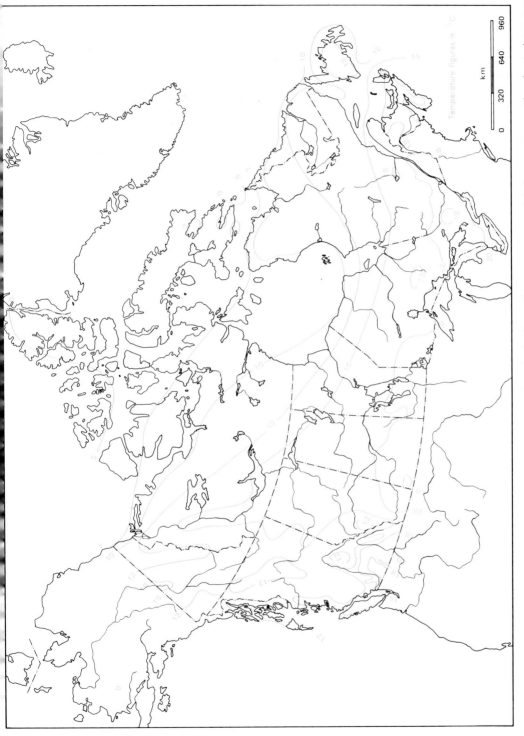

Figure 5.7 AVERAGE JULY TEMPERATURES. Find other places in Canada that have the same average July temperature as the place where you live.

143

However, to compare one place with another, it is necessary to take an average of the average temperatures over a period of years. For example, the long-run average July temperature at Toronto is 22°C. In some years, it could go as high as 25°C, while in others the average July temperature may slip to just below 20°C.

Figures 5.6 and 5.7 show how one can compare average monthly temperatures using special types of maps. In making these maps, the long-run temperature averages were plotted at a large number of stations. Then, lines called *isotherms* were drawn to connect points with equal temperatures. Isotherms show areas of low and high temperature just as contours show areas of low and high land. Even if a place is in between two isotherms you can estimate its January or July average temperature. For example, if a place is halfway between the −15°C isotherm, and the −12°C isotherm, its temperature would be −13.5°C. Because January is usually the coldest month and July the warmest, these maps are good indicators of winter and summer temperatures across the country. Notice that in the interior provinces (Alberta through to the western half of Quebec), the winters become colder towards the north. The moderating effect of large bodies of water is shown along all three ocean coasts. The water in the ocean is warmer than the land in winter, and therefore the coastal areas are milder in winter — particularly along the coast of British Columbia, because the westerly winds blow across the warm Alaska Current before blowing over the land. On the east coast, the Labrador Current is cold, and the winds blow more often from land to water. As a result, the east coast winters are not as mild as those of the west coast. The north coast is not greatly affected by the ocean because the water is very cold and in winter is filled with pack ice. Nevertheless, in Figure 5.6, the −34°C isotherm in the central Territories port shows that temperatures are lower inland rather than along the nearby Arctic and Hudson Bay shorelines.

QUESTIONS

By studying Figure 5.6 answer the following questions about winter climate in Canada:

1. What is the average January temperature in °C where you live? How many degrees higher or lower is this figure than the average for each of the provincial capitals?

2. Give the latitude and the approximate January average temperature for the following places: Whitehorse, Saskatoon, Winnipeg, Chibougamau, and Goose Bay. Is it true that the weather is always colder farther north?

3. Explain why the isotherms are packed so closely together along the west coast. (Your explanation will require your consulting a map of landforms or relief.)

4. Which side of Hudson Bay has the coldest winters? Why?

Large bodies of water, such as Hudson Bay, also modify summer temperatures (Figure 5.7). The reason is that land heats up much more quickly than water, and so, in summer, the surrounding water is cooler than the land. The cooling effect of large bodies of water on the land is greater when most of the winds are onshore. The prevailing westerly winds help to explain why Victoria has cooler summer temperatures than Halifax.

QUESTIONS

Examine Figure 5.7, and answer the following questions:

1. Where are the two warmest places in July?

2. Give the latitude and the average July temperature of Calgary, Dawson Creek, Norman Wells, Moosonee, and Sept Îles. Is it always true that one can get away from summer heat by going north?

3. Explain why the 10°C isotherm loops south around Hudson Bay.

In looking at these maps, we must always remember that they show average temperatures. The average January temperature at Toronto is −4°C. However, in January the temperature at Toronto may drop to as low as −26°C or rise to as high as 10°C during a January thaw.

The effects of local relief on temperature are not well shown in Figures 5.6 and 5.7. The sun's rays do not heat the air as they pass through the atmosphere; instead, the rays heat the ground which in turn heats the air. Because the air is heated by the ground, the lower layers of air are usually warmer than the upper layers. The temperature decreases on the average at a rate of 6.4°C per kilometre of elevation. For example, the temperature at 3000 m in the Rocky Mountains would be about 19°C lower than at sea level. This fact explains why the highest mountains are capped with snow even in summer.

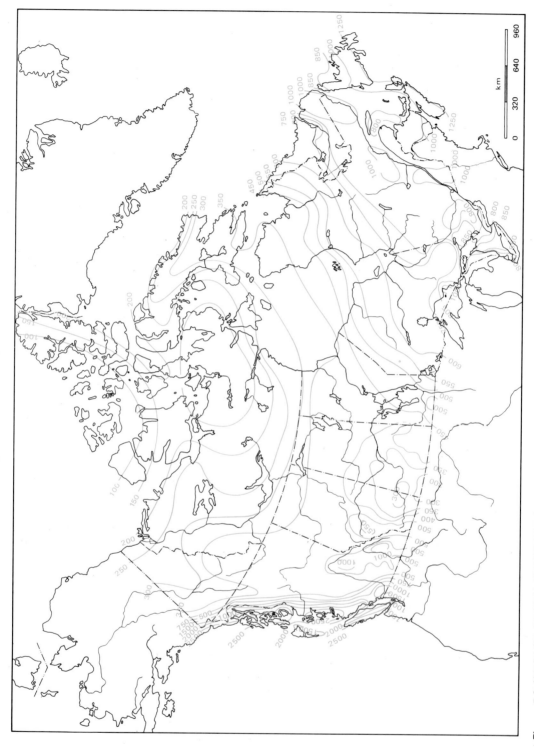

Figure 5.8 AVERAGE ANNUAL PRECIPITATION. The lines connecting points with equal precipitation are called isohyets. Find other places in Canada with the same average annual precipitation (in millimetres) as the place where you live.

Precipitation

Precipitation is totalled for each month, and then, long-run averages are worked out for every station. These monthly averages of total precipitation are often plotted on a graph as in Figure 5.9.

As with temperatures, we must remember what average means. Although the average July precipitation for Toronto is 80 mm, in some years, Toronto gets only a trace of rain in July, while in other years, it gets 100 or 125 mm.

Precipitation is sometimes shown on maps by month or by season. The most common map of precipitation is one that shows average annual precipitation as in Figure 5.8. The lines on this map connecting points of equal precipitation are called *isohyets*.

The most humid part of Canada is along the west coast. Here, the warm, moist air is forced up over the Island and Coast Mountains. As the air rises, it cools and precipitation results on the western slopes. As the air descends the eastern slopes, it is warmed and picks up moisture from the land. Thus, the dry, eastern slopes are said to be in the *rainshadow*.

The most drying wind in Canada is called the *Chinook*. It is a warm wind that comes down the eastern slopes of the Rocky Mountains and spreads out across the Prairies. Sometimes, in the winter and early spring, the Chinook actually makes the snow disappear without leaving any water. In winter, this wind helps cattle to get at the grass. However, in spring, the Chinook does damage by taking up the moisture that is badly needed for crops.

The southern part of the Prairie Provinces is not only in the rain shadow, it also receives less precipitation from Low pressure systems. The centre of a continent is usually dry because it is farthest away from the oceans, which are the major sources of moisture. Thus, it is not surprising to find that one of the driest parts of Canada is the southwestern corner of Saskatchewan and the adjoining corner of Alberta.

It may come as a surprise to you that the Arctic region has little precipitation. The continental Arctic air is cold and dry, and moist Pacific or Gulf air seldom reaches that far. The storm centres (Low maritime pressure systems) that bring precipitation to much of eastern Canada do not often cross the northern parts of Canada.

The northern regions of Canada are not as dry as they appear at first. Because the climate is cooler, there is less evaporation, and so, the precipitation is more effective.

QUESTIONS

Answer the following questions by consulting Figure 5.8:

1. Why are the isohyets on the west coast packed so closely together?

2. By consulting a large-scale relief map of British Columbia, explain why the west coast of Vancouver Island has 2000 mm of rainfall; Victoria has less than 750 mm, and Penticton, in the Okanagan Valley, has only about 250 mm.

3. Give the latitude and the approximate annual precipitation for Ottawa, Moosonee, Churchill, and Chesterfield Inlet; for Thunder Bay, Winnipeg, and Regina; for Toronto, Charlottetown, and Halifax. Make a general statement about the amount of precipitation as you move (a) north, (b) west, and (c) east from Ontario. What would you expect if you moved south?

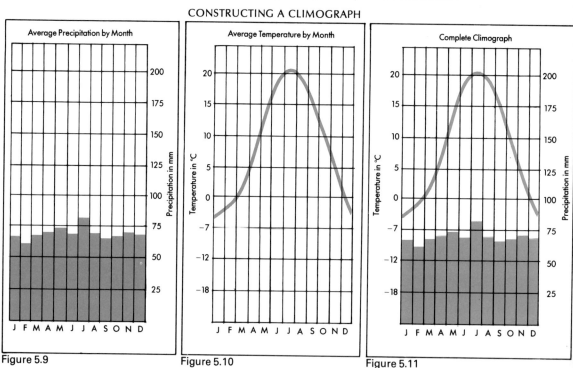

CONSTRUCTING A CLIMOGRAPH

Figure 5.9   Figure 5.10   Figure 5.11

Climographs

Figure 5.9, a bar graph, shows the average monthly precipitation for Toronto. On a similar graph, we can place dots representing the average temperature for each month. These dots can be joined by a line to form a graph like that shown in Figure 5.10. By combining Figures 5.9 and 5.10, we obtain a *climograph* that shows the average monthly temperature and the average monthly precipitation as in Figure 5.11. The temperature scale is placed on the left side of the graph; the precipitation scale on the right.

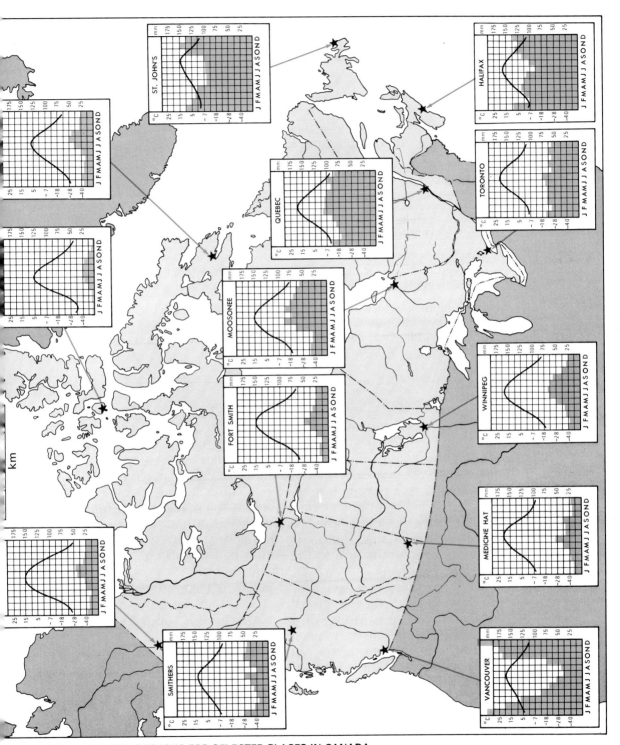

Figure 5.12 CLIMOGRAPHS FOR SELECTED PLACES IN CANADA.

Climographs provide a useful summary of the climate in different places (Figure 5.12). Your answers to the following questions will illustrate this point.

QUESTIONS

1. Compare the annual precipitation of Smithers and Winnipeg. Which place receives the most rainfall in the growing months from May to September?
2. Because of its high annual rainfall, Vancouver is considered to have a very wet climate. In which city would you most likely need an umbrella for July and August: Vancouver, Winnipeg, or Toronto?
3. Which place in Canada has the greatest difference in temperature between the coldest month and the warmest month?
4. Which station shows the fewest months with an average temperature above freezing? Name the station showing no months with an average temperature below freezing. Does this place never have temperatures below 0°C?
5. The months in which the average temperature is below 0°C are most likely to have precipitation occurring in the form of snow. In what months are the people most likely to get snow in Winnipeg? in Toronto? in Halifax?

CLIMATIC REGIONS

Canada has many different kinds of climate. One can find climates in Canada that are similar to those of England, Poland, Finland, and the interiors of Turkey, Romania, and the Antarctic. Locate these places in your atlas to give you an idea of the broad range of Canadian climate.

Every place in Canada has its own peculiar climate. Geographers try to group places with similar climates into climatic regions. If you look at different textbooks and atlases, you will discover that there are different ways of dividing Canada into climatic regions. There is no one "right" way. From the information in the earlier part of this chapter, you may want to try your hand at drawing climatic regions.

One set of climatic regions is given in Figure 5.13. Describe the climate in each of these regions by consulting the other maps in this chapter. These comparisons will be made easier if you trace Figure 5.13 onto a sheet of onionskin paper and then lay it over the other maps. In this way, for each region, you will be able to give details of January and July average temperatures, and annual precipitation. The climatic charts will show you how the precipitation and temperature change from month to month. Additional climatic data is provided at the end of this chapter. Climatic statistics and averages, however, do not provide an accurate impression of what the climate is really like in the various regions, or particularly of how the climate feels to the residents.

In the Arctic and Cold Continental climatic regions, winter temperatures may plunge to −40°C, but it may not feel that cold because the air is dry. However, if there is a wind, the wind-chill factor can make the weather seem colder than the temperature indicates. For example, a temperature of −15°C with a wind of 40 km/h has the same cooling effect on the skin as a temperature of −40°C with no wind. Because of a high incidence of strong winds in the Prairies, the wind chill is often greater there than farther north. Thus Winnipeg, Manitoba, has a higher average wind-chill index than Dawson in the Yukon.

Wind and moisture also greatly affect how hot the summer weather feels. A 25°C temperature in Southern Ontario with a relative humidity of 90 may seem much hotter than a 30°C temperature in the Okanagan Valley with a relative humidity of only 50. In comparing areas which have the same temperature and humidity, hot weather is more tolerable where there is a breeze rather than still air.

In addition to the regional differences in climate, there are differences in the way people perceive and adjust to weather and climate. In the Lower Lakes region the weather is considered cold if the temperature falls to near −18°C, whereas in the Cold Continental region it is not considered cold until the temperature falls below −32°C. Even within the same region, some people enjoy snappy, cold weather; for others, hot humid weather is their cup of tea.

Canadians in the most densely settled part of Canada have pretty well conquered the weather of winter. Central heating in buildings permits people to live, work, and play in comfort, even when dressed in relatively light clothing. Heated automobiles, buses, trains, and planes make it possible to move from place to place without really experiencing the cold temperature. In fact, most buildings and transportation facilities are kept too hot and the air is usually too dry to be healthy. About the only times many urbanites are exposed to winter weather is when they pursue some outdoor recreational activity.

It is ironic that in a country generally considered to have a cold climate, more people suffer from summer heat than from winter cold. In the cities in the Great Lakes–St. Lawrence Lowlands, when the temperature reaches about 30°C and the humidity is high, summer weather is almost unbearable. Offices are often air-conditioned, but factories and most houses and apartments are not. To escape the heat on weekends, city people must travel for several hours in bumper-to-bumper traffic to reach their cottage, a campsite, or a swimming area with clean water.

APPLIED CLIMATOLOGY

A detailed knowledge of weather and climate, and how it differs from place to place is extremely valuable to society. The most obvious application is weather forecasting. Relatively accurate forecasts are important to

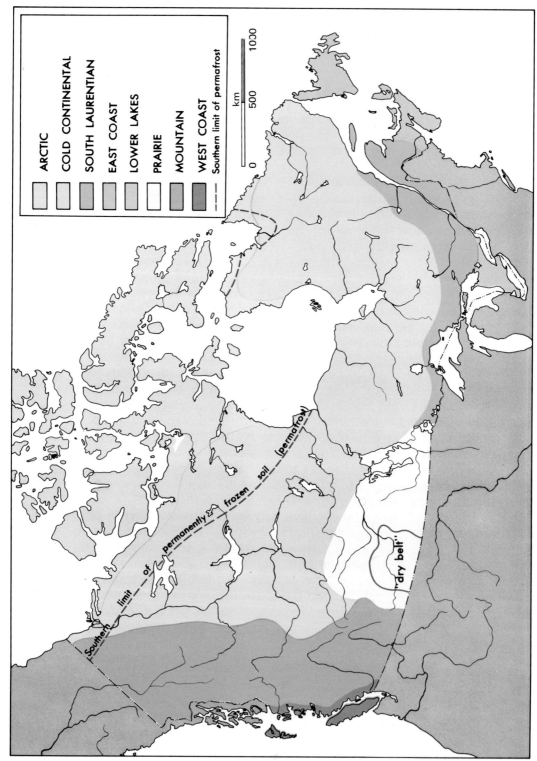

Figure 5.13 CLIMATIC REGIONS OF CANADA. This map shows one way of dividing Canada into climatic regions. Compare this map with climatic region maps shown in other textbooks and atlases.

seamen, air pilots, fishermen, farmers, and outdoor recreationists.

Since many plants are killed when the temperature falls below freezing, information on the average last frost in spring and first frost of fall is important to farmers. They must plant certain tender crops such as corn, soybeans, and vegetables so that they emerge from the ground after most risk of frost is past. However, if some crops are planted too late, they do not ripen before the first frost of fall. The length of the frost-free season gives an indication of the relative risk of frost damage to crops in different parts of Canada (Figure 5.14).

QUESTIONS

1. Which place in Canada has the shortest frost-free season? the longest frost-free season?
2. Why do the isolines in Southern Ontario parallel the shorelines of the Great Lakes?
3. It takes about 100 days free from frost to mature most farm crops. Draw a sketch map showing where the frost-free season is long enough for general farming.

The concept of *degree-days* is even more useful than frost-free days to determine suitability for growing specific crops. Most Canadian crops achieve significant growth only when the average daily temperature is above 5.5°C. The higher and the longer the temperature is above 5.5°C, the greater the growth. If the average daily temperature for one day is 10°C, that constitutes 4.5 degree-days. If the average daily temperature is 10°C for 10 days, that constitutes 45 degree-days. The index of degree-days for growing purposes for a particular place can be calculated by subtracting 5.5°C from the average daily temperature of every day within the growing season, and then adding these differences together.

QUESTIONS

Answer the following questions by referring to Figure 5.15.

1. What place in Canada has the highest index of degree-days above 5.5°C?
2. If crops such as grain corn, tobacco, soybeans and field beans require approximately 2000 degree-days, in which areas of Canada would you expect to find these crops?
3. A minimum of approximately 1100 degree-days is required to mature most farm crops. Draw a sketch map showing where there are enough degree-days for general farming. Compare this map with your sketch map showing areas with 100 frost-free days.

The degree-day concept is also used by the fuel supply industry. It is assumed that when the temperature drops below 18°C heating is

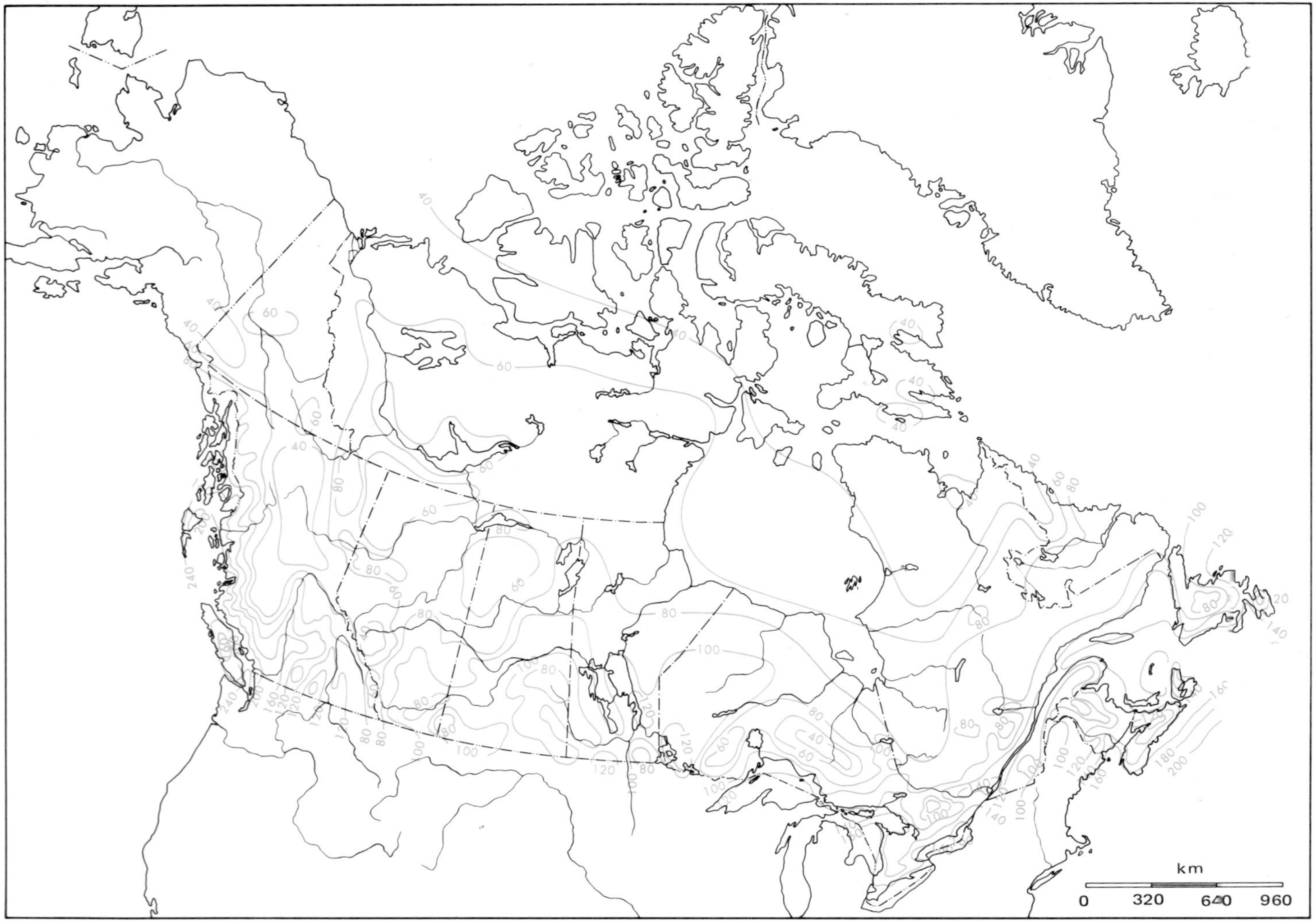

Figure 5.14 AVERAGE ANNUAL FROST-FREE SEASON. The lines on this map join places with the same number of frost-free days.

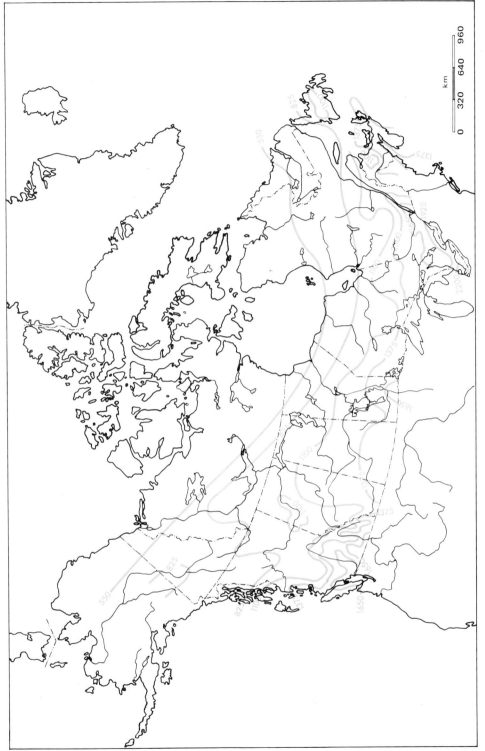

Figure 5.15 AVERAGE ANNUAL DEGREE-DAYS ABOVE 5.5°C. The lines on this map join places with the same number of degree-days above 5.5°C.

required in homes. The amount of heating required is dependent upon the number of degree-days below 18°C. Since the fuel supply industry knows approximately how much fuel is consumed by a household for a given number of degree-days, it can estimate reasonably accurately when fuel oil tanks need to be filled.

Climatologists have devised methods to predict the amount of water runoff or the moisture deficiency in different parts of the country. The method is simple in principle but very complex to work out in practice. The total moisture supply for a place is determined by the amount of precipitation. The amount of moisture needed by the environment is determined by the amount of *evaporation* from the soil and *transpiration* from plants. The term for these two processes combined is *evapotranspiration*. The total amount of evapotranspiration that would occur if the moisture were available is called *potential evapotranspiration*. When the precipitation is greater than the potential evapotranspiration, there is a water surplus or runoff. Huge volumes of water runoff pose flooding problems. When the precipitation is less than the evapotranspiration, moisture is drawn from the soil until there is a water deficit or drought (Figure 5.16). Any moisture deficiency slows the growth of crops, but with a severe water deficit, most vegetation will wilt and die. The average annual water deficit for the Prairie Provinces is shown in Figure 5.17. Even though there is an annual water deficit in most of this region, there is a small amount of water surplus in the winter and early spring when the evapotranspiration is zero or very low. Very detailed studies of water surplus and deficit are required before major flood control or irrigation projects are constructed.

Climatic and meteorological data can also be used to improve the quality of our environment. Information about wind direction and velocity can be used to improve urban design so that people are less exposed to cold winter winds. Industrial parks can be located so that the prevailing winds carry pollutants away from the residential districts of a city.

Atmospheric conditions in some parts of the country are more susceptible to air pollution than others. In areas where the atmosphere tends to be very stable (very little vertical movement of air) and the wind speeds are low, the susceptibility for pollution is high. One geographer has mapped broad zones of air pollution susceptibility in Canada (Figures 5.18 and 5.19). The zones change from winter to summer. It is worth noting that the Arctic has high air pollution susceptibility in both summer and winter. This is another environmental factor that must be considered as Canada contemplates developing the Arctic. Much more of this kind of research is required so that we can plan the location of and determine the pollution controls required for new industrial and urban development. Only if we have adequate knowledge can we protect our fragile atmosphere which is often neglected in environmental management programs.

Figure 5.16 AVERAGE WATER BALANCE, REGINA. The precipitation line shows the amount of water supply. The potential evapotranspiration line shows how much moisture would be used if it were available. After the moisture stored in the soil is used up, there is a water deficit at Regina equal to approximately 175 mm of rainfall (see Figure 5.17).

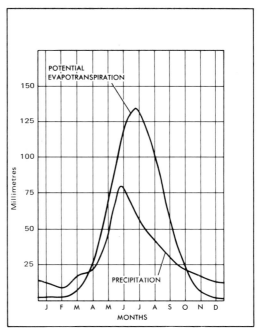

Figure 5.17 AVERAGE ANNUAL WATER DEFICIT IN THE PRAIRIE PROVINCES IN MILLIMETRES. The lines on the map join places with the same amount of average annual water deficit. From similar but more detailed information, the amount of irrigation water needed for certain crops can be calculated.

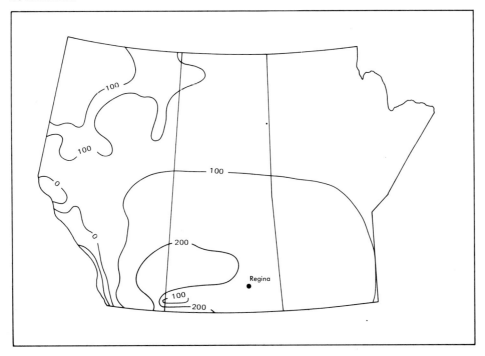

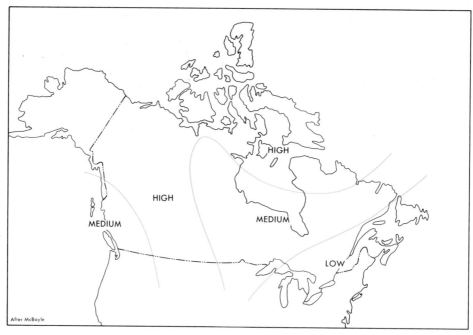

Figure 5.18 AIR POLLUTION SUSCEPTIBILITY IN WINTER.

The degree of air pollution is greatly dependent on weather conditions. Toronto City Hall (left) on a clear day with a brisk wind and (right) on a cloudy, humid day with no wind. In which season of the year from Figures 5.18 and 5.19 would the weather conditions shown in the photographs most likely occur?

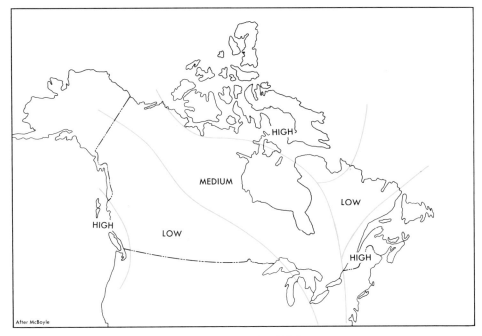

Figure 5.19 AIR POLLUTION SUSCEPTIBILITY IN SUMMER.

THE IMPORTANCE OF CLIMATE TO CANADIANS

The importance of weather and climate to Canadians is suggested by the answers to the following questions.

QUESTIONS

1. Name two popular games or sports activities for each season. Why are they most popular in that season?
2. How do our meals differ from summer to winter? How did the pioneers overcome the lack of fresh fruit and vegetables in winter? How does your family overcome this problem?
3. Name some crops that Canada grows very well because of its climate. Name one crop that is sometimes sown in the fall. Why is sufficient snow necessary for this crop? What other crops benefit from sufficient snow cover?
4. For how many months in the year do farmers in your area keep their cattle in a barn and feed them hay, grain, and other fodder? Find out what this does to the cost of producing beef and dairy products.
5. In what way is the cost of building houses increased in an area that has low winter temperatures?

Cold winters greatly affect transportation in Canada. In this photo, you see the icebreaker *Montcalm* working its way through the ice on the St. Lawrence River. In most winters, ships can reach as far inland as Montreal. Above Montreal, however, the ice on the St. Lawrence is too thick to be cleared by icebreakers.

6. How much does it cost to heat a three-bedroom house in your community for one year?
7. Name, if any, some of the air-conditioned buildings in your community. What would it cost to install an air-conditioning system in your house?
8. Is there serious unemployment in your community in the winter season? If so, why?
9. What is the effect of climate on Canadian transportation costs? (Consider the effects of snow on automobile travel; ice on boat travel; storms and fog on air travel.)
10. In what way does climate affect Canadian natural resources such as forests and wildlife?
11. What are some of the methods used to solve the problems resulting from extremes of weather and climate such as spring frost damage to orchards, floods, drought, freezing of water pipes, and construction problems in areas of permafrost?

In all climatic regions except the West Coast, snow creates great problems for transportation. This photograph shows a snowplow clearing snow on a heavily drifted railroad. Snow removal from highways and city streets costs many millions of dollars each winter.

PROBLEMS AND PROJECTS

1. It is not quite true that no one does anything about the weather, as Mark Twain observed. In fact, there has been considerable experimenting with artificial rainmaking devices. You will find this a very interesting research topic for your class.

    If people could control the weather, what problems as well as economic benefits do you think would result?

2. There are meteorological stations all across Canada. If possible, arrange for your class to visit one of these stations to observe how weather data is collected and analysed.

3. Check through old newspapers to see how often the weather has been a hazard in your area. What is the major hazard of your area? Drought? snowstorms? floods? windstorms? lightning? hail? From the newspaper accounts, can you obtain any indication of the economic costs of these hazards? What measures have been taken to reduce the damage resulting from these hazards?

4. In this chapter only a few climatic factors have been mapped. For many other maps showing different climatic data, see *The National Atlas of Canada* (Ottawa: Energy, Mines and Resources, Canada, 1974).

5. Use the following climatic data to construct climographs. Locate each station on a map of Canada and compare your climographs with those in Figure 5.12. Note that there are differences in temperature and precipitation within the same region.

6. Nitric and sulphuric oxides, given off by industry and coal-burning electric plants, unite with moisture in the air to create what has been called *acid rain*. Explain why Ontario and Quebec suffer from acid rain generated in the United States. Why are lakes in the Canadian Shield more susceptible to poisoning with acid rain than those in the Great Lakes-St. Lawrence Lowlands? Why has a shortage of petroleum increased the severity of the acid rain problem?

7. For some interesting reading on weather and climate and their effects on landscape, consult these references:

    Atmospheric Environment Services Canada: *Knowing the Clouds*; *Knowing Weather (facts and myths)*; *Making the Most of the Forecast*; and *Mapping Weather* (Toronto: Environment Canada, 1981). These four booklets are free and available from your nearest regional centre.

    Canada Forestry Service, *Ecotours* (Ottawa: Environment Canada). These pamphlets on landscape ecology along the Trans-Canada Highway are free and available from Environment Canada.

    Chinook (Brampton: Weather Enterprises). This is an interesting quarterly magazine.

    F.K. Hare and M.K. Thomas, *Climate Canada* (Toronto: Wiley, 1976).

    W.G. Pierce, *Weather and Climate* (Toronto: Holt, Rinehart and Winston, 1976).

Climatic Data

		Jan.	Feb.	Mar.	Apr.	May	June	July	Aug.	Sept.	Oct.	Nov.	Dec.	Yearly
(A) EAST COAST														
Gander, Nfld.	(temp.)	−6.1	−6.3	−3.6	0.8	6.3	11.4	16.5	15.8	11.8	6.3	1.9	−3.4	4.3°C
	(precip.)	94.0	100.8	96.8	85.1	62.5	76.2	77.7	100.8	84.1	95.3	106.9	98.0	1078.2 mm
Saint John, N.B.	(temp.)	−6.5	−5.7	−1.2	4.3	9.5	13.6	16.7	16.8	13.8	9.1	3.7	−3.3	5.9°C
	(precip.)	125.7	114.1	98.3	99.8	102.9	93.7	89.7	99.8	99.8	104.7	145.3	131.8	1305.6 mm
(B) ARCTIC														
Alert, N.W.T.	(temp.)	−32.1	−33.3	−33.0	−24.7	−11.2	−0.6	3.9	0.9	−10.1	−19.7	−26.1	−29.8	−18.0°C
	(precip.)	7.6	5.3	7.1	6.6	10.7	13.5	18.0	27.4	27.9	15.8	8.1	8.1	156.1 mm
Inuvik, N.W.T.	(temp.)	−29.3	−29.4	−23.8	−14.6	−0.8	9.8	13.3	10.3	2.7	−7.2	−20.6	−27.1	−9.7°C
	(precip.)	20.3	10.4	16.5	14.0	17.5	13.0	34.3	46.2	21.1	33.8	14.7	18.5	260.3 mm
(C) COLD CONTINENTAL														
Earlton, Ont.	(temp.)	−16.1	−13.8	−7.5	1.9	9.3	15.2	17.7	16.2	11.4	5.9	−2.2	−11.9	2.2°C
	(precip.)	53.6	45.0	44.5	45.0	63.8	92.0	80.0	82.8	96.8	64.3	69.9	52.8	790.5 mm
Yellowknife, N.W.T.	(temp.)	−28.6	−25.7	−18.6	−7.8	4.0	12.2	16.0	14.1	6.8	−1.2	−14.2	−23.8	−5.6°C
	(precip.)	13.7	12.2	11.7	10.2	14.0	17.3	33.3	36.3	28.2	30.7	23.9	18.5	250.0 mm
(D) SOUTH LAURENTIAN														
Ottawa, Ont.	(temp.)	−10.9	−9.5	−3.1	5.6	12.4	18.2	20.7	19.3	14.6	8.7	1.4	−7.7	5.8°C
	(precip.)	59.9	56.9	61.0	67.6	70.1	72.6	81.3	81.5	78.7	65.8	78.5	77.0	850.9 mm

		Jan	Feb	Mar	Apr	May	Jun	Jul	Aug	Sep	Oct	Nov	Dec	Year
(E) LOWER LAKES London, Ont.	(temp.)	−6.0	−5.6	−0.7	6.6	12.3	18.2	20.5	19.7	15.7	9.9	3.1	−3.6	7.5°C
	(precip.)	76.2	65.3	71.9	78.2	74.9	81.0	81.3	73.4	78.7	74.2	82.8	86.6	924.5 mm
(F) PRAIRIE Saskatoon, Sask.	(temp.)	−18.7	−15.1	−8.7	3.3	10.6	15.4	18.8	17.4	11.3	5.0	−5.8	−14.0	1.6°C
	(precip.)	18.3	18.0	16.8	20.6	34.0	57.4	53.1	45.2	33.0	19.1	18.8	18.3	352.6 mm
Calgary, Alta.	(temp.)	−10.9	−7.4	−4.3	3.3	9.3	13.2	16.5	15.2	10.7	5.7	−2.6	−7.6	3.4°C
	(precip.)	17.0	19.8	20.3	29.5	49.8	91.7	68.3	55.9	35.3	18.8	16.0	14.7	437.1 mm
(G) MOUNTAIN Kamloops, B.C.	(temp.)	−6.0	−1.3	3.6	9.3	14.3	18.0	20.9	19.7	15.0	8.4	1.7	−2.6	8.4°C
	(precip.)	28.7	15.5	8.1	12.5	19.1	36.3	25.9	26.9	20.3	18.5	20.6	28.2	260.6 mm
Prince George, B.C.	(temp.)	−11.8	−6.2	−2.1	3.9	9.4	13.0	14.9	13.7	9.8	4.7	−2.8	−7.6	3.2°C
	(precip.)	59.2	42.9	31.5	29.5	42.2	58.2	57.9	73.4	55.9	61.0	54.9	54.1	620.7 mm
(H) WEST COAST Victoria, B.C.	(temp.)	2.9	4.7	5.8	8.6	11.9	14.5	16.4	16.1	13.9	10.0	6.2	4.2	9.6°C
	(precip.)	146.3	96.8	69.1	44.2	30.5	29.2	18.5	24.9	36.6	87.4	127.5	145.5	856.5 mm

Source: Atmospheric Environmental Services (Canada): 1940-1970 averages.

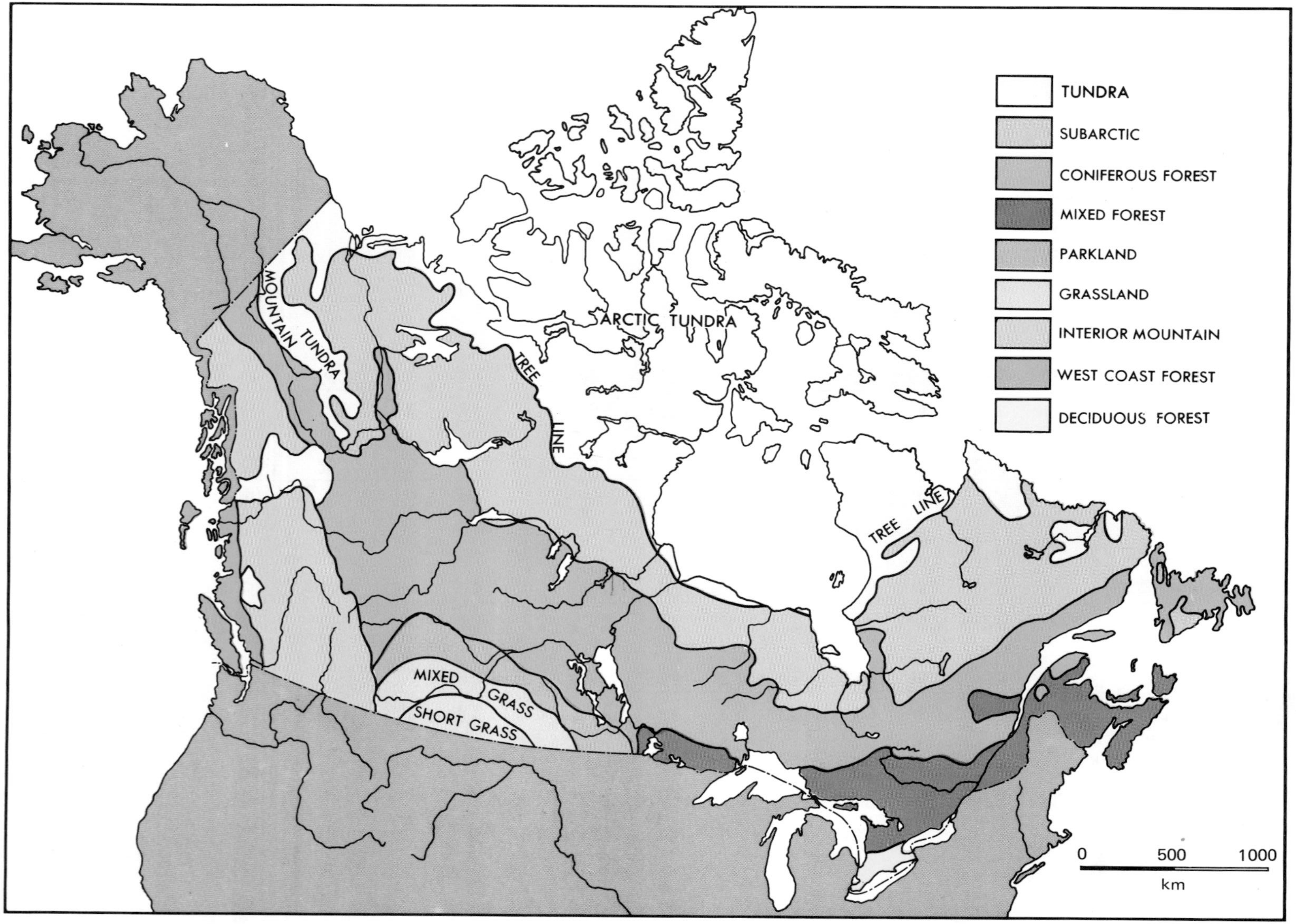

Figure 6.1 NATURAL VEGETATION REGIONS OF CANADA.

6 NATURAL VEGETATION, WILDLIFE, AND SOILS

NATURAL VEGETATION

When the explorers first landed in Canada, they saw the trees, shrubs, and other plants as they had existed for many hundreds of years. Except for clearing some areas of forest by burning, the Indians had done relatively little to change the vegetation that grew naturally. The mixture of plants that grows undisturbed by people is called *natural vegetation*.

In the northern parts of Canada, most of the vegetation looks just as it did hundreds of years ago. However, in the southern parts of Canada there is very little natural vegetation left. People have cut down trees, ploughed the native grasses under, and have introduced many new plants. Even the vegetation of the wooded or wilderness areas, that is now mistakenly called "natural," has been greatly modified by human activity. The remaining woodlots in Southern Ontario are much different now from when the first pioneers arrived. Whereas at one time the giant white pine was common, after the first wave of logging other species such as maple and beech became dominant. Many of the plants that now grow profusely along roadsides and fencerows are not even native to Canada but were unintentionally brought here by the Europeans. Perhaps the vegetation found in cultivated areas should be called *wild vegetation* instead of natural vegetation because, although it is growing wild, it is not natural to the place in which it is found.

Although difficult, it is still possible to get some idea of what the natural vegetation was like in the more densely settled parts of our country. There may still be some old-timers in your community who can remember what the countryside looked like before it was densely settled. Some of the explorers and early pioneers may have written descriptions of the natural vegetation of your local area when they first saw it. In some parts of the country, the original surveyors made detailed notes of the natural vegetation. You may find some of these accounts in your local museum or library; or perhaps your community has a historical society that keeps such records. In the more remote areas of our country, national and provincial parks have been established to preserve some of the natural landscape for present and future generations of Canadians. Although the vegetation in some of these parks has been greatly modified, in others the vegetation is in a relatively natural state.

Even if you cannot find any historical records describing the natural vegetation, you can make some observations about the wild vegetation of your own area. Here are a few ideas for those of you who live in an area that was once wooded.

> LOCAL STUDY
>
> 1. Mark off an area of ten square metres in a wooded park or in a farm woodlot, and try to identify all the trees, shrubs, and other plants in that area. What are the three most noticeable species of trees?
> 2. Note the kinds of trees that have been left standing along fences between farmers' fields.
> 3. What kind of wood has been used in the old barns and houses in your area?
> 4. If there are any stump or rail fences left in your area, find out from what kind of trees they were made.
> 5. Examine woodlots in different parts of your area. Make a chart in which you list the species of trees found in the following different kinds of areas:
>   (a) high and dry land,
>   (b) low and wet land,
>   (c) land with a sandy soil,
>   (d) land with a clay soil.
>   Are there any trees found in only one kind of area? in all four kinds of areas?

Natural Vegetation and Climate

Natural vegetation is affected greatly by the climate, particularly by the winter temperatures and precipitation. Compare Figure 6.1 with Figures 5.6 and 5.8. Which January isotherm corresponds most closely with the southern limit of the Coniferous Forest region? Approximately how many millimetres of precipitation are required to grow trees?

About two-thirds of Canada's forest is made up of *coniferous* trees. *Conifers,* or cone-bearing evergreen trees, such as spruce or pine, can survive very cold temperatures and long periods of drought. The needles of coniferous trees have a kind of "anti-freeze" that prevents them from freezing even in below −18°C weather. The thick skins and the small surface area of the needles result in little moisture being given off by coniferous trees. This fact explains why conifers can survive a long winter without any available moisture. When spring does come, the coniferous trees have green leaves all ready to begin the process of manufacturing food. In this way, coniferous trees make the best use of a short growing season.

The *deciduous* trees, such as maple, oak, and elm, are not as hardy as the conifers. Deciduous trees must get rid of their broad leaves in winter because they give off so much moisture, which cannot be replaced by the roots because the ground is frozen. In the fall, before severe

  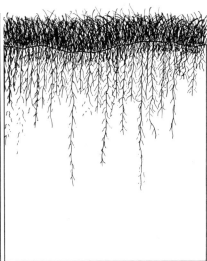

CONIFEROUS TREE   DECIDUOUS TREE   PRAIRIE GRASS

temperatures set in, the deciduous trees withdraw their life-giving sap from the branches and store it in the roots. Then, in the spring, the sap is sent back up to the branches after the risk of very low temperatures is over.

In this way, the deciduous trees are protected from the cold of winter. However, even with this protection, most deciduous trees can grow only in areas with an average January temperature of not lower than −18°C. Although parts of the Prairie Provinces have average January temperatures higher than −18°C, trees do not grow there because of a lack of precipitation. The natural vegetation there is grass.

Grass is particularly suited to dry areas in several ways. It has a surprisingly deep root system so that it can use moisture from quite a depth. Even when the blades of grass die because of drought, the roots remain alive. Then, as soon as water is available, new green shoots are sent up again.

WILDLIFE

Wildlife is greatly affected by both climate and natural vegetation. Some species can survive only in a climate with moderate winters; others prefer long, cold winters and short, cool summers. Some species prefer forest as their habitat, others prefer parkland, and still others, the dry grasslands. Because the natural vegetation provides food and shelter for wildlife, different natural vegetation regions have different wildlife species. For this reason, later in this chapter we have included a brief description of the wildlife found in the major natural vegetation regions of Canada.

Humans have had a great impact on the distribution of wildlife. Many animals such as wolves and bears were eradicated from large areas because they were a menace to people and livestock. Other species have been reduced in numbers through hunting. People have also had an impact on animal populations by destroying the natural vegetation. When forests were cleared, the forest-loving animals decreased in number, while those inhabiting open fields increased. Because many wildlife species are predators, what happens to one species may cause a chain reaction that affects a great number of birds and animals.

Clearing much of the land of its natural vegetation has upset the balance of nature in favour of insects. Thousands of hectares of field crops provided excellent feeding grounds for insects, while many insect-eating birds were reduced in numbers because their forest habitats were being destroyed. As insects reached epidemic proportions, industries began using insecticides, such as D.D.T., to control the insects.

Insecticides kill not only the harmful insects, but also the predatory insects that help control the harmful ones. Birds that consume too many insects exposed to chemical poisons may also become poisoned and die, or their eggs may become infertile. Thus, in the long run, the use of insecticides sometimes leads to even greater populations of harmful insects by destroying their natural enemies. Today, scientists are recommending that insecticides be used sparingly, and at just the right times, so that only the harmful insects are killed. Scientists are also experimenting with biological controls such as the introduction of predatory insects and the use of chemicals that sterilize harmful insects so that they cannot reproduce.

SOILS

Soils are composed of rocky particles such as clay, silt, sand, or gravel, and decayed vegetative matter. The rock portion of the soil is called the *parent material*. The story of how this parent material is weathered from bedrock was told in Chapter 4.

The parent material is not a real soil until it has vegetative matter added to it. Decayed vegetative matter is called *humus*. The humus makes the soil more spongelike, so that it soaks up water more easily. Humus also makes the soil fertile and makes it easier to cultivate for crops.

Different kinds of vegetation provide different amounts and kinds of humus. For that reason, distinctive kinds of soils have developed in the different natural vegetation regions of Canada. This fact explains why the soils map of Canada (Figure 6.2) looks something like the natural vegetation map (Figure 6.1). Since the kind of natural vegetation depends upon the kind of climate, there are also similarities between the soils map and the climate map (Figure 5.13).

Climate has a direct effect upon soils. In a humid climate, the precipitation seeps down through the soil. As it does, it carries dissolved minerals from the top layer of soil to a lower layer. Where there is a great deal of water seeping through the soil, the minerals are carried so deep that

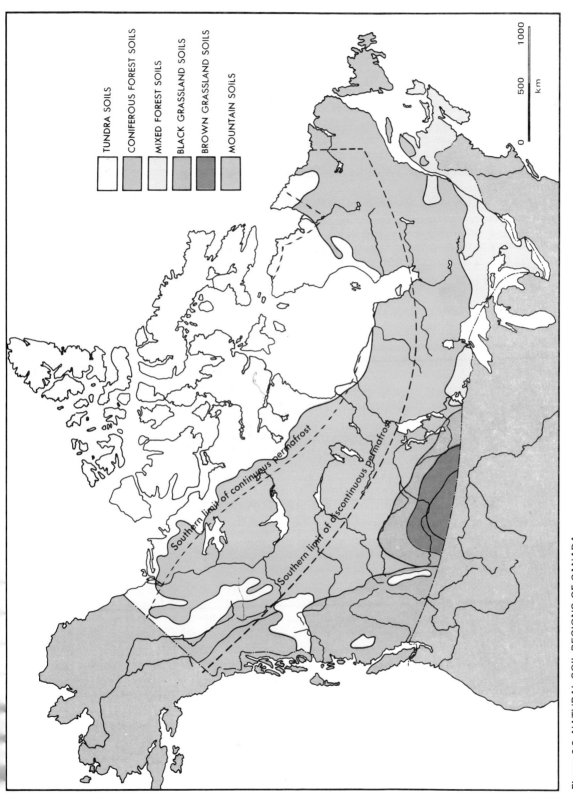

Figure 6.2 NATURAL SOIL REGIONS OF CANADA.

most farm crops cannot use them. Thus, the soil becomes infertile and requires artificial fertilizer in order to grow good crops.

In dry climatic regions, there is less water seeping down through the soil. Thus, the minerals are left near the surface, making a fertile soil. In the very dry, short-grass country, there is much evaporation, which results in moisture moving up through the soil in the same way that coffee will move up through a sugar cube. This upward-moving moisture carries minerals with it, with the result that the dry-land soils have minerals near the surface.

The climate and natural vegetation together cause soils to have distinctive layers. Layering in soils is called a *soil profile*. You can see a soil profile by digging with a spade at the edge of a stream or woods where the soil has not been cultivated. These layers have been named A, B, and C (Figure 6.3). The A layer is called the *topsoil*. It contains humus and is the most important layer for farm crops. The depth and colour of the A layer, or topsoil, depend upon the kind of natural vegetation and climate. The B layer, or *subsoil*, contains minerals carried from the A layer by seeping water. The C layer is the rocky, *parent material*. Only the A layer, or topsoil, is valuable for growing crops. In many parts of the country, much of this valuable topsoil has been lost through misuse of the land.

We said earlier that soils are composed mainly of parent material (weathered rock) and humus. The amount of humus, as well as the depth and colour of the layers of soil, are the factors used for dividing soils into major regions (Figure 6.2). Within each region, however, there are many soil types, depending upon how fine or how coarse the parent material is.

The fineness or coarseness of soil is referred to as its *texture*. A soil developed on a silt or clay parent material is said to have a *fine* texture. A soil formed on a sand or gravel parent material is said to have a *coarse* texture. If the soil is a mixture of clay and sand it is called a *loam*. If the loam has more sand than clay, it is called a *sandy loam*. If there is more clay than sand, it is called a *clay loam*.

If you talk to local farmers, you will likely find that they do not know or even care in what soil region they are located. They will, however, know all about the texture of their own soil, because soil texture is very important to farmers. Some crops, such as hay, grow best on a clay loam because a fine-textured soil holds moisture well. Other crops, such as orchard crops, corn, and tobacco, grow better on sandy or sandy loam soils where excess moisture drains away quickly.

NATURAL VEGETATION REGIONS

Figure 6.1 shows the broad regions of Canada in which the natural vegetation is similar. We mentioned earlier the ways in which the climate, the

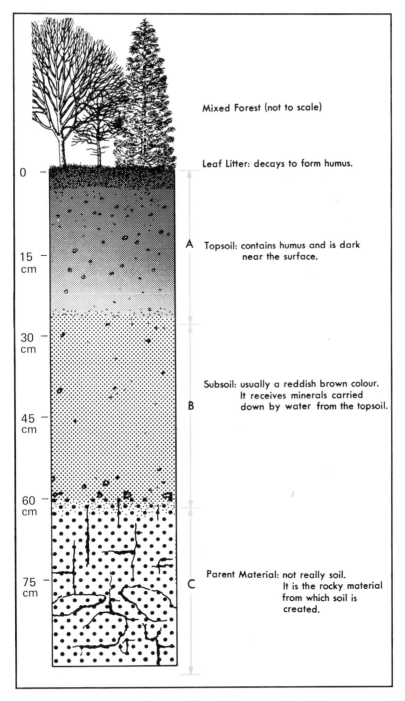

Figure 6.3 A SOIL PROFILE. If you were to dig a pit in an uncultivated area you would be able to see three distinctive layers, known as a soil profile.

natural vegetation, and the soils are related. Because climate has an effect upon vegetation, which in turn has an impact on the kind of soil in any region, it is not surprising that maps showing climatic regions, natural

171

vegetation regions, and soil regions are somewhat similar. You should study the three maps (Figures 5.13, 6.1, and 6.2) until you have a picture of the regions firmly in your mind. If you have difficulty remembering the patterns on these maps, you may wish to trace the maps onto a separate paper so that you can keep them in front of you as you read this section on the natural vegetation regions.

The study of the way plants, animals, birds, and insects interact with one another and their environment is called *ecology*. A community or system of interdependent living things in a particular place is called an *ecosystem*. For example a bog, or a pond, or a river valley could be considered an ecosystem. In order to predict the ecological impact that artificial development will have on the landscape, it is necessary to have detailed knowledge of the ecological relationships in ecosystems across our whole country. To date, the ecological information about our country is very superficial; much remains for biologists, ecologists, geographers, geologists, and other scientists to do.

In the rest of this chapter, we are going to discuss the natural vegetation of the regions shown in Figure 6.1 and briefly describe the related soils and wildlife and some of their ecological relationships. If you ever have the opportunity of travelling from one of these regions to another, you will be able to notice the differences. However, do not expect the changes to occur abruptly, because the regions are not separated by any sharp and clear dividing lines. Instead, there are transition zones between the regions. In these transition zones there is a mixture of the natural vegetation, the soil types, and the wildlife species that are found in the regions on both sides.

Arctic Tundra Region

NATURAL VEGETATION

The Arctic is a treeless landscape. The southern limit of the Arctic natural vegetation region is the *treeline*. North of the treeline, no trees grow because it is too cold and too dry. Only a small part of the annual precipitation can be used by plants because for most of the year the moisture is in a frozen state. This region is the land of continuous permafrost, for

In Canada's northland, the tundra forms a pattern of shrubs, mosses, and lichens over parts of the rocky surface of the land. Numerous plants add colour to the tundra landscape for periods during the short summer. In these photos you can see Broad-leafed Fireweed (red) and Cotton Grass (white).

Ice and snow cover much of the Arctic vegetation for more than eight months of the year. Some animals, like the baby seal in this photo, spend their entire lives in or near the Arctic Ocean where food is more easily available.

Musk oxen browse on tundra vegetation that is partially covered with snow. During the winter, they must paw through the deeper snow to find enough mosses and lichens to eat. Once quite plentiful, today only a few musk oxen are found on the islands of the Canadian Arctic.

here the lower layers of soil never thaw out. In the warmer parts of the Arctic, the top 60 or 90 cm of soil do thaw out, permitting certain plants to grow during the short summer period. In summer, the plants grow very rapidly because of the long hours of daylight.

The Arctic vegetation, called *tundra*, consists of low scrubby shrubs, grasses, mosses, lichens, and Arctic flowers. The vegetation of the glacial drift plains consists mostly of grasses with many flowers such as the Arctic poppy and the Arctic dandelion. These flowers grow low to the ground and mature very rapidly. On the more rocky uplands, shrubs such as ground birch and Labrador tea are found. In areas with little or no soil, there are numerous plants such as lichens and Arctic ferns.

The flowers, fruits, and seeds of many Arctic plants are surprisingly colourful. In late summer and early fall, the tundra landscape is a mass of many different colours.

WILDLIFE

A surprising number of wild animals survive the Arctic climate and thrive on the food supply there. Throughout the Arctic Islands and along the seacoast live numerous polar bears, seals, and walruses. Arctic foxes range over the whole tundra region. In summer, Arctic foxes are brown in colour, but in winter, their fur changes to a pure white colour. They provide the basis for much of the fur industry of the far north. For food, the Arctic fox depends primarily upon small rodents called lemmings. Because the numbers of lemmings vary greatly from year to year, the numbers of Arctic fox also fluctuate greatly.

The caribou is a very important animal for the Inuit, providing them with food and clothing. The caribou that spends most of its time in the tundra region is called the Barren Ground caribou. At the turn of the century, the Barren Ground caribou were very plentiful, but in recent years, their numbers have decreased drastically. There are some Inuit today who have never seen a caribou.

The musk oxen are quite rare, but a few still exist in the Canadian Arctic. They are large, shaggy animals with big feet and long, sharp horns that curve downward. Even on the more northerly islands, they seem able to live by grazing on tundra mosses and lichens. They are able to survive very intense cold and raging blizzards. If a herd is attacked by hunters or animals, the bull musk oxen form a ring with their sharp horns pointing outward. In this way, they protect the cows and calves that stay in the middle of the circle.

There are a few birds that live all year round in the Arctic. They include the great snowy owl, which sometimes ventures into areas farther south if the supply of lemmings runs short in the north. Another permanent resident, the ptarmigan, changes its plumage from a brownish-grey in summer to white in winter.

THE CANADA GOOSE. Several varieties of the Canada Goose spend their summers in a large area of Canada. For many, the favourite nesting grounds are in the more northern parts of Canada where there is an abundance of sedges, bushes, and grasses, and where there are long daylight hours and few human beings to disturb them. In winter they migrate to southern parts of the United States and to Mexico. Only a few Canada geese live all year in the same place or spend their winters in Canada.

The Canada geese are cautious, intelligent, and difficult to hunt. They nest in places where there is a surrounding open view, and feed in wide stretches of water, marshes, or fields where hunters find it difficult to hide. A few members of the flock stand on guard against any possible danger while the remainder of the flock feeds.

Canada geese can be seen by the thousands in the spring and fall migration periods, at a bird sanctuary established by the late Jack Miner at Kingsville, near Windsor, Ontario.

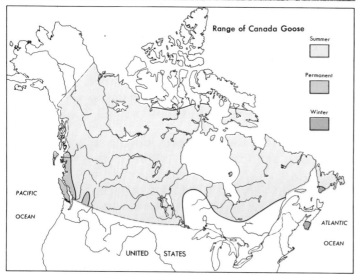

Many birds spend the summer as far north as the Arctic. The Canada goose and the snow goose are among the better known of these. Canada geese are still quite numerous, but the snow geese were slaughtered by the millions until, in 1908, there were only about 3000 left. Strict conservation measures were adopted just in time. Today, there are more than 100 000 of them, all living together in one great flock. They spend the summer nesting in the Arctic, and the winter, along the mid-Atlantic coast of the United States. In the spring and fall, they stop at a place called Cap Tourmente on the St. Lawrence River about 40 km from Quebec City.

SOILS

The soils that develop north of the treeline are given the same name as the natural vegetation—tundra. Because tundra vegetation is sparse and grows slowly, there is little humus in the tundra soils. In the long winter season, the soil-making processes stop altogether. For these two reasons, the tundra soils are very thin and poor. As a result of the cold climate and poor soils, there is no agricultural activity at all in the Arctic tundra.

Subarctic Region

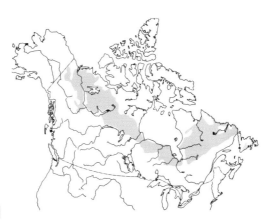

NATURAL VEGETATION

Between the Arctic tundra and the coniferous forest, there is a transition zone of scattered coniferous trees and shrubs mixed with tundra vegetation. This region is called the Subarctic. The most common coniferous trees of this transition zone are spruce and larch. There are also some white birch, one of the few deciduous trees that can survive such a cold climate.

Because of the short growing season and the small amount of precipitation, the growth of trees is very slow. As a result, most of the trees are small, stunted, and of no commercial use. For this reason, the area is known as the "land of the little sticks." The rocky areas, having no trees and very little other vegetation, are called *barrens*.

The Subarctic region has many lakes and swamps. Swampy areas, or bogs, that have filled in with peat are called *muskegs*. Spruce and tamarack trees grow densely around the edges of these muskegs. Towards the centres, there are many shrubs and cranberries. These muskegs have been a transportation barrier to summer travel in the north. Only in winter, when they are frozen over, is it possible to cross the muskegs.

WILDLIFE

Some of the wildlife of the Arctic is also found in the Subarctic region. The caribou, lemming, and great snowy owl are all common. There are also very many animals in the Subarctic that are usually associated with the Coniferous Forest region to the south. In summary, it may be said that the Subarctic is a transitional zone of wildlife, just as it is a transitional zone between the natural vegetation of the Tundra and Coniferous Forest regions.

SOILS

The soils of the Subarctic are similar to those in the Coniferous Forest region and will be discussed in the following section.

FRAGILE ECOSYSTEMS

Both the Arctic Tundra and Subarctic regions contain what ecologists call fragile ecosystems. This means that even a minor unnatural disturbance or interference will have grave consequences. Tundra vegetation that is destroyed by running a bulldozer over it will not grow back for decades. A road, unless properly constructed, will cause the permafrost to melt, leaving a water-filled depression in the summer. Because of the cold climate, garbage that would decay in a few years in southern Canada lasts for decades in the Arctic. Birds often concentrate their nesting in limited

Muskeg along the southern edge of the Subarctic forest. Shrubs and grasses thrive around the outside of the muskeg and each year the vegetation gradually grows in toward the centre of the muskeg. You can see patches of coniferous forest beside the muskeg. Note the dead trees (lower left) that were killed by a rise in the water level.

areas and have a short season in which to raise their young. If people disturb these nesting areas during the hatching and rearing season, the populations may be seriously reduced. Some animals, such as the caribou, must migrate to find sufficient food. However, barriers such as pipelines can greatly affect these animals. Because of the fragility of this northern environment, the Canadian Government is conducting ecological studies and formulating rules and regulations for any company that wishes to explore for, develop, or transport mineral resources in the Arctic and Subarctic regions.

Coniferous Forest Region

NATURAL VEGETATION

The Coniferous Forest region extends in a broad arc from Alaska to Newfoundland. It is the second largest continuously forested area in the world; second only to the great northern forests of Siberia.

It is the Coniferous Forest region that has made Canada famous as a supplier of pulp and paper. The most numerous trees are spruce, balsam fir, jack pine, birch, tamarack, and aspen. Because only a few tree species are found to predominate in any one area, it is easier to develop the forest resources commercially in this region than it is in areas where there is a wider variety of trees.

In addition to pulp and paper, the Coniferous Forest region is an important source of lumber. However, the more accessible stands of timber have already been exploited. There is much more forest land that could be used for lumber, but for the present it costs too much to get to this lumber, to cut it, and to transport it to markets.

WILDLIFE

The Coniferous Forest region abounds with wildlife. Members of the deer family—including woodland caribou, white-tailed deer, and moose—are

A stand of primarily black spruce and jack pine in the Coniferous Forest region north of Thunder Bay. In this photo, a forest harvesting operation is underway. The jack pine is used for lumber; the black spruce and other conifers for pulp. After it is clear, the area will be reseeded. It takes between 80 and 100 years for the forest to become mature again.

found here. The moose is the largest member of the deer family, sometimes with a mass of as much as 900 kg. The bull moose is rather aggressive and is known to have charged people, automobiles, and even trains.

There are too many species of wild animals in this region to name all of them. A few of the more common ones are the black bear, wolf, red fox, mink, otter, muskrat, and beaver. There are also many birds, including many types of wild ducks and large numbers of birds of prey such as hawks, owls, and eagles.

SOILS

The soils of the Coniferous Forest and Subarctic regions are not very fertile because coniferous trees do not provide much humus. The soils are also very acid and this property increases the amount of minerals dissolved from the topsoil and carried down into the subsoil.

The lower part of the topsoil (A layer) is grey in colour and the subsoil (B layer) is reddish-brown. This colouring results from the reddish-brown minerals having been carried down from the A layer to the B layer. Because of the grey colour in the topsoil, soil scientists call these soils *podzols*, which is a name derived from a Russian word meaning "ash grey." The soils in the southern margins of the Coniferous Forest region are more fertile than those of the Subarctic.

THE WHITE-TAILED DEER. The white-tailed deer is so named because of the pure white undersurface of its tail, which stands erect when the deer is frightened. The white-tailed deer is found as far south as Lake Erie and as far north as Reindeer Lake. It feeds on buds and twigs of deciduous trees and shrubs. It generally lives at the edge of the forest where there is more undergrowth and in areas where the forest has been cut and there are many shrubs and young trees. It also likes grazing on winter wheat, and sometimes is numerous enough to be a hazard to farmers.

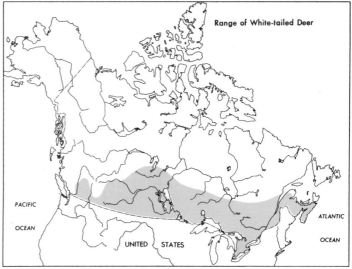

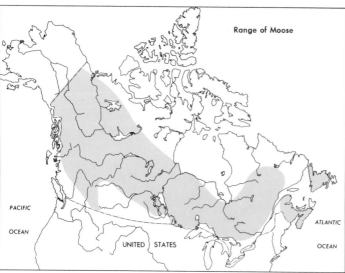

THE MOOSE. Although the moose is most numerous in the Coniferous Forest region, it is also found in the Mixed Forest, Interior Mountain, and southern parts of the Subarctic regions. The moose prefers to live near lakes, streams or bogs.

The moose is well adapted to its environment. Its long legs help it to wade through deep snow, and its hooves provide support in walking through muskeg. In spite of its great size, it can move silently and swiftly, and is a powerful swimmer. The moose's poor sight is compensated for by its keen senses of hearing and smell.

The moose feeds on tree leaves, shrubs, grasses, water lilies, and other water plants in the summer. In the winter it eats the twigs of trees and shrubs, and the bark of some kinds of trees.

Mixed Forest Region

NATURAL VEGETATION

This region extends from the Manitoba-Ontario boundary to Nova Scotia. In it are found scores of species of trees, both coniferous and deciduous. It is really a transition zone between the Coniferous Forest to the north and the Deciduous Forest to the south. The predominant coniferous trees are pine, hemlock, cedar, fir, and spruce. These are often called the softwoods. The most important hardwood deciduous trees are the sugar maple and silver maple, beech, birch, oak, elm, and ash. The stately, umbrella-shaped elm is beginning to disappear because of the Dutch Elm disease which is spreading rapidly through this region.

In pioneer times, the white pine was a very important species. It was used for the masts of ships and also used to make excellent lumber. It was shipped to Europe in great quantities. Unfortunately, a lack of conservation measures has resulted in a scarcity of white pine in the region at the present time.

In this region, there are also countless numbers of shrubs and other plants. The forest floors are scattered with dainty wild flowers, most of which bloom early in the spring before the leaves of the trees come out to shade the ground from the sun. Wild flowers are abundant only where we have not interfered too much with nature. In many farm woodlots, cattle have eaten or trampled the wild flowers. In public parks, many of the wild flowers have been destroyed by thoughtless people who pick them. The beautiful white trillium, for instance, dies if picked with its leaves.

Most of the Great Lakes-St. Lawrence Lowlands and the areas with good soils in the Maritime Provinces have been cleared of forests and turned into farmland. Some of the poorer land that has been used for cattle pasture is almost overgrown with hawthorn bushes, wild raspberries, and brambles. Along the roadsides and fences, many weeds grow tall and healthy. Among the more common roadside weeds are goldenrod, wild carrot, and blue chicory.

WILDLIFE

Many of the same species of wildlife were originally found in the Mixed Forest region as in the Coniferous Forest region. The moose was found

Here is an autumn scene in the Mixed Forest region of Ontario. The coloured leaves of deciduous trees stand out in contrast to the dark green of the coniferous trees.

These flowers are white trilliums, the floral emblem of Ontario. If the trillium is picked with its leaves, the whole plant dies.

THE BLACK BEAR. The black bear is found in all parts of Canada where there are large tracts of forested land. Although black bears are entirely black except for a brown patch on the nose, the occasional one has a white spot on its chest. Some varieties of black bears are brown instead of black.

Although black bears are classified as flesh-eating animals they can eat many kinds of food because their teeth are suited for both grinding and tearing. Their favourite foods are small animals as well as fish, grubs, bird's eggs, honey, berries, nuts, and leaves and roots of certain plants.

Black bears spend much of the winter sleeping in caves, hollow trees or holes dug in the earth. When the weather is cold they may not stir out of their shelter for months at a time. The cubs are born during the winter.

Black bears are very timid and will usually run away from a human being. However, in parks or camping grounds they become quite tame. They seldom attack people, but can be very dangerous if wounded or when they feel their young are in danger.

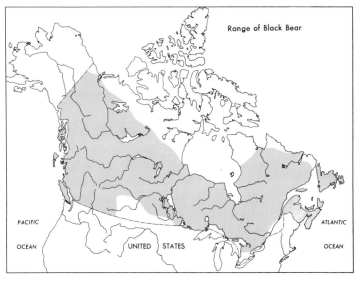

only in the northern edges of the Mixed Forest region, but the white-tailed deer were numerous throughout the region. The clearing of a large part of the area for farming greatly reduced the numbers of deer, bears, wolves, bobcats, beaver, and other forest dwellers. Throughout the region, however, there are still many small animals such as red and black squirrels, chipmunks, jack rabbits, cottontail rabbits, and many others too numerous to mention. Birdlife is also abundant, ranging in size from tiny songbirds to the blue heron that has a wingspan equal to that of a large eagle.

The clearing of forest reduced the numbers of some wildlife species but increased the populations of others. For example, the woodchuck (ground hog) has multiplied to the point where it is a nuisance, and field mice are much more numerous than when the area was all wooded. Open fields have also attracted birds such as the meadowlark and the bobolink.

SOILS

The soils of an area with mixed deciduous and coniferous forests are more fertile than the podzols of the coniferous forest. The topsoil is deeper and contains more humus. Not as many minerals have been carried away from the A layer. As a result, the topsoil is not as grey in colour, but has a more brownish colour. Soil scientists call these soils *brown* or *grey-brown podzols*. When artificial fertilizer is added, these soils make very good farmland. The most fertile soils are found in the most southerly parts where the deciduous trees are more numerous.

Deciduous Forest Region

NATURAL VEGETATION

The only pure deciduous forest in Canada is found in the most southerly part of the Ontario peninsula. This forest is a small part of a broad deciduous forest region that extends into the United States. Compare the northern limit of the Deciduous Forest region with the 20°C July isotherm (Figure 5.7) and the −4°C January isotherm (Figure 5.6). What you observe indicates that deciduous trees like warm summers and moderate winters.

The Deciduous Forest region has the same hardwood trees as the Mixed Forest region. In addition, there are some deciduous trees that are commonly found in much milder climates. These include the chestnut, hickory, sycamore, black walnut, and the tulip tree. The sweet chestnut trees have been almost completely wiped out by a disease called chestnut blight. There are very few evergreens, except pines, which grow on the very sandy soils.

In early pioneer times, this region was almost completely wooded, with the exception of limited areas cleared by the Indians for agricultural purposes. Today, the entire area has been cleared for farming except for farm woodlots and natural parks.

WILDLIFE

The wildlife of this region is similar to that of the Mixed Forest region. Because the flyways of many migrating birds cross this area, many varieties of birds are seen only in the spring and fall. Bird watchers often congregate at points along the Lake Erie shoreline to observe birds during the migration periods.

SOILS

The soils of this region are also very much like those of the Mixed Forest region. They are classified as grey-brown podzols but are less acid and more fertile than the soils farther to the north.

Grassland Region

NATURAL VEGETATION

The grasslands are found where there is not enough moisture for trees. The grasslands are usually called prairies; this is how the Prairie Provinces get their name. A comparison of the above sketch map with a political map will show you that the Prairie Provinces are really misnamed, because at least three-quarters of the area of the Prairie Provinces is covered with forest.

SHORT GRASS PRAIRIE

In the drier parts of the prairies short grass grows in bunches often mixed with sagebrush (the taller, lighter green plant). Note that the vegetation is so sparse that the soil shows through in many places. This photo was taken in the spring when the grass is green. By midsummer the whole landscape usually takes on a brown hue. Grain stored in the elevators (background) is grown on flatter land beyond the skyline.

ANTELOPE

The antelope is an animal commonly found in this short grass country.

The short grass country has a surprising number of colourful flowers such as the prickly pear cactus and the prairie crocus.

PRICKLY PEAR CACTUS

PRAIRIE CROCUS

The common gopher of the Prairies is a member of the squirrel family that burrows into the soft earth to make a home. Part of the gopher's diet is composed of the grass that is found close to the gopher's burrow.

The driest part of the prairie is in the south around the Alberta-Saskatchewan boundary. This area is often called the short-grass country. It is also called *steppe,* which is derived from the Russian word meaning "grassland."

The natural vegetation of the short-grass or steppe country consists mostly of bunches of several types of short grasses with some sagebrush and small cactus. Bare earth is visible between the bunches of grass. Where the grass has been overgrazed by cattle, there is more sagebrush and cactus, and considerable wind erosion is in evidence.

One of the more useful grasses is commonly called buffalo grass. It starts growing late in the spring but keeps on growing well into the summer drought. When rain does come after a long period of drought, buffalo grass comes back to life and grows more quickly than other grasses. In the past, it provided the bison (popularly called the buffalo) with excellent feed.

Bison (commonly known as buffalo) on a reserve can hardly be called wildlife because they can no longer roam freely over the prairies. These bison show us what the largest and, in the past, a very common animal on the prairies is like.

In midsummer, when most of the grass has turned to a light brown colour, a trace of green can be seen at the edges of roads where rain has run off the pavement. There is also a green fringe around natural ponds, called *sloughs*. Most streams dry up in summer. In the resulting dry streambeds there is enough moisture available for a few small trees, shrubs, and green grass.

Of the flowers growing on the short-grass prairie, the most showy flower is that of the prickly pear cactus. The wild crocus, wild rose, dwarf phlox, and dwarf goldenrod are all common flowers. The semitropical yucca is found in some of the southern dry streambeds, called *coulees*.

Surrounding the dry, short-grass country is an area of both short and long grasses. With slightly more precipitation, the grass covers almost all of the ground. Towards the moister margins of this area, most of the grass is long and stays green for a much longer period of time.

In 1860, Captain Palliser described these grasslands as an arid infertile zone, surrounded by a fertile belt with darker soils and trees. To this day, the dry area outlined by Palliser is called Palliser's Triangle. Ironically, however, the "arid infertile zone" became an important grain and livestock producing area.

WILDLIFE

The most common animals of the grasslands are the pocket gopher, the Richardson's ground squirrel, and the prairie dog. These animals are similar to one another and are often collectively called "gophers." Because they eat both grass and young grain crops, they cause considerable damage and force farmers to use poison, gas, and guns in attempts to reduce their numbers. "Gophers" also have a number of natural enemies, including the badger, which digs right into the ground after the "gophers", and also several types of hawks and owls.

When settlers first arrived on the prairies, there were countless bison, but by the beginning of this century, hunters had reduced their numbers to the point where the bison seemed headed for extinction. The bison have made a comeback on reserves set aside for them, but they will never again be seen roaming over all the prairies because they would ruin croplands and would compete with cattle for pasture.

Now that the bison have gone, the largest wild animals on the short-grass plains are the mule deer and the antelope. They do not exist in large enough numbers to be a hazard to farmers.

Because of the many sloughs found in the prairies, many varieties of wild fowl nest in the area.

SOILS

The grassland soils are a brown colour and very high in mineral content. The soils of the very dry, short-grass country are a very light brown colour.

THE BEAVER. The beaver is found in every province of Canada as well as in the western part of the Northwest Territories. The beaver possesses many features that are adapted to its environment. Its front paws are used to plaster mud on the dams or lodge. Its front feet are equipped with long and strong claws adapted to digging. Its trowel-shaped and flat tail, covered with leathery scales with a few coarse hairs in between, is used as a rudder in swimming; its hind feet with large supporting surface and thick, blunt toenails provide propulsion. The beaver has small nostrils and ears that can be closed under water. Above all, the beaver possesses an acute sense of smell, which helps protect it from its enemies.

The beaver depends mainly on the barks of green herbs and young trees for its food. In summer, its favourite foods include aquatic plants, grasses, sedges, rushes and the like. In winter food is stored near the beaver's home and in as deep water as possible.

Many beaver houses are merely burrows in stream banks, but others are lodges built in the pond or on an adjacent shore. Most lodges are 4.5 m in diameter and 1.8 m high with single living compartments 1.2 or 1.5 m in diameter and 0.6 m high. Dams are built across streams or river beds to create ponds deep enough to free the bottom from freezing during the coldest winters and to provide storage for winter food supply. Dams as high as 5.5 m have been discovered.

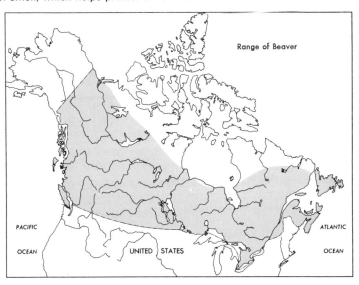

The drought in the light brown zone causes an upward movement of water through the soil, resulting in mineral salts being brought near or up to the surface. In some places, the ground is white with these minerals. The light brown soils are not very good for crops because they lack humus and are too rich in mineral salts. Also, there is not enough precipitation for most crops. For these reasons, much of the area has been left in grass for cattle grazing.

The soils of the zone of mixed long and short grass are a dark brown colour because of the greater amount of humus. The dark brown soils also contain fewer mineral salts. Much wheat and other grain crops are grown on the dark brown soils.

Parkland Region

NATURAL VEGETATION

The grasslands are surrounded by an arc of land where the natural vegetation is long grass interspersed with clumps of trees. This Parkland region is a transition zone between the dry prairies and the more humid Coniferous Forest region to the north. Some coniferous trees are found along the northern edge of the zone as well as along some of the river valleys. The grass is much taller and thicker than in the Grassland region.

Along the river courses in the Manitoba Lowland, a ribbon of forest developed, consisting of elm, Manitoba maple, basswood, green ash and willow. It is probable that the absence of trees found by settlers in the rest of the Manitoba Lowland was a result of too much water instead of a deficiency of moisture as in the rest of the prairies. The poor drainage resulting from fine-textured soils and a very level surface may have resulted in soils too saturated for trees.

In the balance of the Parkland region the clumps of trees consist mostly of aspen and cottonwood. These clumps are sometimes called bluffs in western books and movies.

WILDLIFE

This region is also a wildlife transition zone. Species common to both the Coniferous Forest region and the Grassland region are found in the Parkland region.

SOILS

The long blades and deep roots of the tall grass found in the Parkland region have provided a large amount of humus making the soil very black. Because of its blackness, this soil is called a *chernozem* which comes from the Russian word meaning "black earth." The medium amount of precipitation in the region has resulted in very little mineral material being carried away from the topsoil. Nor are there too many mineral salts as in the light brown zone.

The black soil region is an excellent area for growing wheat. There is enough precipitation to produce high yields of wheat but not enough to reduce the fertility of the soil. Wheat can be grown on the black soils for many years without the addition of fertilizer.

Interior Mountain Region

NATURAL VEGETATION

There are many different kinds of vegetation in the Western Cordillera. The lower mountain slopes are covered with coniferous trees. Some mountain peaks rise so high that it is too cold for tree growth. Meadows, similar to those of the Arctic and other tundra vegetation areas, are found above the treeline. Some of these alpine meadows are covered with beautiful wild flowers in the late spring and early summer. In some places, the mountains rise so steeply that no trees can gain a foothold, even at lower elevations. Despite this wide variety, the Interior Mountain region

Grass vegetation and ponderosa pines on the interior plateaus of British Columbia. This kind of land is used as open range for cattle grazing.

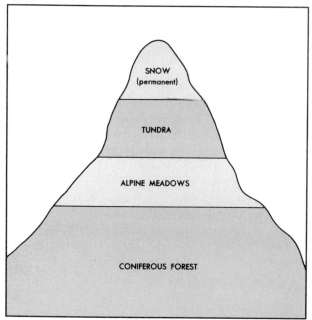

The type of vegetation changes with altitude on mountain slopes of the Western Cordillera.

is primarily a coniferous forest region.

The forest vegetation of interior British Columbia ranges from scattered pines on the upper slopes of the dry grassland valleys, to dense stands of coniferous forest on windward slopes, similar to those of the Coniferous Forest region. The southern half of the Interior Mountain region has a greater range of species and larger trees than in the north. In the south, the Douglas fir is the most common species; in the north the balsam fir and spruces are more abundant. In the interior, red cedar is found extensively only in the Columbia Mountains.

The valleys of the interior plateaus receive relatively little precipitation because they are in the rain shadow of mountain ranges. Some of the southern valleys are so dry that the natural vegetation consists of short grass and semiarid plants, very similar to the driest part of the prairie Grassland region. A few of the southern interior valleys have been irrigated and are used for orchard crops and vegetables. Both the valleys and uplands in the southern interior plateaus are extensively used for grazing cattle.

WILDLIFE

Because so much of the Western Cordillera is still in its natural wooded state, wildlife is abundant. Bears are commonly seen even along the highways. There are several types of deer and both mountain goats and sheep. Other animals include several varieties of the wildcat family. Birdlife is also abundant.

SOILS

Just as there are many different kinds of natural vegetation, so there are many different kinds of soils in the Western Cordillera. The forested mountain slopes have coniferous forest soils. The dry valleys and plateaus of the interior have grassland soils ranging from black to light brown. In the high mountains, above the treeline, tundra soils are found.

ROCKY MOUNTAIN SHEEP. Rocky Mountain sheep are found in the Rocky Mountains and other parts of the Western Cordillera. They eat mainly grass, but also graze on leaves and twigs of trees and shrubs. In summer they migrate to the higher parts of the mountains; in winter to lower south-facing slopes where the snow is not so deep; in spring they are frequently found in the lower forested slopes.

Rocky Mountain sheep have very keen sight. They are quick moving, very sure-footed and can climb precipices with seeming ease.

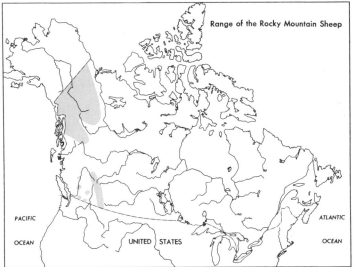

West Coast Forest Region

The West Coast Forest region also has mountain vegetation but has been shown separately on the natural vegetation map because of the lush tree growth and the importance of lumbering in this area.

Coniferous trees in the West Coast Forest region grow tall and very close together. The tall, straight trunks of Douglas fir, red cedar, and hemlock trees make excellent lumber, and logging companies do not have to move far to obtain all the timber they want.

The West Coast Forest region contains most of British Columbia's remaining mature stands of western hemlock, red cedar, Sitka spruce, and Douglas fir. Although the hemlock and cedar are the most dominant, Douglas fir is abundant in the drier and warmer parts of the south. Sitka spruce is most abundant on the lower slopes where precipitation and soil moisture are high.

The generally mild, wet climate of most of the coastal region promotes rapid tree growth. Although the heaviest timber stands of large mature trees in the more accessible areas have been cut over in this century, individual trees in excess of 60 m in height and 1 m in diameter are still common. Occasionally one will find a giant Douglas fir which reaches close to 90 m in height and has a diameter of 1 to 2 m.

The wildlife and soils of this region are similar to those found in the rest of the Cordillera.

PROBLEMS AND PROJECTS

1. Illustrated articles on Canadian vegetation and wildlife are found in the journals *Nature Canada* and *Canadian Geographic*. In back issues, find feature articles about each of the regions discussed in this chapter.

2. For detailed discussions of ecological problems in the Canadian North and the conservation of wildlife in Canada, consult the following books:

    T.R. Berger, *Northern Frontier, Northern Homeland* (Ottawa: Supply and Services Canada, 1977).

    R. Byran, *Much is Taken, Much Remains* (North Scituate: Duxbury, 1973). See Part 4, "The Status and Conservation of Wildlife in Canada".

    D.H. Pimlott et al. (eds.), *Arctic Alternatives* (Ottawa: Canadian Arctic Resources Committee, 1973).

3. What birds, animals, and insects are considered of greatest value in your area? Which are considered to be a nuisance? What steps are being taken to control them?

4. Along a roadside or beside a ditch, dig a hole about a metre deep. Try to identify the following layers: A layer (topsoil), B layer (subsoil), C layer (parent material). Describe the colour and texture of the material in each layer.

5. Many of the scientific names for different kinds of soils are taken from Russian words because Russian scientists have been very active in soil science. Using an atlas, compare the kinds of soils found in Canada with those in the Soviet Union. What soil region in both countries is important for wheat growing?

6. Consult Canada Land Inventory (CLI) reports for your region which map the land's capability for forest and wildlife production. How does the actual production compare with the potential? These CLI reports are available from: Canada Map Office, Surveys and Mapping Branch, Energy, Mines and Resources Canada, Ottawa, K1A 0E9. Also available at Canadian government bookstores.

7 POPULATION

In the last three chapters, we have been describing the physical geography of Canada—its landforms, climate, natural vegetation, wildlife, and soils. In this chapter, we are going to discuss one of the most important geographical elements of all—population. It is the people of a country who give it its distinctive personality and character. It is the people who make a country either divided and weak, or united and strong. It is the people more than the natural resources that make a country rich or poor, because resources are of little use to a country unless there are people who know how to develop them.

A great amount of information about the population of your community and of the whole country is found in the *Census of Canada*. In Canada, a census is taken every five years. Census takers visit every household in the country and record information about the number of people, their age, sex, occupation, and so on. This information is then summarized and published in census volumes that can be found in most public libraries. A summary of Canadian population statistics is given in the *Canada Year Book* published by Statistics Canada. Another good source of population information is your local municipal assessment office. Many municipalities take a population census every year.

LOCAL STUDY

If you have carried out the local studies suggested in earlier chapters, you should have a good appreciation of the physical geography of your local area. You now should find out all you can about the nature and distribution of the local population. In later chapters you will study how the population has made use of the natural environment of your area. The following suggestions will help you to gather information about the population of your local area.

1. Obtain the population totals for your municipality for every ten years back to 1901. Plot these totals and years on a graph and connect the points with a line. (Place the population scale on the left side and the years across the bottom. See Figure 7.1.) Can you explain any sharp rises or falls in population?

2. Draw a population distribution map of your local area using dots. The number of people you make each dot represent will depend on the density of population and the scale of your map. Before you start drawing your map, you should look at several dot maps in atlases and geography textbooks. Figure 7.6 in this book will provide you with one example.

   To obtain the information needed for this map, you will have to do some field work. For instance, in a town or city, you can count the number of dwelling units in all houses and apartments in each block. Then, by estimating the number of persons in a family, you can calculate the population for each block. In a rural area, a topographic map is very useful because it shows the location of every house. However, unless the map is very recent, you will have to check to find how many new houses have been built and how many farmsteads have been abandoned.

3. From the distribution map you drew, calculate the density of population per square kilometre. Compare this with the generalized population density for your region as shown in Figure 7.3.

THE POPULATION OF CANADA

Population Growth

At the time of Confederation the population of the area that is now Canada was 3.5 million. By 1981, the population had increased to slightly over 24 million (Figure 7.1). The two factors determining this population growth have been *natural increase* (excess of births over deaths) and *net migration* (excess of immigrants over emigrants).

In total, natural increase has been the most important factor in Canada's population growth. In a little over a century, the country gained approximately 16 million people by natural increase and only 3 million by net migration. In recent decades, the birthrate has had a greater effect on natural increase than the deathrate. The latter declined from approximately 10 deaths per thousand people in the 1930's to a little under 8 deaths per thousand in the 1970's. During the same decades, the births per thousand people ranged from an average of 21 in the 1930's to 28 in the 1950's. By 1971, the birthrate of Canada had dropped to approximately 17 and throughout the 1970's it has remained relatively stable, at an average of about 15 births per thousand people. At the same time, the

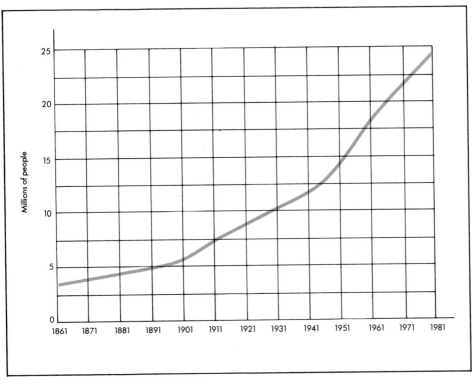

Figure 7.1 POPULATION GROWTH OF CANADA, 1861-1981. Between 1881 and 1981, Canada's population increased sixfold. Since World War II, it has doubled.

birthrate in Quebec—one of the highest in the 1950's—dropped to 14 in the mid-1970's, the lowest of all provinces. The reasons for fluctuating birthrates are very complex, but include the number of women of childbearing age, the age at which people marry, and the size of families. These reflect a number of factors, including attitudes toward birth control.

Had Canada been able to retain all of its immigrants since Confederation, its current population would be well over 30 million. However, in over a hundred years, while immigration has amounted to 10 million people, about 7 million people have emigrated from Canada, leaving a net migration of 3 million. Over the years, much of the emigration has resulted from the movement of both Canadian-born and foreign-born people from Canada to the United States. This movement was very great in the early decades when job opportunities were much greater and more varied in the United States and many immigrants used Canada as a stepping-stone to get to that country. In recent decades smaller numbers of Canadians have been emigrating to the United States, while the number of American immigrants to Canada has been increasing. This increase reflects changes in United States immigration laws and the attractiveness of Canadian living conditions to some American residents.

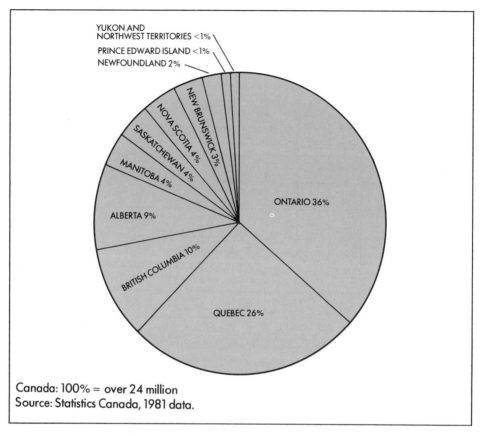

Figure 7.2 POPULATION BY PROVINCES AND TERRITORIES. Even with the rapid increases over the last quarter century, the distribution of Canada's population across the country has remained much the same, as you can see in this graph.

Population Distribution

The population of Canada is not equally distributed among the ten provinces and the territories (Figure 7.2). Ontario and Quebec together have more than three-fifths of the Canadian population. The Western Provinces make up over one-quarter, while the Atlantic Provinces have about one-tenth of Canada's population. The 70 000 people in the northern territories compose only 1/350 of the Canadian population.

Ontario and Quebec have the greatest population because these are the provinces where most of the great factories and office buildings are located. People live where they can find work. In general, primary industries such as forestry, mining, and agriculture do not employ as many people as the manufacturing and service industries in the cities and towns.

If Canada's more than 24 000 000 people were spread evenly across the country, there would be only 2.5 persons per square kilometre (Figure 7.3). In some northern parts of Canada, the population density is much less than 2 persons per square kilometre. In fact, most of the northern areas of Canada are not settled at all. Because of these vast, empty spaces, Canada is considered to be a sparsely populated country. The population density map shows that this conclusion is not true of many parts of southern Canada. In fact, the highly urbanized areas around Montreal, Toronto, and Vancouver have some of the highest population densities in the world, with densities running into thousands of people per square kilometre. About 90 percent of Canada's population occupies a narrow strip of land that comprises less than 10 percent of the country's area.

It is interesting to try to explain the distribution of population in Canada. As we discovered in Chapter 3, the Great Lakes-St. Lawrence waterway made it easy for the explorers, fur traders, and settlers to penetrate to the interior of the continent. The cheap transportation provided by this waterway has continued to contribute to the population growth of the Great Lakes-St. Lawrence Lowlands to this day. Other factors that help to explain the distribution of population are the availability of flat land, good soil, adequate precipitation, moderate winters, availability of power, availability of key mineral resources like coal and iron, and access to other densely populated areas by several types of transportation.

Answers to the following questions will guide you in explaining the population distribution pattern of Canada. You will need to consult a number of maps in your atlas to answer these questions.

QUESTIONS

1. Why is most of the population of the Atlantic Provinces found either along the coast or in the valleys?
2. Why is there a medium population density around Lac Saint-Jean and the upper part of the Saguenay River in Quebec?
3. On a sketch map of Ontario, draw and label the following lines:
   (a) the division between the most densely populated area and the area north of it with a medium population density;
   (b) the southern edge of the Canadian Shield; and
   (c) the January average temperature of −7°C.

   Use the sketch map to explain how landforms and climate have helped affect the distribution of population in Ontario.
4. What transportation and market advantages does the Great Lakes-St. Lawrence Lowlands have over other parts of the country?

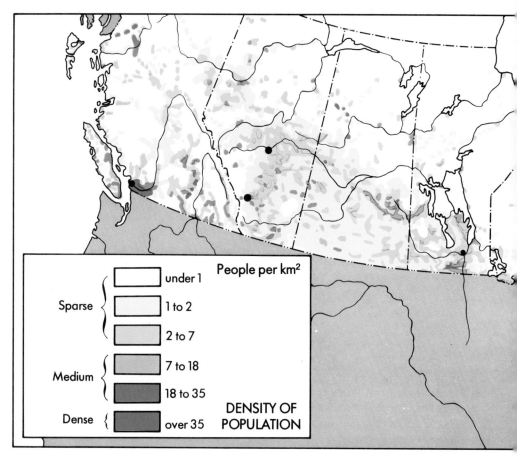

Figure 7.3 POPULATION DENSITY OF CANADA. Can you give reasons for the distribution of population density as shown on this map?

5. Explain why there is a concentration of population in the area around the Ontario-Quebec border about midway between Georgian Bay and James Bay.

6. What are the physical geographical factors that have kept the densest population of the Prairie Provinces in the south? In which Prairie Province does the settled area extend farthest north? Why?

7. Explain the clustering of population in British Columbia.

8. In what ways have settlement patterns been influenced by road, rail, shipping, and air transportation facilities? Give examples.

Ethnic Origins

The Indians and Inuit, native to Canada for thousands of years, make up only about one and a half percent of the population. The rest of

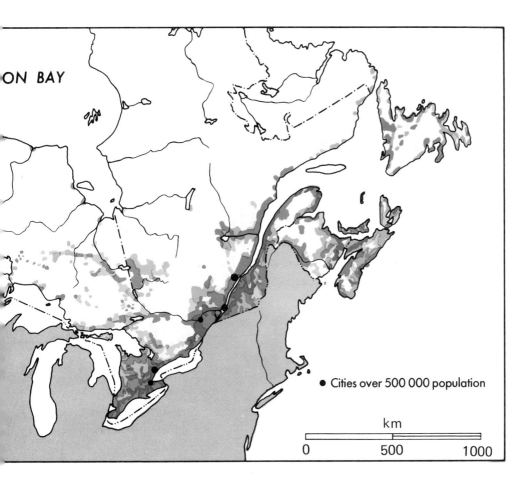

Canada's population has descended from immigrants who have migrated relatively recently—some came in the seventeenth and eighteenth centuries; others, in the nineteenth century; and still more, in the twentieth century.

About six million people, or 30 percent of Canada's population, are descendants of the early French colonists. Most of these French Canadians live in Quebec (77%), although many live in Ontario (11%) and in New Brunswick (5%). The French Canadians in the Province of Quebec have retained their French language and customs, and in fact, many people in Quebec speak only French.

Approximately half of the Canadian population is of British origin. People of British descent are found all across Canada. It may surprise you to find that there are more people of British descent in Quebec than in seven of the other provinces. The largest proportion of British descendants, however, is found in Ontario and British Columbia (Figure 7.4).

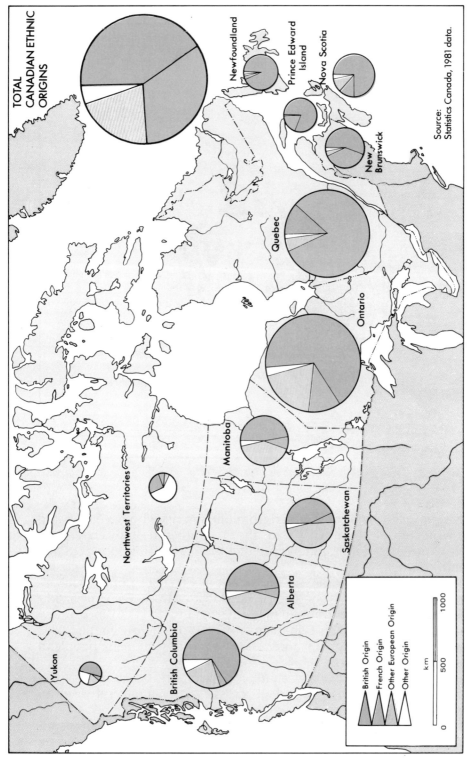

Figure 7.4 ETHNIC ORIGINS OF CANADIAN POPULATION. (The differing sizes of the circles indicate the relative amount of the Canadian population in each province or territory.) In which provinces or territories are more than half the people of British origin? In which provinces are more than one-third of the people of French origin? In which provinces are more than one-third of the people of European origin other than British or French? Although three-quarters of the people of Nova Scotia are of British origin and only about one-seventh of the people of Quebec are of British origin, there are more British descendants in Quebec than in Nova Scotia. Can you explain how this can be?

The balance of the Canadian population has many different ethnic origins. Next in numerical importance to the British and French are the Germans, the Italians, the Ukrainians, the Netherlanders, the Scandinavians, and the Polish people. People of various European origins are found in every province. However, a number of national groups have tended to settle in groups or "colonies". The Ukrainians have been attracted to the Prairie Provinces where the land and climate is similar to the Ukraine in the Soviet Union. Many Netherlanders have settled on the flat plains of southwestern Ontario where they can carry out such intensive farming activities as the growing of vegetables and other specialized crops.

Immigrants have come to Canada in three major waves: the two decades following Confederation; from the turn of the century to 1930; and the post World War II period from 1945 to the early 1960's. In the thirty-five-year period following World War II, immigrants from the United Kingdom accounted for 24 percent of all new Canadians. Another 45 percent have come from Europe other than the U.K., mainly from Western Europe, Italy, West Germany, and the Netherlands (Figure 7.5). Since the mid-1960's, the number of immigrants from the United States has increased significantly. In 1971, for the first time, the United States replaced the U.K. as the major source of immigration to Canada. However, now the United Kingdom once again contributes the largest number of immigrants each year.

Figure 7.5 POST WORLD WAR II IMMIGRATION TO CANADA.

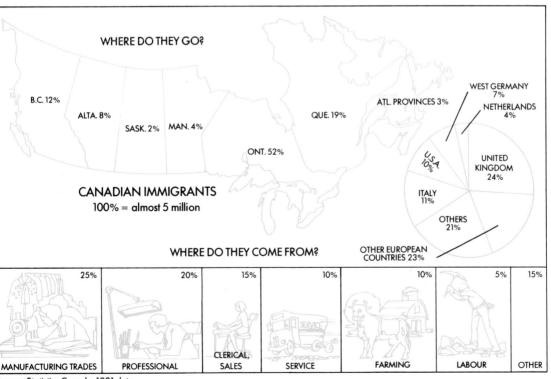

Source: Statistics Canada, 1981 data.

In the 1970's, Canada developed a new immigration policy. It does not favour or discriminate against any race or nationality, but permits entry to Canada to people who have the education and training that makes finding jobs in Canada easy for them; they must be able to support themselves. As a result, immigrants are coming from an even wider range of countries than ever and Canada's population is becoming increasingly cosmopolitan.

In the post World War II period, the majority of new Canadians have settled in the large cities in Ontario and Quebec (Figure 7.5). These cities, and particularly Toronto and Montreal, offer more job opportunities and a chance to settle among friends and relatives while learning the language and customs of a new country.

Most Canadians, other than those of French descent, have adopted English as their first language. There is a great deal of intermarrying between ethnic groups, and thus, many Canadians have a mixture of several nationalities in their background.

We are involved in what is sometimes called "the great Canadian experiment." We are trying to build a strong, unified nation across the northern part of the North American continent despite physical barriers and cultural differences. To succeed, Canada needs the support of all Canadians, regardless of their ethnic origins.

SOME MINORITY GROUPS

There are several small groups of people in Canada who, because of their own special culture or way of life, face unique problems with the rest of Canadian society or with their environment. In the following pages, we are going to discuss three of these minority groups: the Indians, the Inuit, and the Old Order Mennonites.

The Indians and Inuit are minority groups that should be of concern to all Canadians because the non-natives have caused many problems for both of these peoples. The Old Order Mennonites have been chosen because they are a very interesting and important group of people about which the authors have considerable information.

Frame shacks such as this provide poor shelter for thousands of Indians who live on reserves too stony, swampy, or infertile to farm and who cannot find jobs off the reserve. Few of these shacks have the modern services of running water, electricity, or oil or gas heating. In some instances, the people must walk two or three kilometres for water. With conditions such as these, as well as a serious lack of doctors, it is not surprising that the infant mortality rate is high.

Artists and artisans among Canada's native peoples produce beautiful works and useful products. In the photographs are (top) a raven painted on a building in 'Ksan, British Columbia, and (bottom) a skin being stretched to make clothing. The Cowichans in British Columbia are just one group that have formed a co-operative for manufacturing and selling goods made on the reservations. The Cowichans produce wool and thick sweaters, scarves, and mittens; the distinctive wool is made from reservation animals, and local women knit the unique patterns. These products are world famous and many are exported annually to Japan. Native art fetches good prices across Canada and the federal government now has strict regulations about what kinds of native art can leave the country.

The Indians

When Europeans first came to North America, there were about 200 000 Indians of various tribes in what is now Canada (Figure 3.3). This number was gradually reduced by the introduction of new European diseases, by intertribal wars that were often incited by the Europeans, and by the non-natives' wholesale slaughter of game animals and destruction of the habitats of much wildlife. In the twentieth century the Indian population

Many young Indians are finding employment off the reserves. Some are leaving to become Canadian citizens without reserve privileges and to work in construction, in forests, mines, and factories. These young Indians are being trained on oil rigs in Alberta.

began to increase again. By 1981, the Indian population had reached more than 300 000, and is currently increasing at the rapid rate of about three percent a year. In addition, there are at least as many people with Indian blood who are not classified by the government as Indians. These include Indians who have given up their Indian status and the offspring of Indian women who married non-Indians. The largest group of people with mixed Indian and non-Indian ancestry are the Métis, most of whom live in western Canada. (See Figure 7.6.)

Early in the settlement of North America the British recognized Indian rights to land and began making agreements called treaties with the various Indian tribes. With the British North America Act (1867), the administration of Indians came under the Canadian Federal Government, which continued the practice of signing treaties in which the Indians surrendered their interest in the land in exchange for special reserves. About half of the Canadian Indians today live on more than 2000 reserves scattered across Canada. The land set aside for reserves was often poor agricultural land that non-Indians did not want, or was too small to provide the forest and wildlife resources to support the Indian population. The Indian treaties gave the Indians certain special privileges and benefits. They did not pay taxes on Indian reserve land, were guaranteed annuities of agreed amounts (usually about $5.00 per person per year), were given livestock, machinery and tools, and were promised educational facilities.

Serious poverty is evident on the Indian reserves. The resource base of the reserves is inadequate, and usually there are insufficient jobs in the surrounding area. Consequently, the housing is usually substandard and overcrowded and the children are shabbily dressed. The education provided on the reserve is inadequate to prepare the young people for jobs in the "outside world." When Indian children attend off-reserve schools, they have to live away from home, and often face racial prejudice. Also, many Indians are opposed to too much of non-Indian education for fear that the Indian race will lose its cultural heritage, language, and social customs.

Despite the problems noted above, many young Indians receive sufficient education and training to obtain good off-reserve jobs. Some Indians have developed expertise which is greatly in demand in Canadian society. For example, wherever high buildings or bridges are being constructed of structural steel, you are likely to find Mohawks from the Caughnawaga Reserve near Montreal. Some of the Indians who take off-reserve jobs retain their connections with the reserve, sometimes working for only part of the year. Others renounce their treaty rights and become Canadian citizens without reserve privileges.

Most of the Indians who do not live on reserves live in small settlements in the northern parts of Canada. The plight of these people, including the Métis, is sometimes worse than that of the reserve Indians. They can no longer make their living from hunting, trapping and fishing, and alternative steady employment is not available. Since they are not eligible for the assistance received by treaty Indians, they must rely on occasional employment, unemployment insurance, and welfare payments. Their housing and health conditions are substandard or worse.

In general it is fair to say that the Canadian Indians are a minority group suffering from serious social and economic disadvantages. Only three percent of Indians reach a Grade 12 level of education and it is estimated that at least 35 000 Indian adults can neither read nor write. The infant mortality rate is much higher than the average for Canada and the life expectancy of an Indian is only 36 years, compared to 62 years for the average Canadian.

Over the years, the Canadian Government has introduced a number of programs to improve the welfare of the Indians. A special housing program has helped provide many Indians with better shelter. Greater emphasis has been given to education, so that today virtually all elementary-school-age children are in school. New schools have been built and teachers are better qualified. The government encourages Indians to attend provincial public schools and even university by paying tuition fees and other costs. The health and welfare services to Indians have been greatly improved in recent years. On some reserves new industries and business enterprises are being established with the help of government loans.

A basic problem with government programs has been that they have made the Indians dependent on non-Indians for their livelihood. This treatment of Indians as though they were dependent children is termed *paternalism*. To combat this paternalism, in the 1960's, the Canadian Government proposed that reserves be phased out, that Indians be given equal status with other Canadians, and that the government provide them with the assistance necessary to become integrated into Canadian society. While this proposal would have wiped out paternalism, it also would have wiped out the Indians' culture, heritage, way of life and, ultimately, their very identity. The Indians reacted against this proposal of integration so vociferously that the government was forced to withdraw it.

Recently Indians and Inuit have organized themselves into associations called brotherhoods that are negotiating with the federal and provincial governments for financial assistance that will permit them to help improve the living conditions of their own people. Through these brotherhoods they are also applying legal and political pressure to force the provincial and federal governments to recognize their claims to land in the northern parts of Canada. Their legal claim is based on the Royal Proclamation issued in 1763 that designated all lands northwest of the Atlantic watershed as hunting grounds for the Indians. If non-Indians wanted to buy land for settlement, it had to be purchased from the Indians. The Nishka tribe in British Columbia took its claim all the way to the Supreme Court of Canada, where a tie vote was registered ("The Calder Case," 1973). At a later date, the Canadian Government appointed an Indian Claims Commissioner and provided funds to native groups to help them research their rights and claims, poll their people, and come up with specific claims.

By the early 1980's, fourteen major native land claims had been accepted by the federal government for negotiation. Ottawa is concentrating on settling first with five native groups in the northern territories where mineral and pipeline development plans make it urgent to come to an agreement soon. These five groups are as follows (see accompanying map):

–COPE: Committee for Original People's Entitlement, 2500 Inuit in the western Arctic, has reached agreement with the federal government for control over 96 000 km² of land and over $100 million in compensation.
–ITC: Inuit Tapirisat of Canada, 12 500 Inuit in the eastern Arctic, has asked for 65 000 km² of land and control of subsurface mineral rights. The ITC also wants the eastern Arctic to be created as a separate Inuit province called Nunavut.
–CYI: Council for Yukon Indians, 5000 Métis and non-status Indians in the Yukon, originally claimed 70 percent of the Yukon landmass and $670 million. The CYI is in process of revising its claims.

– Dene: about 10 000 status and non-status Indians in the Northwest Territories, many of them in the Mackenzie Delta where their claim overlaps those of COPE and the Métis. The Dene claim is still undefined but includes the concept of a separate Dene nation.
– Métis: about 5000 originally included in the Dene claim who later broke away to make their own claim. The federal government wishes to make a single settlement for both groups because their traditional hunting areas overlap.

The native peoples in the James Bay region, where Hydro-Quebec constructed a huge hydroelectric project, reached an agreement that gave them a three-part settlement (see Chapter 12 and Figure 12.7).

The Inuit

There are only about 20 000 Inuit scattered across the Canadian Arctic (Figure 7.6). The Inuit are a hardy race who have learned to survive in a very harsh environment. The winters of the Arctic are cold, long, and dark, and there are no trees to provide building materials. Berries seldom ripen because the summer is too short, and most of the birds migrate from the region for eight months of the year. The sea freezes over in the winter so that fishing and sealing have to be done through the ice. The caribou herds wander over a broad area in search of food, making it difficult to know when and where they might be found.

How could anyone feed and clothe a family in such an environment? What sorts of houses could be built to shelter families? How could homes be heated? How could they be lit during the months when the sun does not rise above the horizon? How could enough food be found and stored for the long winter? How could enough variety of food be stored to keep a family healthy?

The late Dr. Diamond Jenness, famous Canadian anthropologist who lived with and studied the Canadian Indians and Inuit for many years, wrote the following account of how the Inuit of earlier times adapted to their environment.

Some of us have seen them with our own eyes. I have stalked the wild caribou with Eskimos who were armed only with bows and arrows. I have watched them practise those clever tricks to which every primitive hunter must resort if his family is not to die of starvation. In northernmost Alaska, I have helped Eskimos to build small cabins from the driftwood which the Mackenzie River carries down to the Arctic Ocean and to insulate the rude dwellings with clods of earth. Farther east, where driftwood is lacking, I have learned from them to erect a dome-shaped house of snow. The igloo melts away under the rays of the spring sun, and even in midwinter drips water on your head if too many warm-blooded visitors crowd into your home.

In my own snow-hut I could fry bacon, and boil rice and oatmeal over a kerosene-burning stove. The only food many of my Eskimo friends had ever tasted was the meat of the caribou, seal, and other wild game that they killed, and the fish they caught in the lakes and in the sea. Sometimes, through lack of fuel, they ate their food raw, but generally, they cooked it in stone pots heated by stone lamps that burned the animals' fat or blubber. A trained economist would have observed how they practised in their kitchen the same division of labour as we do. It was the men who manufactured the pots and the lamps, but only the women knew how to use them efficiently. A ten-year-old girl could trim a lamp to perfection, but whenever I dared to disturb its wick I invariably smoked up the whole dwelling.

As in the kitchen, so too in the art of tailoring, women displayed a marked superiority over men. Of all the world's peoples outside Europe and parts of Asia, only the inventive Eskimos ever made separate coats and trousers tailored to fit their wearers. The ten-year-old girl who learned to keep the lamp smokeless, and to preserve all the flavour of caribou tongues by boiling them with the tips upward, could cut out and stitch a perfectly-fitting suit of caribou fur, artistically adorned with inset patterns. The man who attempted to make or mend his own trousers looked like a hobo. We need not wonder, then, that before the white man entered the Arctic, no one ever heard of an Eskimo youth, or an Eskimo maid, reaching the age of eighteen unmarried. Spring is very short in the Arctic, and children, like young birds, must take flight earlier than they do in southern climes. By sixteen at the very latest, the youth was a trained hunter who made his own tools and weapons and could build his own home. The girl was a trained cook and seamstress, ready to help build that home and preside over it with all the poise and dignity of her mother.*

*From an address given at Wilfrid Laurier University, 1962. (Eskimos are now known as the Inuit, their own word meaning "The People".)

Before the fur-trade period, the locations of the Inuit living sites were determined by the availability of wildlife and fish that were used for food, clothing, fuel, and many other things. Thus the Inuit led a nomadic life moving from camp to camp along the coastal areas and throughout the barren lands. With the coming of the fur trade, many trading posts were established along the coasts where ships could visit to deliver goods and to pick up furs. In these tiny posts, the trader would be joined in permanent residence by a missionary and a member of the Royal Canadian Mounted Police. The only native people encouraged to live in these tiny

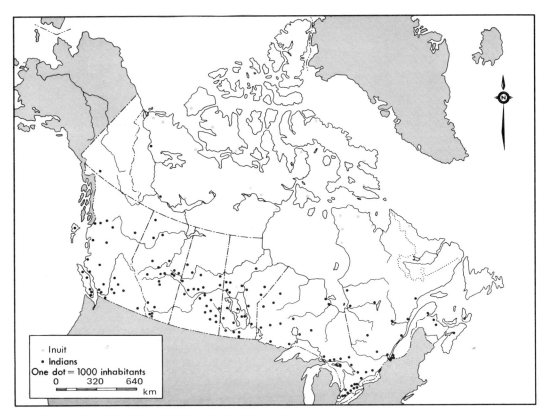

Figure 7.6   INDIANS AND INUIT IN CANADA

settlements were those who served as assistants to the trader, policeman or missionary.

In the first half of this century the Inuit became increasingly dependent upon trapping furs for a living. The successful trappers had to spend more of their time at trapping and less at hunting. This increased their dependence upon trading posts and stores for food and other goods. It also changed their areas of residence, since the areas for successful trapping (particularly for the white fox) were not always the best for hunting caribou, or for fishing and sealing. Thus, the Inuit dispersed from their traditional hunting and fishing areas and moved into areas that had been deserted for centuries.

The Inuit specialization in trapping and dependence upon selling furs for their livelihood had disastrous consequences for them in the economic depression of the 1930's. The sudden severe drop in fur prices caused great hardships among those Inuit who had become accustomed to a new standard of living. Many of them had turned from their nomadic ways and had built frame houses heated with coal or fuel oil. Many owned power boats, and they had become accustomed to non-native clothing and food and a variety of other store goods. Because, on the world market, the prices of luxury items such as furs dropped more than those of the more common products, the Inuit suddenly discovered that their furs would no longer pay for their flour, tea, bacon, clothing, fuel, ammunition, and other modern tools and gadgets. Without these things

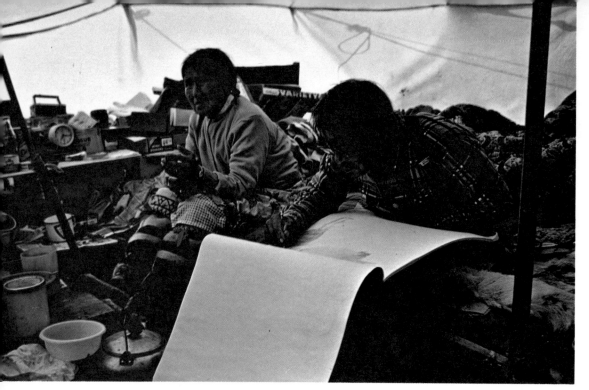

An Inuit at a drawing inside his summer home. Note the many non-native items in the photograph that this Inuit has purchased.

they could not survive. Some Inuit attempted to go back to their old ways, but the younger generation had lost some of the traditional Inuit skills and no longer cared for the former ways of life. As a consequence, many Inuit lived in poverty and became increasingly dependent upon government pensions and welfare payments.

The construction of the DEW line brought another wave of modernization to the Inuit. Most able-bodied Inuit men between the ages of sixteen and sixty in the western Arctic were employed at one time or another on the radar line when it was being built near the coast of the mainland. They received high wages and grew to enjoy the standard of housing and all the other amenities of non-natives. They began buying expensive boats, outboard motors, and many other luxuries.

When construction on the DEW line ended, many Inuit were laid off. The Inuit who obtained maintenance jobs were usually those who had gone to school and could speak English well. Until this time, there had been great resistance among many Inuit parents to sending their children to school because they thought that the education would spoil them for a life of trapping and hunting. But now most parents want their children to go to school so that they can get good jobs. In order for children to go to school, it is necessary to send them away to board or to move the whole family to the large settlements for the whole winter. If an Inuit family lives in a settlement for the winter it cannot trap and hunt as effectively, and thus becomes more dependent than ever on jobs, unemployment insurance and government pensions and welfare.

Top: Inuit paintings and soapstone and whalebone sculptures are in much demand across Canada and on international markets. Many Inuit artists have formed co-operatives to complete their works and to sell them. These photographs show (left) one such co-operative workshop in Cape Dorset and (right) an Inuit pulling a print of a painting by Kenojuak, a famous artist who is also well known for her sculpture. Usually, Inuit art depicts natural objects and animals of the Arctic.

Lower: The Inuit in this photograph is working on a northern construction project. Unfortunately, once the construction phase of development is over, many Inuit become unemployed.

The current wave of petroleum exploration and exploitation is again creating jobs, but when the construction stage is over, the Inuit will again be unemployed. In the process of developing the petroleum resources, more of the Inuit hunting and trapping environment will be destroyed and even more Inuit will be living in fewer and larger settlements.

Until World War II, native affairs in the north were mainly in the hands of the church, the Hudson's Bay Company, and the R.C.M.P. In the 1940's the Canadian Government took over direct responsibility for Inuit affairs, providing facilities and services that touched on every aspect of native life. The Inuit's material standard of living has been greatly improved since this government intervention. Housing standards are vastly superior to what they were. Famine is a thing of the past. A comprehensive health program has greatly reduced the death rate and modern educational facilities are available in the major settlements.

However, these improvements have been made at the expense of the Inuit's self-sufficiency. Only about 25 percent of the Inuit men have full-time employment. A small minority still make a living from hunting, fishing and trapping; the majority do some hunting, fishing and trapping and supplement their income with part-time jobs, the sale of handicrafts, and various government assistance programs. Within this century the Inuit have changed from a proud independent race, well-adjusted to their harsh environment, to a people completely dependent on non-native economy. Unfortunately, along with their loss of independence, they are also losing their language, mythology, and social customs. How to reverse these trends in the future is a major Canadian problem.

The Old Order Mennonites

A number of ethnic and religious minority groups in Canada have settled in colonies and, by retaining most of their traditional beliefs and ways of life, have helped to make Canada a more interesting country. A few of these groups are the Doukhobors, the Hutterites, and the Old Order Mennonites.

The Old Order Mennonites are a religious group with strict religious regulations concerning the kinds of clothing they can wear, the kinds of pleasures they can enjoy, and the modern conveniences they can use. A large number of Mennonites have broken away from observing the strict rules of their religion to become Reformed Mennonites, who generally live in cities. Those Mennonites who still follow the older, stricter religious rules are called Old Order Mennonites, and they generally live on farms.

Large Mennonite settlements are found in Manitoba, British Columbia, and Ontario. One of the largest groups of Old Order Mennonites is found in the Regional Municipality of Waterloo (formerly Waterloo County) in the centre of southwestern Ontario. The original Mennonites came from Germany to Pennsylvania and then made the long trek to Ontario in covered wagons called Conestoga wagons.

The Mennonites chose their new land carefully. Without the assistance of soil scientists or soil reports, they decided to settle in a part of the Waterloo region that has some of the best farmland in Ontario. With good land and hard work, the Mennonites soon became prosperous farmers. Successful farmers needed grist mills, lumber mills, blacksmith

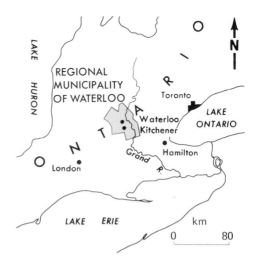

shops, harness shops, general stores, and other services, which led to the establishment of towns and villages; thus, the Mennonites made an important contribution to the development of the Waterloo region.

If you should visit the Waterloo region today, you would see Old Order Mennonites who dress and live very much as they did when they came to the area in the early 1800's. Most of them still drive their horses and buggies. Those who use cars choose a black colour and often paint the chrome black. The most conservative groups do not use tractors, electricity, or telephones and have no curtains on their windows. The Old Order Mennonites still use a kind of German dialect called "Pennsylvania Dutch."

Most of the clothing of the Old Order Mennonites is homemade. The women wear long, plain, loose-fitting dresses. When it is cool out-of-doors, they wear long black coats over which they place a large black shawl. All Old Order Mennonite women wear a white net cap known as a prayer veil, over which they wear large black bonnets. In some groups, this veil is worn all the time, while in other groups, it is worn only at church meetings. The men's clothing is also very conservative. Both work clothes and "dress" clothes are black and no ties are worn. No zippers

Many Old Order Mennonites travel by horse and buggy in the Waterloo region. Signs are posted beside the roads they frequently use to warn motorists to drive carefully. The buggies are equipped with battery-powered lights so that the Old Order Mennonites can drive at night. On the left of the close-up photo you can see something of how the Old Order Mennonites dress. They sell a number of homemade and home-grown products at the Kitchener and other local farmers' markets.

A white, frame meeting house or church of the Old Order Mennonites. Note the horses and buggies between the church cemetery and the meeting house, and the German dialect on the tombstone.

are used, and the most conservative groups use hooks-and-eyes to fasten their clothing. In one group of Mennonites, called the Amish, the men wear beards.

Although Old Order Mennonites do not buy luxury goods, their appreciation of beauty can be seen in the attractive needlework designs on cushions, mats, and quilts. Often a large group of women will spend afternoons at quilting bees making beautiful patchwork quilts. Neatly arranged flower beds and whitewashed fences demonstrate the Mennonites' love of simple beauty. The women also have a very good reputation for cooking and baking.

The churches of the Old Order Mennonites are white frame buildings called meeting houses. Attendance at church is very regular. The men sit on one side of the church, the women on the other. Some meeting houses are used by two groups of Mennonites, with one group driving cars, and the other driving horses and buggies. The car drivers use the meeting house on one Sunday, and the horse-and-buggy drivers use it the next Sunday.

The average Old Order Mennonite family in the Waterloo region has six to eight children. The children go to regular public schools, but they are not allowed to go to school after they are fourteen. At an early age, the boys have to work hard on the farm, while the girls learn to sew, to do housework, and to look after the younger children. In some families everyone has to help prepare products and to sell them at one of the farmers' markets in the Waterloo region. These markets are famous for Mennonite specialties such as cooked cheese, sauerkraut, summer sausage, and homemade soap.

The Old Order Mennonites do not believe in buying insurance. If someone's barn burns down, all the neighbours get together to build a new one at what they call a barn-raising bee. They not only help one another in a number of ways like this, but they also help other people who are not Mennonites. Once, after a tornado left a path of destruction in Southern Ontario, the Mennonites of the Waterloo region formed work-bees to help people to rebuild their houses and barns. In their willingness to help one another, the Mennonites have set a good example for other Canadians.

Although the Old Order Mennonites do not use modern farm equipment, they are excellent farmers. They may take longer to plough, cultivate, and sow their crops by horsepower rather than tractor power, but their harvests are just as bountiful as those of their non-Mennonite neighbours. By rotating crops properly and using barnyard manure and other fertilizers, they have maintained the fertility of their soils. They are also good at raising livestock and at growing fruits and vegetables. Some of them avoid the loss of selling to a middleman by selling their own produce at farmers' markets.

The production costs of the Old Order Mennonites are lower than those of other farmers because they do not use expensive equipment.

This photograph shows the farmers' market that opened in Kitchener in 1976. Local farmers bring products that they have grown or made to be sold here; as you can see, very few Old Order Mennonites come to this newer market, which is part of Kitchener's urban renewal program.

Their cost of living is very low because they lead a simple life without many of the luxuries and entertainments that most other Canadians enjoy. A Mennonite farmer seldom has any hired labour expenses because his whole family is required to help out with the various farm jobs. In general, the Mennonites of the Waterloo region are a prosperous group, and are an asset to their community and their country.

The Old Order Mennonites are having difficulty in resisting modern trends. In the Waterloo region, the cities and towns have grown close to the Mennonite rural areas. There is not room to keep all the young people on the home farm or another farm in the neighbourhood, so some of the young people are taking jobs in the city, where they learn to enjoy modern city life, and leave the old ways of their religion. One way of preserving their culture and simple way of life would be to move to a rural area far from urban influences. However, with present-day communication and transportation, there are few farm areas left in Canada that have not already been reached by modern ways of living.

It is interesting to note that in formulating a land-use plan, the Regional Municipality of Waterloo decided to restrict urban development from growing farther into the Mennonite area. Also the Provincial Government has agreed to build special underpasses in a new expressway so that the Mennonites will not be forced to travel many kilometres by horse and buggy or to cross the highway by way of a cloverleaf. An annual Maple Syrup Festival at Elmira, in the heart of Mennonite country, attracts tens of thousands of people from across the province. These are a few of the indications of the interest and respect that other Canadians have for this minority group with its unique religious and cultural practices.

PROBLEMS AND PROJECTS

1. Write a letter to the community-development officer of the nearest Indian reserve. Ask for a classroom visit, possibly with an Indian visitor.

2. Through your local public library obtain National Film Board movies on minority groups in Canada.

3. Arrange to have a recent immigrant to Canada speak to your class.

4. Have some of your class go through old issues of a local newspaper for "human interest" stories involving minority groups in your community.

5. Copy the graph in Figure 7.1, leaving enough room at the right to extend the population line to the year 2000. Project the population growth at the 1971-1981 rate. What is your projected population for the year 2000?

6. How large do you think Canada's population should be? Consult Science Council of Canada Report No. 25, *Population, Technology and Resources* and *A Population Policy for Canada*, published by the Conservation Council of Ontario and the Family Planning Federation of Canada. Consider the following factors:
   (a) How adequate is Canada's resource base?
   (b) What standard of living do Canadians want to have?
   (c) What impact will a much larger population have on the quality of our environment?
   (d) Can additional population be directed to the more sparsely populated parts of Canada?
   (e) In what way would a larger population help Canada's economy?

7. For some very interesting reading on the nature of Canada's population, see A. Skeoch and Tony Smith, *Canadians and Their Society* (Toronto: McClelland and Stewart, 1973). Accounts of minority groups may be found in the following books: K.F. Dudley, *An Arctic Settlement: Pangnirtung* (Toronto: Ginn, 1972); S. Wilson, *Opasquiak: The Pas Indian Reserve* (Toronto: Holt, Rinehart and Winston, 1973). Descriptions of provincial population characteristics may be found in D.C. Bennett and R.P. Mogen, *Alberta: A People and a Province* (Toronto: Fitzhenry and Whiteside, 1975); L.R. Knight, *Manitoba: A People and a Province* (Toronto: Fitzhenry and Whiteside, 1977); and C. Taylor, *Population and Canada* (Toronto: Guidance Centre, 1977).

URBAN DEVELOPMENT. *Above*. Don Mills and Eglinton, Toronto. *Below*. The same area about ten years later.

8 SETTLEMENT AND LAND-USE PATTERNS

Over the years, the people who have settled Canada have changed the face of the land. Paths have been cut through the wilderness for highways and railways. Vast areas of natural vegetation have been cleared; trees and grass have been replaced by cultivated crops. Wild animals that once stalked the forests or ranged the plains have been replaced by domesticated animals.

In the more densely settled areas, people have greatly changed the appearance of the natural landscape by the things they have built. In the *rural* areas of Canada, the survey system, the road pattern, the type of farm fence, the nature of farmsteads, and the appearance of villages give character to the countryside. In the towns and cities, the natural landscape is scarcely visible. The *urban* landscape is a constructed environment of factories, stores, houses, streets, schools, and so forth. The settled area, including both rural and urban, is often called the *cultural landscape*. The Canadian cultural landscape has distinctive settlement and land-use patterns that differ from region to region.

RURAL SETTLEMENT

LOCAL STUDY

The best way to learn about the rural, cultural landscape in your area is to drive slowly along the back roads. A topographic map will help you find your way. It will also show you the road pattern and the precise location of houses, barns, schools, and churches.

The following questions will give you some idea of the things you should look for the next time you drive through farm country. You may find it interesting to try to answer the following questions from memory and then check your answers in the field. When doing field work, you will find that sketch maps and drawings are a very useful form of note taking. Photos are a great help as well.

1. How far apart are the main roads and crossroads? Compare the width and condition of the main roads with the crossroads. On which roads are most of the farmsteads located? How much land is enclosed in each block surrounded by the roads? (This question applies only in areas with a regular road grid.)

2. How large are the farms? Describe the size and shape of the fields. What proportion of the land is under cultivation? If there is a woodlot, where is it usually found? Can you explain why?

3. Describe and make sketches of the different types of fences. Can you explain why the different kinds of fences are used?

4. How far are the farmsteads from the road?

5. Make sketch maps of several farmsteads, showing where the different buildings are located.

6. Note the location of the farmsteads in relation to the terrain. Are they on the tops or on the sides of hills? Are the barns built *into* the sides of hills? If so, what advantage is there in building this way?

7. Of what materials are the houses made? Are there any particularly interesting architectural styles?

8. What proportion of the barns are painted?

9. How large are most of the rural villages? How far apart are they?

10. What kinds of stores are found in the villages?

11. What kinds of special services do the villages provide for the rural people?

12. How far apart are the rural schools and churches? Are there any abandoned schools or churches? If so, where are the new ones being built?

13. Try to find out how many rural houses there are from which people commute daily to the nearest town or city.

14. How many abandoned farmsteads are there?

Some Rural Settlement Patterns

The answers to the above questions will be different for the different regions of Canada. We are not going to describe in detail the cultural landscape of all parts of Canada, but we will take a brief look at several settlement patterns that are quite different from one another.

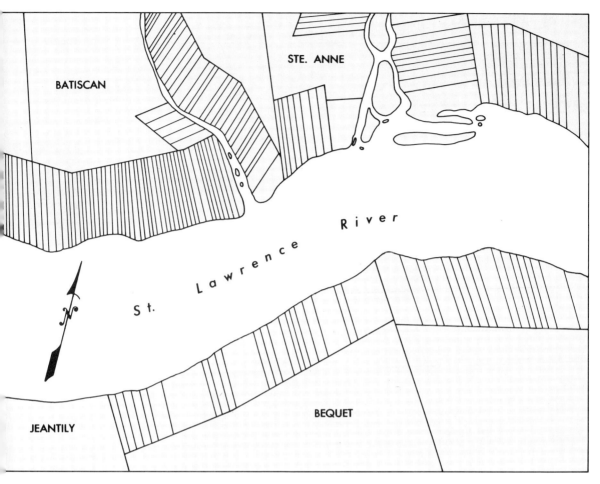

Figure 8.1 FRENCH STRIP FARMS. The French used the river as a road. Their farms fronted on the river and went back in long narrow strips. This pattern of farm settlement can still be seen in Quebec today.

RURAL QUEBEC

Most of the early settlement in Quebec took place along the rivers. The *habitants,* as the French settlers were called, used the river for a road. Their farms fronted on the river and went back in long, narrow strips. Narrow frontage meant lower taxes. The long, narrow farms also gave each farmer a small amount of good farmland along the shore and some rocky forested land farther back (Figure 8.1).

Later, roads were built back from and parallel to the rivers. Here, too, by force of habit, the settlers laid out long, narrow strip farms with long, narrow fields. Because the farms were so narrow, the farmsteads were very close to one another. Even today, in some parts of Quebec, the farm houses are so close to one another that the country road looks like a continuous village (Figure 8.2).

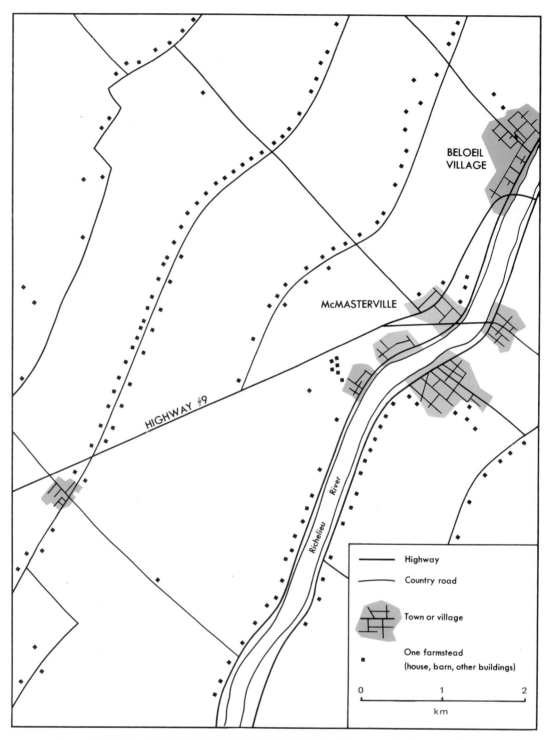

Figure 8.2 RURAL SETTLEMENT PATTERNS IN QUEBEC. Note how the farmsteads are strung along the river and the roads that parallel the river.

In a typical Quebec village, the houses are strung out along a main road. The church is the tallest and most prominent building. Because the surrounding farmland has been divided by fences into long, narrow fields, the farmstead were built close together along the road leading into the village. Thus, it is very difficult to tell exactly where the village ends and the open countryside begins.

Many villages in Quebec seem to be centred on the church. You will find large Roman Catholic churches with towering steeples even in the smallest of villages—a reflection of the great importance of the church to the life of the people of Quebec. The houses in Quebec villages are generally smaller than in most other parts of Canada, despite the fact that traditionally French-Canadians had large families. The houses are usually built very close to the road, and sidewalks are uncommon. Frame houses are much more popular than brick.

On a typical Quebec farmstead, there are a number of buildings, constructed entirely of wood; these include a large barn, several smaller sheds, and a small farmhouse. Two farmsteads are shown in the foreground of the photograph and each farmstead is connected to the country road by a separate farm lane. The farm houses are of two different architectural styles. Home A has a mansard roof design while home B possesses a French-style roof, also very common in rural areas of Quebec. At C, a snake or wooden rail fence leads to the back part of the farm. The village in the middle ground is Baie St. Paul, on the north shore of the St. Lawrence River.

There are many other unique characteristics of the Quebec cultural landscape. Outdoor crosses and religious statues are much more numerous than in other parts of Canada. Differences in housing architecture are immediately noticed by visitors from other provinces. For instance, the mansard roof and the outdoor staircase are seldom found outside of Quebec.

The settled part of rural Quebec is organized into municipal areas called counties. Each county is subdivided into a number of irregularly-shaped municipalities called townships, or parishes. Like the farms, many of the counties are long and narrow and front on the St. Lawrence River.

SETTLEMENT IN SOUTHERN ONTARIO

In Southern Ontario more regular rural settlement patterns were created than in Quebec. Although several survey systems were used in Southern

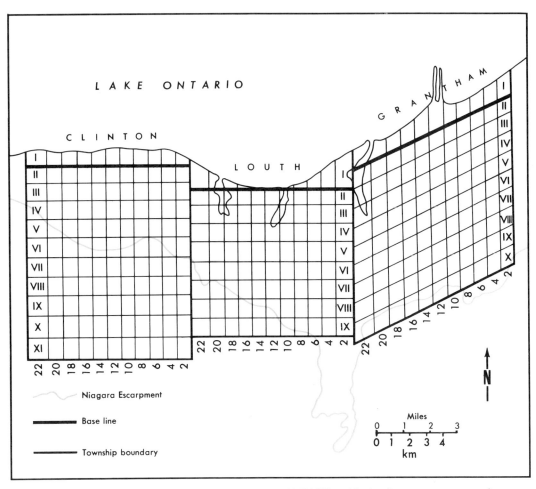

Figure 8.3 TOWNSHIP SURVEY IN SOUTHERN ONTARIO. All the lines represent roads. Those running east-west parallel to the base line are called concession roads; those running north-south are called sideroads. The strips between the concession roads are called concessions and are identified with Roman numerals. Each concession is divided into lots that are numbered with Arabic numerals from right to left. In Figure 8.3, each block bounded by roads contains two lots. It should be noted that the terms concession and lot are not used in all townships in Southern Ontario.

Ontario, the concession and lot system was the most common. The base line for a township was laid out parallel to the shorelines of one of the Great Lakes, and roads were then built parallel to the base line and approximately one mile apart. These were usually called concession roads and the strips between them were called concessions. Crossroads, usually called side roads, were then laid out perpendicular to the concession roads at approximately the same distance apart as the concession roads. This pattern formed equal units of land, each approximately one square mile in size.

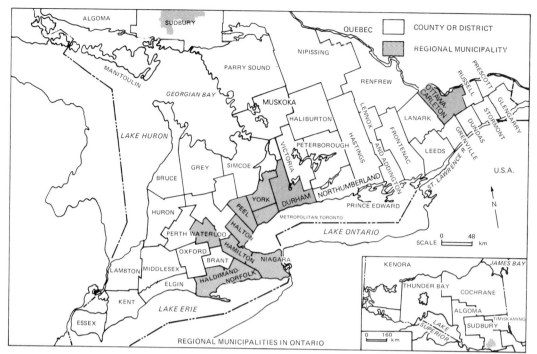

Figure 8.4 COUNTIES, DISTRICTS, AND REGIONAL GOVERNMENTS IN ONTARIO. Because urbanization required new forms of government, the cities and townships of some original counties were reorganized into regional governments (in grey). In the cases of Haldimand-Norfolk and Niagara, two counties became a single municipality. Districts are large, less populated areas of government found mainly in Northern Ontario.

Because the townships were surveyed independently of each other, each township had a different base line, and therefore, the concession roads ran at different angles and did not always meet at the boundaries between townships. This road pattern is well illustrated by the survey pattern of three townships in the Niagara Peninsula (Figure 8.3)

The townships were organized into counties and all counties and townships were given names. The counties and townships along the lakes tended to be rectangular and regular in shape. However, by the time the survey reached the centre of the Ontario peninsula, the shapes became very irregular. The resulting pattern of counties—and the later districts and regional municipalities—in Southern Ontario looks something like the crazy-pattern quilts the pioneer women used to make (Figure 8.4).

In most parts of the province, each block bounded by roads was divided into about six to eight farms of approximately 100 acres each. The farmsteads were located at equal distances from each other along the main concession roads.

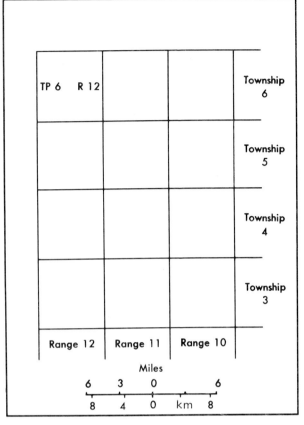

Figure 8.5 TOWNSHIP SURVEY IN WESTERN CANADA. In the upper left-hand corner, the township would be referred to as Township 6, Range 12 (Tp.6,R.12). Each township is six miles square.

Each township usually has several very small hamlets and at least one village or town to provide the services for the surrounding farmers. Most of the farmers move to these villages and towns when they retire. Many of the villages have little industry and have remained approximately the same size for many years. Some of the villages are actually declining in population. On the other hand, some villages have grown into towns, and some towns have become cities because of industrial development.

PRAIRIE SETTLEMENT PATTERN

Most of the farmland in the Prairie Provinces was surveyed, and the road pattern established, before the area was settled. Because of the relatively level land and the perfectly straight Canada-United States boundary, most of the land was divided into square blocks called townships, which measured six miles on each side. The townships in the Prairie Provinces were not named, as in most of eastern Canada, nor were they grouped into counties. The townships were arranged in rows running north and south. Each row of townships was referred to as a *range*, and each range was given a number. Each township within the range was also numbered

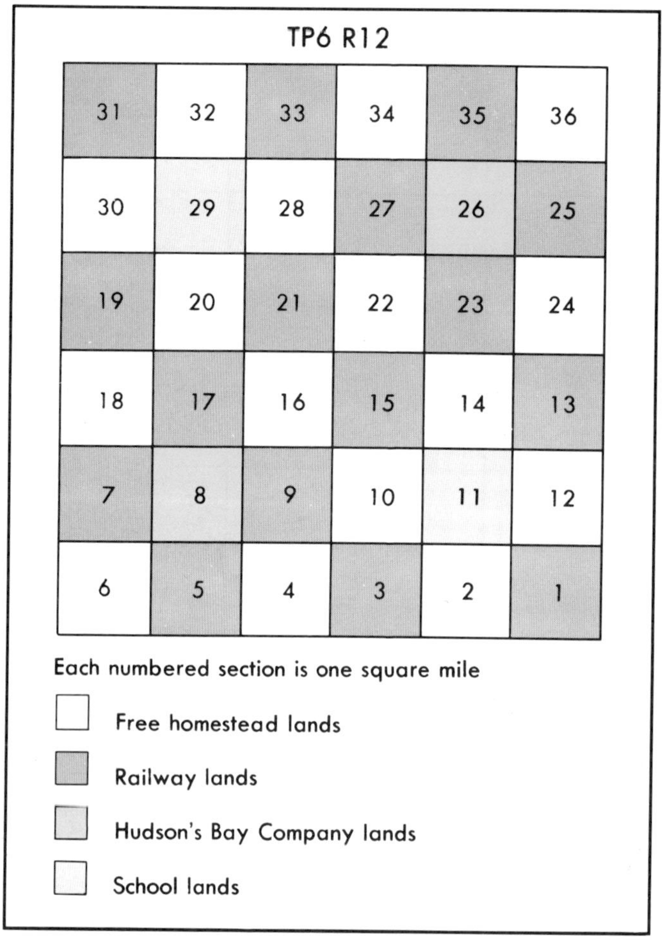

Figure 8.6 SUBDIVISION OF LAND WITHIN A PRAIRIE TOWNSHIP (Tp.6, R. 12). Townships in the Prairies like Tp.6, R. 12 are divided into 36 one-square-mile units of land called sections; each section contains 640 acres of land.

(Figure 8.5). Each township was then divided into 36 one-square-mile units called sections. These sections were further divided into quarter-sections of 160 acres each. Although roads were not built around every section, enough roads were built to make it possible to settle all the land.

In the late 1800's and early 1900's, quarter-sections were granted free to the settlers or homesteaders. But not all of the land was given to these homesteaders. Usually, every other section was given to the railway company, and when the settlers needed more land, they could buy it from the railway company. This system was the government's way of helping to pay for railroad construction. In a similar manner, two sections in each township were reserved as school lands, and when these sections were sold, the money could be used to build schools. In many townships, the Hudson's Bay Company was also granted one or two sections (Figure 8.6).

In the past few decades, the rural farm settlement has been thinning out in the Prairie Provinces. Modern farm equipment is not economical to use unless the farmer has a very large area, and thus, as farms come

When seen from the air, a prairie village looks isolated by the large, open fields surrounding it. The tall elevator at (A) dominates all other buildings such as the few stores and the homes of retired farmers. Prairie towns are often located at the point where a county road (B) and a railway line (C) cross each other. You can see that fences do not enclose the grain fields nor are trees very common, except for the few around each farmstead. Most of the farm buildings are sheds (D) used for grain storage and they look like small cabins. A larger structure such as (E) serves as an implement shed. (See two coloured photos of prairie farmsteads on page 121.)

up for sale, neighbouring farmers usually buy them. It is quite common today for one farmer to own one or two sections of land. Sometimes, the farmer does not live on the farm at all but lives in a nearby city or town.

Fences do not dominate the farm landscape in the Prairies as they do in eastern Canada. In the grain-growing areas where no cattle are grazed, there is little need for fences. Even in the cattle-raising areas, fences are few and far between. In some areas, the cattle are free to roam over what is called open range.

The most striking features of the cultural landscape of the grain-growing areas of the Prairies are the towering elevators in which farmers store their grain. Almost every village and town in the grain-growing districts has at least one, and often several, elevators, which can be seen on the horizon for many kilometres. Almost all elevators are built and look the same. Thus, from a distance at least, almost all Prairie villages and towns look much alike.

In recent years, farmers have been storing grain in metal, cylindrical bins that look like short, stubby silos. A series of the bins are usually built in a straight line. These bins have added a new element to the landscape of the Prairies.

A number of Prairie villages are declining in importance. At one time, they were essential as service centres for the surrounding farmers. However, with better roads and automobiles, farm people are willing to drive farther to the larger towns for their supplies. Also, the greater use of trucks means that elevators need no longer be located so close to the farm. As more farmers truck more grain for greater distances, there is less need for branch railroads. If a branch railroad is closed, the villages on that line suffer an even greater decline in business and population.

Thus, the rural farm and village population is decreasing while the major towns and cities are growing in size. If this trend continues for a period of time, the cultural landscape of the Prairies will be quite different from what it has been in the past.

VARIATIONS IN SETTLEMENT PATTERNS IN MANITOBA

The early French settlers in Manitoba established farms along some of the major rivers. They used the river as a road and divided the land into long narrow strip farms just as they had done in Quebec. This river-lot strip farm development is still evident today and can be seen on topographic maps of southern Manitoba.

The Mennonite farm villages in Manitoba provide another interesting variation to the settlement pattern of the Prairies. Instead of scattering the farmsteads along the section roads, the Mennonites congregated their houses and barns in villages, as they were accustomed to doing in Russia.

They pooled their quarter-sections so that all could share equally from the different qualities of land. The meadows that were least valuable for crops were used as common pasture for everyone. The remainder of the land was divided into narrow strips, and each farmer was given several strips of different quality land at different distances from the village (Figure 8.7).

The villages were located near creeks from which the Mennonite farmers obtained their water supply. The length of the village street depended upon the number of farmsteaders (usually between 20 and 30), but few villages were longer than a half mile. The streets between the two rows of farmsteads were usually over 100 feet wide. These wide streets provided plenty of room for turning implements, piling firewood, and hitching horses. In the early days, the buildings were constructed of logs, but later, sawn lumber was used. The house was always attached to the barn.

Before the turn of the century, there were about 100 Mennonite villages in the Manitoba Lowland along the United States border. Today, there are only a few villages left, and in these, only a few of the original

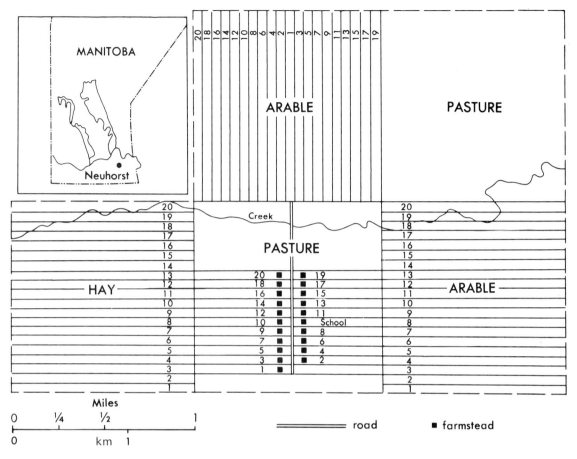

Figure 8.7 NEUHORST: AN HISTORIC MENNONITE VILLAGE IN SOUTHERN MANITOBA. The Mennonite farmer who lived on farmstead 16 worked the fields marked with that number. He could also make use of the common pastureland to graze his livestock.

In the Mennonite districts of Manitoba, the barns are usually attached to the houses. Farmsteads like these add variety to the cultural landscape of the Prairies.

buildings remain. The land-holding system has also changed. The remnants of these villages can be seen on the east side of the Red River between Highway 14 and the United States border.

The villages broke up for a number of reasons. The system of individual narrow strips of land was not suitable for Canadian commercial wheat growing because the long trip to the distant fields consumed too much time each day, and the narrow grass bands between the farm strips became infested with weeds and insects. Some Mennonites became impatient with the uneconomic way of farming. They demanded their quarter-section from the pooled land and built their house and barn on their own land. Others followed suit, and soon, the villages had broken up.

The Mennonite settlements in Manitoba had no influence on the settlement patterns of the rest of the Prairies. The Mennonites came after the square-mile survey had been completed. Some people believe that if the Mennonites had settled on the Prairies before the survey, the settlement pattern would have been one of farm villages, as in Europe, instead of the scattered farmsteads with which we are familiar.

A LAND OF CITIES

For many years, Canada was considered to be a rural country, and was known for its products from farm, mine, and forest. Indeed, at the turn of this century, Canada *was* a rural country, with almost two-thirds of its population living in rural areas.

Today, Canada may be called a land of cities, since about eighty percent of all Canadians live in cities or large towns. In addition, many people, classified by the census as rural, are really city folk living in the country. Although they live in homes in the country, these people travel to the city for their work, shopping, and entertainment; thus, they may rightly be classified as urban instead of rural.

The shift of population from rural to urban is a result of technological change. With mechanized equipment, it now takes fewer people in the primary resource industries (mining, fishing, lumbering, farming) to support very large populations. The modern industrial society, with its high standard of living and demand for many goods and services, creates numerous job opportunities in the towns and cities.

In addition to the shift from rural to urban, there is also a trend toward larger cities. Many villages and towns and some small cities are either growing very slowly or actually declining in size. The people are moving to the larger cities where there are more job opportunities and a different style of life. In 1981, almost 45 percent of all Canadians lived in the nine largest cities. Even more striking is the statement that one out of every three Canadians lives in our three largest metropolitan centres of Montreal, Toronto, and Vancouver. It has been predicted by urban researchers that this trend will reverse slightly and that by the end of this century only

40 percent of the Canadian population will live in the nine largest cities, and one out of every four Canadians will live in the three largest metropolises. Some western cities will grow at an even faster rate (see accompanying table).

Large Metropolitan Populations

City	Population (in thousands)	
	Approximate 1981	Projected 2001
Toronto	2999	3319
Montreal	2828	2928
Vancouver	1268	1328
Ottawa-Hull	718	923
Edmonton	657	895
Calgary	593	875
Winnipeg	585	678
Quebec City	576	683
Hamilton	542	569
Total	10 767	12 198
% Canadian Population	44%	40%

Source: Statistics Canada, 1981 and estimates for 2001. For the most recent figures, consult the latest edition of the *Canadian Almanac and Directory*.

Urban Geography—A Study of Cities

Since most Canadians live in cities, and since cities are so important to the growth and development of our country, many Canadian geographers are specializing in the study of urban geography and the problems related to cities. The following questions are a few of the ones for which geographers try to find answers.

QUESTIONS

1. Why do cities exist? What goods are manufactured in cities? What other services do cities provide? From how far do people come to cities of different sizes for shopping, entertainment, and other services? Why do some cities become important for specific functions? Why do most cities grow? Why do some grow faster than others?

2. Why are cities located where they are? How far apart are the major cities? How far apart are the small cities, towns, and villages? How do the functions of various sizes of cities differ? Why? In what way are cities dependent upon one another?

3. In what shapes and directions are the cities growing? What are the major uses of land in cities? Where are these land uses located? Why? How and why are the land use patterns changing?

4. What are the major problems facing cities? What can be done to solve the problems?

In Montreal harbour, western wheat stored in elevators (A) is loaded aboard ships destined for overseas ports. Warehouses (B) receive imported manufactured goods that are then put into railway boxcars (C) to be sent to central and western Canada.

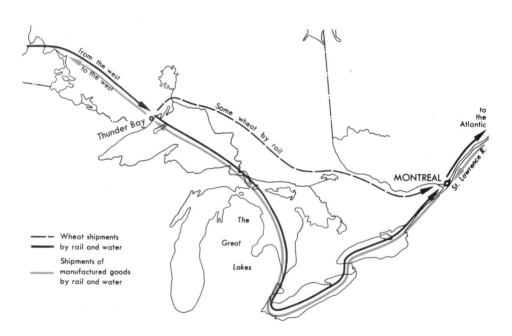

Thunder Bay, like Montreal, is a major port that functions as a break-of-bulk centre. Huge elevator facilities for grain storage line the waterfront. Boxcars full of wheat from the west are unloaded into the elevators all year round. During the shipping season, freighters carry the grain eastward on the Great Lakes.

There is not enough room in this book to answer the previous questions about all Canadian cities. Instead, we are going to discuss the topics suggested by these questions in a general way so that you may apply the ideas to a study of the cities in any Canadian region.

The Origin and Growth of Cities

Historically, cities grew out of the production of farm surpluses and the desire to trade. So long as farmers produced only as much as their families could consume, built their own shelter, and wore homemade clothes, there was little need for cities. When farmers started producing more than they could consume and wished to exchange surplus produce for city-made goods and services, the need for trading centres was established.

In pioneer agricultural areas, the farmers needed a central place to provide them with a market for their produce, a general store, a blacksmith shop, a harness shop, a gristmill, and a sawmill. As farm production grew, business in the town increased, and there was a demand for hotels, law offices, banks, and other services. These new services employed more people who, in turn, required more personal services from barbers, tailors, butchers, and bakers, and the demand grew for even more specialized stores and services. The growing population provided a larger labour supply for factories that manufactured goods for the surrounding country. If the factories were successful, they expanded and created more jobs

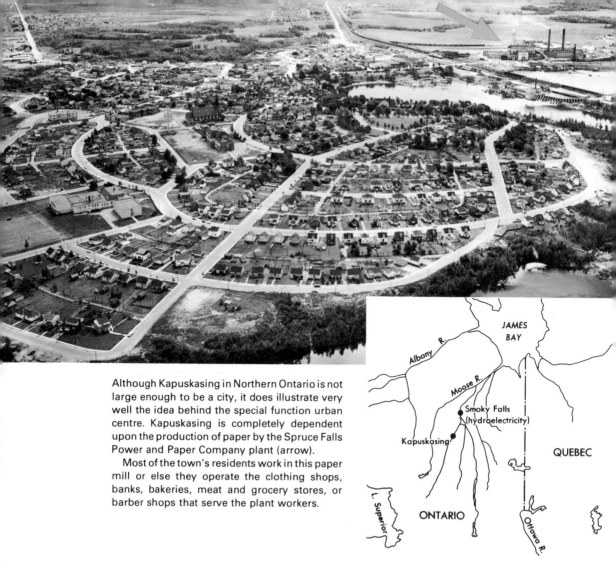

Although Kapuskasing in Northern Ontario is not large enough to be a city, it does illustrate very well the idea behind the special function urban centre. Kapuskasing is completely dependent upon the production of paper by the Spruce Falls Power and Paper Company plant (arrow).

Most of the town's residents work in this paper mill or else they operate the clothing shops, banks, bakeries, meat and grocery stores, or barber shops that serve the plant workers.

which attracted still more people to the town. As the town grew to become a city, it provided special recreational facilities such as theatres, parks, libraries, and museums, as well as special schools and colleges.

Not all cities began as trading centres for farm areas. Some cities were located where cargo had to be moved from one type of transportation to another. For example, at a port city, export cargo is transferred from rail cars and from trucks to ships, and the reverse occurs with import cargoes. Places where cargo is moved from one form of transportation to another are called *break-of-bulk* places. Break-of-bulk places require many other types of urban services such as churches, schools, banks, warehouses, hotels, stores, restaurants, and so on. Available labour and city services attract factories, and a "snowball" effect takes place just as with the trade centres we discussed earlier.

Actually, a port city is a special kind of trade centre. A port's trade area is called a *hinterland* (land behind) and may even include the greater portion of the country. The hinterland of Montreal, for example, extends as far west as the Prairies because western wheat is exported through

Montreal and because imports to the Prairies often enter Canada by way of Montreal. Montreal is a break-of-bulk place in two ways. Cargo is transferred from railroads to ships, and also, from lake freighters to ocean ships.

Thunder Bay (formerly the twin cities of Port Arthur and Fort William, amalgamated in 1970 as one municipality) is a good example of a port city. It is located at the Canadian head of navigation on Lake Superior. Since water provides the cheapest transportation for bulky cargo, Prairie wheat is taken by rail to Thunder Bay and is then shipped by boat to eastern Canada or overseas. Likewise, manufactured goods from eastern Canada are shipped as far as possible by boat and are then transferred to railroads at Thunder Bay. Because the Great Lakes are closed to shipping in winter, great storage facilities have been built at Thunder Bay. The need for winter storage explains the many huge elevators found at this lakehead city, in spite of the fact that there is no wheat produced in the surrounding area.

Still other cities have grown up to serve special purposes. They are often dependent upon a single resource or upon the manufacture of a single product. Timmins, for example, is a mining town, Arvida is an important aluminum production centre, and Oshawa has an economy that depends primarily upon the automobile industry. Once started, a specialized city acts as a nucleus for similar or related activities. Arvida's cheap hydroelectric power used for aluminum smelting has attracted other manufacturing industries. Oshawa has attracted a number of factories that manufacture parts for the automobile industry. As with all cities, a whole range of commercial and service activities develop as the city grows.

We have discussed three general reasons for the origin of cities:

(1) central places providing a market and services for the surrounding trade areas;
(2) transport cities providing break-of-bulk and related services;
(3) special function cities providing one special service such as the mining or manufacturing of one product.

While the origin of cities can be explained by one of the above factors, further growth of cities results from a combination of these factors. In the case of many cities, it is difficult to determine what is the most important growth factor. It is clear that there is always a "snowball" effect in urban growth because growth seems to attract even more growth. For example, many manufacturing industries like to locate in or near large urban centres where there is a large labour supply and a large market for their products. Some gravitate towards existing concentrations of industry because they require the products of, or supply products to, other manufacturing plants. A large population, regardless of its employment, attracts commercial, financial and personal services, which in turn employ more people, who require more goods and services.

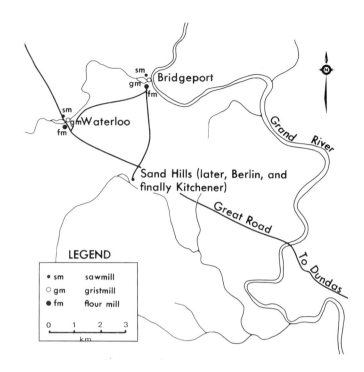

LOCAL STUDY

As a project, make an historical study to find out how and why a city or town in your region got started, and then, find out how and why it continued to grow. If you are lucky, you will find that some local histories have already been written and are in the public library. Sometimes, the local Historical Society or Women's Institute may be able to supply unpublished historical records. From old newspapers and from interviews with senior citizens, you may be able to gather other valuable information.

In doing this study, you will have to keep in mind that you are writing an historical geography, not a history. History is concerned with everything that happened in the past. Historical geography deals only with the factors that led to the present geography. The following study of the growth of Kitchener-Waterloo in Ontario will provide a guide for the kind of information you should try to collect.

HISTORY OF THE GROWTH OF THE CITIES IN THE WATERLOO REGION

The prosperity of the Mennonite farmers in Waterloo County led to the establishment of a number of sawmills, gristmills, and flour mills on small tributaries of the Grand River where waterpower could be developed. Blacksmith shops and general stores followed, and soon, the three villages of Bridgeport, Waterloo and Sand Hills were established. Sand Hills was later called Berlin — a name which was changed to Kitchener during World War I.

The surrounding area provided excellent resources for further industrial development of these settlements. The farmland was excellent for grain growing, and there were heavy stands of timber including white pine, cedar, hemlock, oak, cherry, and black walnut. The first mills were erected to provide lumber, grist, and flour for the local people. Very soon, these mills were shipping flour and lumber by oxcart to Dundas by way of the Great Road. It is interesting to note that Dundas was more important than Hamilton at this time.

In the 1840's and 1850's, breweries and distilleries were established by German settlers. These industries provided another market for the grain of the local farmers. Furniture factories were also established, using the lumber from the local sawmills, and meat-processing plants provided a market for the livestock of local farmers.

In the 1840's, steam power was first used in the area. This new power overcame the problem of limited supplies of waterpower, and new factories were attracted to the area. The construction of the Grand Trunk Railway in 1856, connecting Kitchener with Hamilton and Toronto, gave further encouragement to industrial development. Because the railroad went through Kitchener, that community developed more rapidly than the other towns and it became the dominant city in the region.

Figure 8.8 CITIES AND TOWNS OF FORMER WATERLOO REGION. By 1972, the communities of Kitchener-Waterloo and Hespeler-Preston-Galt had grown to form one urban area.

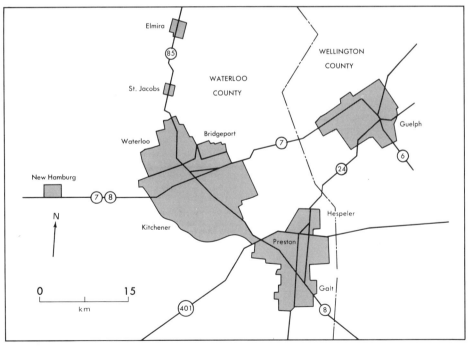

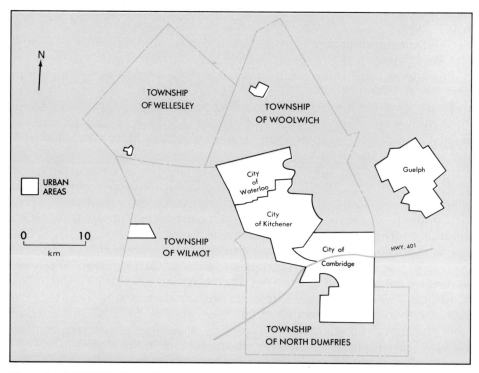

Figure 8.9 THE REGIONAL MUNICIPALITY OF WATERLOO. Since 1973, there has been a regional government over the whole area to administer regional affairs and to plan urban development so that a high quality environment will result. Despite its closeness and common interests, Guelph decided not to be included in the new regional government.

After the coming of the railroad the industrial development no longer depended on local raw materials. By 1912, Kitchener had more than 100 factories manufacturing such articles as buttons, leather goods, clothing, shoes, and rubber products. By the 1920's, Kitchener had become Canada's leading producer of rubber products and furniture, and its shoes, clothing, and meat products were marketed nationally. The twin city of Waterloo had become famous for its breweries, distillery, and life insurance companies.

Diversified industrial development has continued in Kitchener up to the present. A wave of new manufacturing industries followed the opening of the four-lane Macdonald-Cartier Freeway (Highway 401). This highway reduced travel time between Kitchener and Toronto to one-and-a-half hours. Huge new rubber, electronics, and auto-parts plants have been located in an industrial park near the highway. Because of Kitchener's good transportation connections with the major automobile manufacturing cities (Windsor, St. Thomas, Oshawa), when the Canada-United States auto pact led to increased Canadian auto production, a large number of auto-parts companies were built in Kitchener. On a lesser scale, the same kind of diversified industrial development occurred in the surrounding cities and towns: Preston, Hespeler, Galt, and Guelph. The great diversification of industry has led to a very stable job market in the area. It has one of the lowest unemployment records in Canada.

Along with the industrial development came the growth of commercial, financial, cultural and personal services. Waterloo became the home of two universities, another university was located in nearby Guelph, and a community college was built in Kitchener. The rapid growth of these educational institutions in the 1960's and 1970's greatly stimulated the local construction industry, noticeably increased the business of local merchants, and helped to enrich the cultural activities in the communities.

A Central Business District (C.B.D.) developed in each of the cities and towns, with the largest one, of course, in the largest city—Kitchener. To prevent the decay of the C.B.D., as has occurred in some other major North American cities, in the early 1970's Kitchener embarked on a major urban renewal program. The major concepts included taking all traffic off the main street, and turning the street into a shopping mall; building parking garages on a one-way road that rings the C.B.D.; and building pedestrian ways that lead to a park on one side of the C.B.D., and on the other side to the civic centre which contains the library, concert hall and government buildings. The famous Kitchener farmers' market has been retained in a new building right in the centre of the C.B.D. The public urban renewal program has stimulated many store owners either to redevelop or renovate their properties so that they can remain competitive in a bustling new downtown. The enthusiasm has spread to the other cities and towns in the area which have begun their own urban renewal programs on a more modest scale.

By the mid 1960's Kitchener-Waterloo and Hespeler-Preston-Galt had grown so much that they were fast becoming one urban area (Figure 8.8). Intermunicipal problems in urban planning and in co-ordinating services such as water, sewers, major roads, public transportation, policing, and welfare, were becoming greater every year. After several years of study by the Provincial Government and local municipalities, in 1973 the Regional Municipality of Waterloo was formed. It includes all the municipalities within the old Waterloo County but excludes Guelph, which is in the adjacent County of Wellington. Hespeler, Preston and Galt were amalgamated to become the City of Cambridge, a name derived from the first settlement in that area, Cambridge Mills. Bridgeport was amalgamated with Kitchener, but the City of Waterloo remained an independent municipality. Towns were amalgamated with the surrounding rural areas to form new township municipalities (Figure 8.9).

In this new local government organization, the local municipalities look after purely local affairs and the Regional Government administers matters that affect the whole region such as regional planning, roads, parks, water supply, police, and health and welfare services. The Regional Government has prepared a master plan of development that is designed to preserve a healthy economy and create a high quality environment for the people of the region. A unique aspect of the Waterloo Region Official Plan is a program to protect environmentally sensitive areas, based on recommendations from academic and other professional ecologists.

Locations of Cities

QUESTIONS

In an atlas, locate the Canadian cities with populations of 100 000 or more and answer the following questions about their locations (see accompanying table of city populations).

1. What proportion of the cities are within 240 km of the United States border? What is the urban locational advantage of being near the U.S.A.?

2. What proportion of the cities are within 240 km of either Montreal or Toronto? Why do smaller cities tend to cluster around the large metropolitan centres?

3. What proportion of the cities are located within Canada's major agricultural regions (Figure 9.3)? In what way is agricultural production an urban locational factor?

4. Which of the cities are located on navigable waterways? Which industries in these cities are primarily dependent on water transportation? Why? What locational advantages are there, besides transportation, in locating a city on a major body of water?

5. Which cities are seaports (excluding those on the Great Lakes)? Is the tonnage of cargo going through the ports proportional to the size of the cities?

6. Which cities are on the four-lane highway connecting Montreal and Windsor?

7. Which cities are on the mainline of either the Canadian Pacific or the Canadian National Railways?

8. Which cities are serviced by Air Canada or Canadian Pacific Airlines?

9. Which cities are within 80 km of the oil and natural gas pipelines originating in Alberta?

10. Which cities have you named most often in answering the above questions? Are these the largest cities?

11. On the basis of the answers to these questions, state several general principles of the location and distribution of cities.

We mentioned before that in the early days of any agricultural region, a large number of crossroad hamlets and villages developed which served as trading centres for the farm population. Those with the best locations

Canadian Cities over 100 000 Population

City	(in thousands)
Toronto	2999
Montreal	2828
Vancouver	1268
Ottawa-Hull	718
Edmonton	657
Calgary	593
Winnipeg	585
Quebec City	560
Hamilton	542
Mississauga	310
St. Catharines	304
Kitchener	289
London	284
Halifax	278
Windsor	246
Victoria	234
Brampton	165
Regina	164
St. John's	155
Oshawa	154
Saskatoon	154
Sudbury	150
Chicoutimi	135
Thunder Bay	122
Saint John	114
Trois-Rivières	112
Burlington	105

Note: population figures include the whole urban area around a central city, called the metropolitan area or greater city. More recent figures than these 1981 estimates can be found in the latest edition of the *Canadian Almanac and Directory*.

usually grew into towns and cities while others grew very little or even disappeared. In time one city outgrew all the others until it was several times the size of the next largest city. This largest city is usually called a metropolitan centre, and provides a wide range of specialized services for the entire region. In the same region there are usually several other major cities, a large number of smaller cities, and an even larger number of small towns and villages. The smaller the centre, the smaller is its market area and the fewer are the services it provides.

QUESTIONS

The *distribution patterns* of cities of different sizes are well illustrated in the settled parts of southern Canada. Using a road map of your province that shows the location of various cities, towns, and villages, describe these patterns. (Official road maps are available from government departments of transportation or tourism.)

1. How many metropolitan centres and cities with over 10 000 people are there in your province?

2. In what way are these urban centres distributed (arranged) across the province? Explain why this distribution pattern has occurred.

3. On the map, draw circles with a radius of about 16 km, using the location of all towns between 1000 and 10 000 population as centres for the circles. Then find the average number of villages under 1000 population within each circle. Are any large parts of your province not within your circles? Can you account for this?

Good examples of metropolitan centres in Canada are Vancouver, Winnipeg, Toronto, and Montreal. These metropolitan centres have no other city within several hundred kilometres of them that can compete in size, importance and in the variety of goods and services provided. Edmonton may also be called a metropolitan centre, but it has some competition from Calgary, less than 320 km to the south. In a sense, the metropolitan function in Alberta is split between the two cities.

These metropolitan centres attract people from several hundreds of kilometres for special shopping and entertainment. They have a great variety of stores, specialty shops, and restaurants. They are the homes of professional sports teams, attract major theatre and musical productions, and their newspapers have subscribers over a large region. Although not all of these metropolitan centres are provincial capitals, they are all financial capitals of their regions. The major financial institutions such as banks, trust companies, insurance companies, and the headquarters of large manufacturing companies operating in the region are located in these metropolitan centres.

These metropolitan centres influence vast areas that often cross provincial boundaries (Figure 8.10). The shape of the tributary areas of metropolitan centres depends upon the direction of the major transportation routes and the distance from competing metropolitan centres. It is interesting to note that no major metropolitan centres have developed in several of the provinces. Regina, the political capital of Saskatchewan, is overshadowed by Edmonton and Winnipeg. The branch offices of companies developing resources in northern Saskatchewan are usually located in Edmonton or Winnipeg.

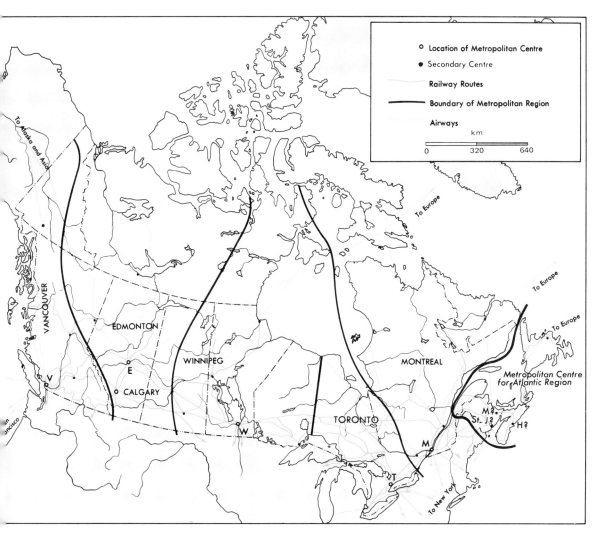

Figure 8.10 REGIONS OF THE METROPOLITAN CENTRES OF CANADA. Give the advantages of using these urban-centred regions as a new set of provinces to replace the current ones.

In the Atlantic Provinces, there is no one city that overshadows all the others and that serves as a regional capital for the whole area. This fact may be partly explained by the small population of the region, and the physical barriers separating the provinces.

The two Canadian metropolitan centres of Toronto and Montreal overshadow all others in both their size and their influence over the economy of the country. With each of their populations over two million, they are more than twice the size of the next largest city, Vancouver. In addition to the impact on the surrounding tributary areas (Figure 8.10), Toronto and Montreal are the financial centres of the whole country. In these cities are located the head offices of most of the major banks, insurance and loan companies, and manufacturing and mining companies. The two major Canadian stock exchanges, where shares in the major Canadian companies are bought and sold, are located in Montreal and Toronto. The two cities account for approximately two-thirds of all cheques cashed in Canada, which indicates the magnitude of financial activity.

There has been friendly rivalry between Toronto and Montreal to see which city would be the most important financial centre in Canada. Since about 1950, Toronto has been in the lead. There are more head offices of banks, and insurance and loan companies in Toronto, and more stocks are sold each year at the Toronto Stock Exchange than at the Montreal Stock Exchange. About 70 percent of the head offices of Canadian mining companies are located in Toronto.

Sites of Towns and Cities

While the general location of towns and cities is heavily dependent upon transportation routes, the precise location or *site* of a city is dependent upon a number of factors. For example, there should be reasonably flat and well-drained land and, preferably, enough earth over the bedrock to permit the excavation work for the foundations of buildings. A supply of pure water is required, as well as a stream or lake into which sewage (treated or untreated) can be discharged. Cities should not, of course, be located on river floodplains because of the threat of floods. Winnipeg provides a good example of the flood damage that results, and costly flood control projects that are required, when cities are built on floodplains.

LOCAL STUDY

Describe the site of your hometown or city. In doing so, consider the following questions.

1. Name the type and describe the nature of the landform on which the original settlement was made.
2. How good is the natural drainage of the land? Does all of the land slope in one direction? Do pumping stations have to be used in the sewage system?
3. How deep is the bedrock? Are there any problems in finding good support for the foundations of buildings?
4. What is the source of the water supply? Is there ample supply for future growth? Is the water hard or soft? How much does it have to be purified before it can be used for human consumption?
5. Where does the city sewage go? If septic tanks are used, is the soil suitable for this method of sewage disposal? If there is a sewage system, is the lake or stream sufficiently large to take the flow of treated sewage without ruining the water for other purposes?
6. Are valleys, gullies, or steep slopes a hindrance to transportation?

7. Is there any flood threat to the city? If so, how often has there been flood damage since 1900? What flood-control measures have been taken? What building restrictions are there on the floodplain?

8. Is the future development of the city in any way hindered by the nature of the site?

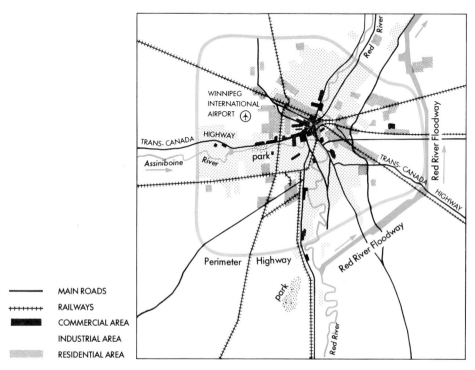

Figure 8.11 LAND USES IN WINNIPEG. From a central site close to where the Assiniboine and Red rivers meet, land uses in Winnipeg have expanded outward. Explain how these rivers and other factors have influenced the location of the larger commercial, industrial, and residential sections of Winnipeg. Because of frequent severe flooding along the banks of the Red River, a floodway channel was built to divert much of the floodwater around Winnipeg.

Uses of Land in Cities

Every city has its own peculiar geography of land uses. The exact locations of commercial, industrial, residential, and other urban land uses are determined by many factors including landform features, transportation routes, land ownership, and the whims of individuals. However, there are some general patterns of land use that apply to the land-use geography of most cities. Study Figure 8.11 showing land use in Winnipeg.

COMMERCIAL LAND USES

Every city has a core of commercial and business land uses. This "downtown" area, which contains major department stores, many specialty shops, theatres, restaurants, and office and government buildings is called the Central Business District, or C.B.D. for short. It has usually developed on the original site of the city, and has retained its commercial and business pre-eminence because it has good transportation access to all parts of the city and the surrounding region.

Land prices in the C.B.D. are very high, therefore it pays to expand vertically instead of horizontally. Thus, the C.B.D.'s in the largest cities are characterized by tall buildings. When space becomes very expensive, even automobiles are stored in multiple-level parking garages, instead of space-consuming surface lots.

The C.B.D. is really the heart of the city. Although not many people live in the C.B.D., it has a dense daytime population. Thousands of people pour into the C.B.D. every morning for work and rush out again at night, and thousands more flock into the C.B.D. each day for shopping, entertainment, or other business. Although there are other shopping centres in large cities, the C.B.D. has the widest variety of goods and the best range of quality and prices.

Many of the C.B.D.'s in the older Canadian cities have had difficulty modernizing enough to meet the competition of suburban shopping plazas. The streets of the downtown areas are often too narrow for the flood of cars, and there is insufficient parking space. Sometimes the buildings are old, dilapidated, and overcrowded, and there is a lack of open space and green area that would make the C.B.D. attractive to shoppers. Cities are combatting these problems in a number of ways. The federal and provincial governments have given a number of cities financial assistance to redevelop obsolete buildings and streets in their C.B.D.'s. Modern shopping facilities, wider streets, parks, and parking garages are being provided. In some cases, streets have been turned into shopping malls from which cars are excluded. These malls are decorated with trees, shrubs and flowers and are provided with outdoor cafés and park benches to make them more attractive and convenient for shoppers. In many cities the public transit system is being improved in order to discourage so many automobiles from entering the C.B.D., thus reducing the congestion and air pollution resulting from the heavy traffic.

Although the C.B.D. is the most important commercial district of a city, various types of commercial land uses are located in other parts of the city. There are usually stores strung out along the major streets which fan out from the C.B.D. There are also major shopping plazas located in the suburbs. In the very large cities some of these suburban shopping centres almost rival the C.B.D. for range and variety of goods provided. However, normally the suburban plaza caters to weekly and even monthly family shopping, whereas in the C.B.D. very specialized goods and services are found. In addition, each neighbourhood has a shopping centre or corner store where some daily needs of the family can be supplied.

INDUSTRIAL LAND USES

In the early stages of a city's growth, industry is usually located near the downtown area. From there, it grows out along the waterfront or the railroads serving the city. With the growing importance of truck transportation the location of industries has more recently been greatly influenced by major highways.

Today, industrial plants require large amounts of land. The new industrial buildings are one-storey plants surrounded by large landscaped areas. It is common for a large industry to buy a 40 ha site for its building.

Because there is little room for industrial development in the built-up parts of the cities, industry is going to the suburbs. The modern tendency is for a city to develop an *industrial park* in an area well served by highways and railroads. The city buys the land, services it with roads, water, and sewers, and then sells or rents it to industrial companies.

Industries are important to cities because they provide employment and large amounts of money in real estate taxes to help pay for costly streets, parks, schools, police protection, and the many other services that are provided by a city. For this reason, all cities promote industrial development and often overzone land for industrial purposes, or induce industry to establish in less than ideal industrial locations.

RESIDENTIAL LAND USES

Residential land uses are found in all parts of the city. There are usually apartments right in the downtown area where land is scarce and expensive. There is housing surrounding the C.B.D. and throughout the rest of the city right to its suburbs. Housing takes up more city area than any other land use.

The first residential districts of a city developed close to the downtown area. As the city grew, some commercial land users moved into this residential area ringing the C.B.D. Those who could afford better housing built new homes farther away from the downtown area.

The ring of older housing around the downtown area that is gradually being taken over for commercial land use is usually called the *transition zone*. In this zone, many of the old houses are converted to small apartment units or boarding rooms. The houses are very close to one another and there is usually insufficient play space for children. People who cannot afford to rent better housing, live in this area. The owners of the houses often live in the better residential districts and often may not repair and paint their houses in the transition zone.

Because of these conditions, the transition zone around the C.B.D. is frequently described as a blighted or slum area. We hear more about

The C.B.D. of Montreal. Land values in the centre of large cities are very high, and so high-rise buildings are erected to make the best use of space. The very tall towers are office buildings. The headquarters of some important national companies are located in Montreal.

Part of the expressway system in Toronto. In large cities, traffic becomes so congested that multilane expressways with complicated cloverleaf interchanges have to be built. However, it is not long until even the expressways become clogged with traffic.

The Montreal Métro. The subway systems in Toronto and Montreal were built to provide city-dwellers with efficient rapid transit and to reduce the number of cars in the C.B.D.

The transition zone around the C.B.D. in Montreal. In this area, you would find a mixture of commercial and residential land uses. The houses are usually in poor condition.

Inferior housing in Toronto. Almost all cities have areas where homes are jammed up against one another leaving little open space. In many cases, these crowded homes are firetraps. The worst of the housing areas are called slums. Do you have any slum areas in your city or town?

Right: An ordinary housing subdivision. Almost every growing town and city in Canada has subdivisions similar to the one in this photograph. The houses are modern and well-kept, but the similarity of the subdivisions becomes monotonous. Also, subdivisions like this provide very little space for privacy or play areas in the backyards.

Below: High-rise apartments in Toronto. More and more, Canadians are taking to living in apartment buildings. Apartment buildings make it possible for many people to live in a small area of land and still have adequate open space.

Below, right: Well-designed townhouses (sometimes called row houses) provide an alternative to ordinary subdivisions with single lots. In this development, the fronts of the housing units are staggered to provide privacy.

slums in large cities such as Montreal and Toronto, but slum conditions occur in almost every city. With financial help from the federal and provincial governments, some cities are tearing down slums and are building low-rental, public housing. The city builds the houses or apartments, and then rents them at low rates to families with low incomes.

Because of the scarcity and high prices of single residential dwellings, in the larger cities the trend has been for more and more people to live in townhouses or other multiple housing such as walk-up or high-rise apartments. Although most people seem to prefer to own a single detached house, some people prefer to rent an apartment or townhouse where they have no responsibility to maintain the property, and which permits them more flexibility in moving from one place to another. Some people prefer to live in an apartment in the central part of the city so that they are close to their work and other attractions of the C.B.D. and thus do not have to spend time commuting.

Growth Patterns

From a central starting point, cities tend to grow in a pattern of concentric rings with fingers of development along the major transportation routes. The development of Toronto demonstrates this pattern well (Figure 8.12). The same general pattern is exhibited by other major cities.

In discussing land uses, we mentioned that commercial, industrial, and residential developments were moving to the suburbs. In many cases, these developments are actually going beyond the suburbs and out into the country. Lower land prices, lower taxes, and the availability of land are attracting urban uses into the rural municipalities that surround the cities. Houses, wholesale businesses, light industry, service stations, and billboards form ribbons of urban land use along the major roads leading from the city. Houses are scattered throughout a broad area surrounding the city, and there is sometimes an urban housing subdivision separated from the city by many kilometres of farmland. This kind of scattered urban development around the urban fringes is called *urban sprawl*. It is most obvious around the large cities, although there is some sprawl around all cities and towns. A good example of urban sprawl is found in the Fraser Delta area south of Vancouver. Ribbons of housing line all the rural roads, and scattered housing subdivisions extend over hundreds of square kilometres.

Urban sprawl causes many problems. It uses up far more farmland than is necessary. It forces people to drive long distances to work, and as a result, it helps to clog the highways with automobiles. Scattered urban developments with wide spaces between them are very expensive to service with sewers and water. The rapid increase of population in the rural municipality requires the municipality to provide a whole range of urban-type services including modern schools. This demand causes increased taxes, not only for the urban people who have moved to the country, but also for the farmers. The rural government is organized to handle rural problems, not urban ones and, often, there is a lack of proper land-use planning.

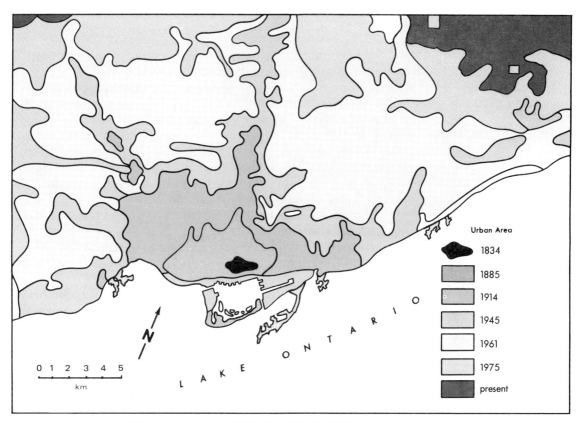

Figure 8.12 GROWTH PATTERN OF METROPOLITAN TORONTO. Notice the way the city grew in rings around the centre. Fingers of development also followed the major transportation routes. The areas between these fingers were filled in later. New fingers of growth outside of Metropolitan Toronto are also following the major roads, such as highways 400, 401, and the Queen Elizabeth Way. Compare the pattern with other cities that you know.

Urban and Regional Planning

The extent of the urban sprawl in Canada's industrial heartland is shown in the degree of urbanization map of the Windsor to Quebec City corridor (Figure 8.13). A township becomes semi-urban when more than fifty percent of its population is urban. The semi-urban townships are not solidly built up, but have enough urban land uses that urban services such as water and sewers are needed. If current trends continue, most of the semi-urban townships will have become urban by the turn of the century.

From Figure 8.13, note that the urban areas around the western end of Lake Ontario are actually growing into one another to form what geographers call a *conurbation*. This Lake Ontario conurbation, sometimes called the "Golden Horseshoe", has a population of about four million. If current trends continue, by the year 2000 as many as eight million people could be living in the area.

The Lake Ontario conurbation demonstrates many of the problems that result from rapid urban growth with inadequate planning. Although a number of individual municipalities have been planning their growth for many years (not always with sufficient foresight), others have not. Moreover, there has been no effective co-ordination of development for the entire region. In the early 1970's, the government of Ontario co-operated with the municipalities in most of the area to create a development concept for the Toronto-Centred Region (Figure 8.14). However, opposition by some municipalities and many developers led to the abandonment of this regional concept. Only a couple of the elements of the plan have been retained, such as the Parkway Belt West and the land-use plan for the Niagara Escarpment.

The lack of effective regional planning has led to a number of land-use and environmental problems in the Golden Horseshoe:

(1) Much of the Toronto waterfront is still occupied by transportation facilities and industry, thus denying the people of its use for recreational purposes.
(2) The entire Lake Ontario waterfront, from Oshawa to the Niagara River, has an incredibly small amount of public parkland.
(3) Industrial wastes and domestic sewage have been insufficiently treated; they pollute local streams and Lake Ontario.
(4) Only a small proportion of the scenic Niagara Escarpment has been set aside for public use.
(5) As a result of the above facts, hundreds of thousands of automobiles clog the highways every summer weekend as people head out to the northern lakes for recreation.
(6) Too much stress on highways and expressways has led to long traffic jams.
(7) The heavy use of automobiles, combined with inadequate pollution control for industries, as well as heating and thermal electricity plants, has led to air pollution that, under certain weather conditions, reaches levels dangerous to public health.
(8) Because of the low density urban sprawl pattern of growth, valuable farmland such as the orchard land between Toronto and Hamilton has disappeared and the Niagara Fruit Belt between Hamilton and Niagara Falls is being seriously threatened.
(9) If the Lake Ontario conurbation continued to grow unchecked and uncontrolled, by the year 2000 the entire area would be either urbanized, under the process of being urbanized, or under the influence of urbanization. Such massive urbanization would lead to an inferior living environment and a high cost of providing essential urban services.

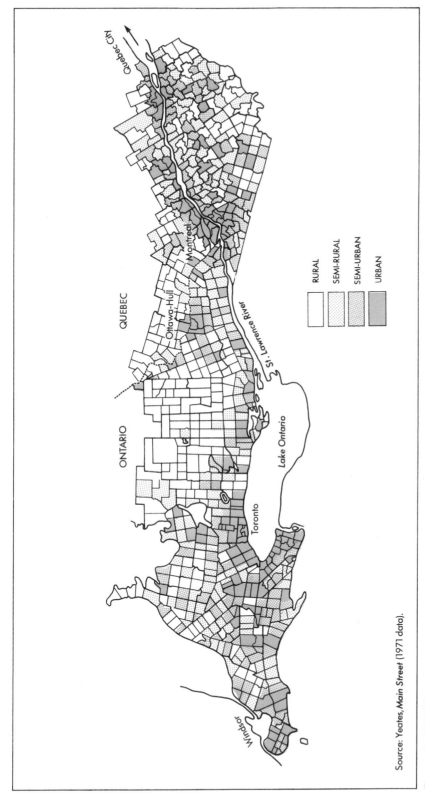

Figure 8.13 DEGREE OF URBANIZATION IN WINDSOR TO QUEBEC CITY CORRIDOR. Can you explain the corridor pattern formed by the concentration of cities along or near the lower Great Lakes and the St. Lawrence River? Why do you suppose there are no major cities along the shores of Lake Huron and Georgian Bay? The solid blue areas of this map are not completely built-up, but are so highly developed that they are likely to become totally urbanized in the future.

Source: Yeates, *Main Street* (1971 data).

261

Despite the lack of effective regional planning for the Golden Horseshoe, some positive steps have been taken to improve the urban environment of Toronto and surrounding area. Redevelopment projects in downtown Toronto include a new city hall, concert halls, restaurants, office towers, shopping centres, and recreational facilities along the waterfront. These developments have made the Toronto C.B.D. an attractive place that draws many visitors to the area and has encouraged the renovation of many old downtown homes.

To reduce automobile traffic in the Toronto C.B.D., a subway system has been built, a fast commuter GO-Train service for Oshawa, Burlington, Richmond Hill, and Georgetown has been established, and plans for the Spadina Expressway have been modified.

Figure 8.14 DEVELOPMENT CONCEPT FOR THE TORONTO-CENTRED REGION. For planning purposes in the early 1970's, the Toronto-Centred region was divided into three zones. Zone I was to be densely urbanized but the individual cities were to be separated by parkway belts. Zone II was designated for very little urban growth; this agricultural land and open space was to restrict urban growth to the north. Zone III was designated for agriculture and open space with specified urban growth centres. Most of this plan has now been abandoned.

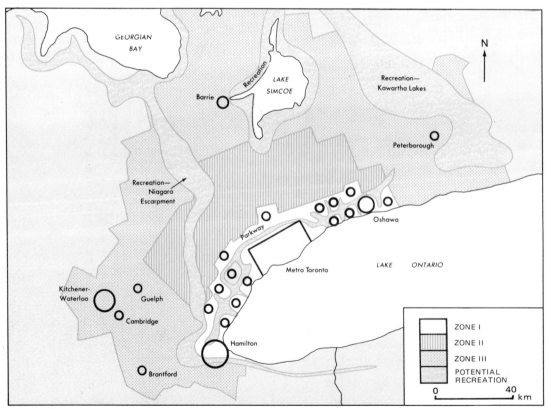

Local governments in the Toronto area have been reorganized into regional governments (similar to that of Metro Toronto) and each of these has prepared a land-use plan that has been approved by the provincial government (Figure 8.4). A major goal of each official plan is to reduce urban sprawl and preserve agricultural land and as much other open space as possible. Only time will tell whether these are effective planning approaches.

LOCAL STUDY

1. On a street map of your hometown or city, colour the land uses according to the following legend: commercial—red, industrial—blue, residential—yellow, park—green. Although you probably have a general idea as to where these land uses are, you will have to do some field work to help you draw precise boundaries. If you live in a very large city you may wish to do only one part of it.

2. Describe, in words, the land-use patterns shown on the map you drew. Are these land-use patterns like the normal patterns described earlier in this chapter?

3. How healthy is the C.B.D., or downtown, of your hometown or city? Are major, new department stores locating in the C.B.D. or in suburban shopping plazas? Are there any old parts of the C.B.D. that should be redeveloped? Does your city have an urban renewal plan?

4. What kind of industry is moving to your city? Find out why recent industries moved to your city. Where is industry locating within the city? Why?

5. What proportion of housing units built last year was apartments or row housing? In what part of the city are apartments and row houses being built?

6. Try to obtain a copy of the master or official plan and of the zoning bylaws for your city. Compare these planning maps with your land-use map.

7. By checking your local newspaper over a period of time, identify the major planning problems and issues in your city.

8. In what ways have increased energy costs changed the design of housing and the layout of subdivisions in your city?

Some Canadian Cities

On the following pages you will find photos, sketch location maps, and some information about a selected group of Canadian cities. Study these photos and answer any questions asked about them. Discuss in class the importance of different kinds of transportation to these cities. Find photos of several other Canadian cities and draw a sketch location map and write a short account about each.

Vancouver, British Columbia, population 1 268 000, is located on a deep fiord (Burrard Inlet), on a rolling upland near the flat delta of the Fraser River. Vancouver is the metropolitan centre of British Columbia and Canada's most important west coast port.

In the photograph, note the spacious harbour, the docking facilities, and the tall buildings of the C.B.D. Explain why Vancouver and the surrounding area has become important for pulp, paper, and wood products and oil refining. Why is urban sprawl a very serious problem in the Vancouver area?

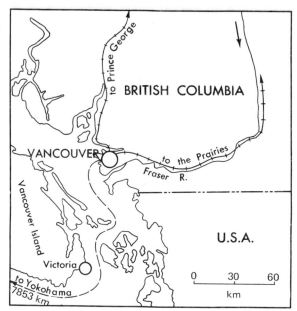

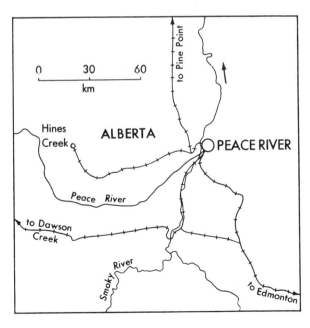

Peace River, Alberta, population 6000, is a service centre for the surrounding Peace River District's agriculture activity (mixed farming with wheat specialty). Peace River is about midway between Edmonton and Pine Point (N.W.T.) and is connected to them by both rail and highway.

As seen from the air, the town of Peace River has an unusually long and narrow shape. Why has the growth of the town taken place in this way? Do you think Peace River will ever become a large city? Why?

Edmonton, Alberta, population 657 000, is located on the flat banks of the North Saskatchewan River. Much of the city's growth started in 1912 when the first railway bridge was built across the river. Edmonton today is the focal point of several railways and roads that serve northern Alberta and the Northwest Territories. It is also a major break-of-bulk centre on the Yellowhead Pass route through the Rockies between Winnipeg and Vancouver. Among the important industries of Edmonton are petroleum refining, meat packing, chemical processing, metal fabrication, and building materials.

In the photograph, note the concentration of new office towers in the C.B.D. and the parkland along the river.

Why has Edmonton developed as the supply and distribution centre, but not the administrative and operational headquarters, for the petroleum industry? Other than petroleum, what resources are found in Edmonton's hinterland that help make it the most northerly large city in North America?

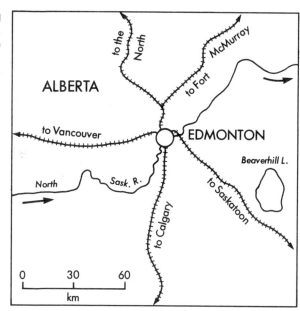

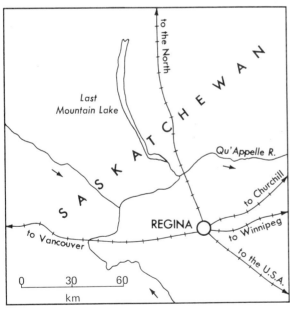

Regina, Saskatchewan, population 164 000, is located on a flat, treeless plain across which winds Wascana Creek. Regina is the capital city of Saskatchewan and is also the major commercial and manufacturing centre of the province. Its main industries —manufacturing farm machinery, oil refining, meat packing, and food processing—are related to the resources and economic activities of its trade area.

In the photograph, note the rectangular grid pattern of streets. Why does Regina have fewer tall buildings than larger cities do?

Why has Regina not become as large a city as Edmonton or Winnipeg? What is the next largest city in Saskatchewan? Why was it easy to lay out a grid street pattern for Regina? What is the predominant type of agriculture in the area surrounding Regina?

Winnipeg, Manitoba, population 585 000, is located at the junction of the Assiniboine and Red rivers. Because it is located on the floodplains of these rivers, Winnipeg suffers from periodic flooding. Winnipeg is the oldest and largest city in the Prairie Provinces. In the early part of this century Winnipeg was clearly the predominant metropolitan centre of the Prairie Provinces, but in more recent years Edmonton and Calgary have been growing much more rapidly.

In 1972 the City of Winnipeg and all of the surrounding suburban municipalities were amalgamated into one city under one local government. This is in contrast to the Toronto situation where there is a metropolitan government and a separate local government for each municipality.

The number and height of tall buildings is usually a good indication of the size of a city. Compare the height (each window is one storey) of buildings in Winnipeg with those in Saint John and Vancouver.

In what ways is Winnipeg the "Gateway to the West"? What are Winnipeg's major industries? What geographical advantage does Winnipeg have for these industries? Explain the origin of the extensive flat plain surrounding Winnipeg.

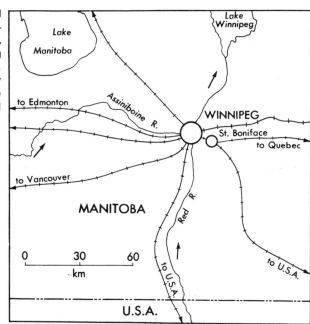

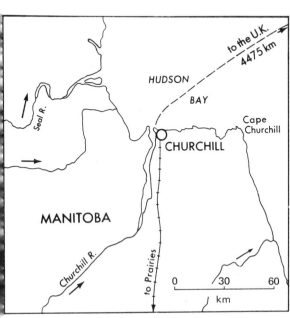

Churchill, Manitoba, population 2200, is located at the mouth of the Churchill River. However, it is not the river but the Hudson Bay Railway (completed in 1929) that has made it possible for Churchill to become a port. The shipping distance for Prairie grain going to Europe is about 1600 km shorter through Churchill than through Montreal. However, because less cargo is imported to Canada through Churchill, many trains return to the Prairies with empty cars. Churchill's importance as a port is greatly reduced by the short shipping season of 88 days (July 22 to October 20).

Why does the lack of imports through Churchill add to the cost of transporting grain to Churchill? The Hudson Bay Railway connects Churchill to what Prairie city?

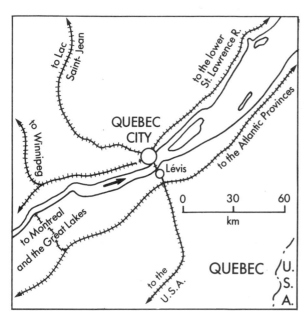

Quebec City, Quebec, population 576 000, is located at the point where the St. Lawrence River broadens out to form an estuary. In the Algonkian language, Quebec means "where the river narrows". The high cliff on which "Upper Town" is located provided Quebec with a commanding position as a fortress. "Lower Town" is crowded between the cliff and the river.

Quebec City, the capital of the Province of Quebec, is an important manufacturing centre and an important port. Although it is not as important a port as Montreal, many large ocean liners stop at Quebec City because the St. Lawrence River is not deep enough for them to proceed to Montreal. With the development of improved icebreakers, Quebec has become an all-year-round port.

Because of its "walled city", old-world charm, and historic buildings, Quebec has become an important tourist centre. It is estimated that tourists bring as much money to Quebec as its largest industries. Note the famous Château Frontenac that caters to many tourists.

Saint John, New Brunswick, population 114 000, is the oldest incorporated city in Canada. It is located on the sheltered estuary of the Saint John River. Saint John has excellent rail facilities. It is the eastern terminus of the CPR and has rail connections with Boston and Halifax. It has some of the most modern docking facilities for container cargo and for super oil tankers. Saint John has been designated as one of the major growth centres in the regional development program for the Atlantic Provinces.

The photograph shows the Saint John harbour with its many facilities: large berthing space for ships, grain elevators, and railroad marshalling yards. Most of Saint John's industries front on the Saint John River.

Explain why the following industries developed in Saint John: sugar refining, oil refining, pulp and paper, and the manufacturing of railway cars. From where do the raw materials come and to where do the finished products go?

Why is a great amount of urban renewal required in Saint John?

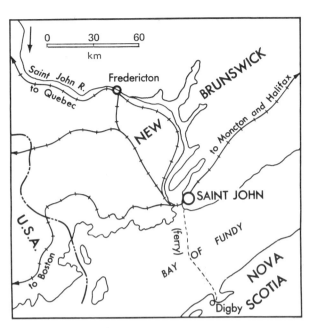

Halifax, Nova Scotia, population 278 000, is located on a small peninsula part way along a drowned valley about 10 km inland from the sea. The city was founded in 1749 as a strategic military site because of its defensive position from a fort, known as the Citadel, on top of a low hill overlooking the harbour. Later, during the two world wars, Halifax developed into a major naval base to help defend Canada's east coast.

Today, Halifax, the capital city of Nova Scotia, is the chief seaport of the Atlantic Provinces. Including Dartmouth across the harbour, it has also become the major ice-free winter port for eastern and central Canada. Surrounding the harbour shoreline are many shipyards, grain elevators, oil refineries, warehouses, and ocean terminals. In the early 1970's, a special container terminal, called Halterm, was built south of the main harbour to handle prepacked freight sealed in large, metal, all-weather boxes. Using Halifax as a focal point, these containers can be transported interchangeably by ship, train, or truck between Europe and Central Canada.

In the photograph, note the container area near the C.B.D. and the numerous areas of parkland.

Why has Halifax assumed the role of leading transAtlantic shipping port over the other east coast harbours? What are the vital road and rail links that serve the Halifax-Dartmouth parts of the harbour? How has the location of this port created urban growth problems for both cities?

PROBLEMS AND PROJECTS

1. Newspaper distribution gives a good idea of the influence of a town or city on its surrounding area. See if a local newspaper company can provide you with a map of its distribution area.
2. From your own local observations, estimate how large a village or town has to be before it can support the following: a weekly newspaper, a drugstore, a dentist, a department store.
3. Topographic maps of all Canadian cities (scale, 1:50 000) and military map plans of the larger Canadian cities (scale, 1:25 000) are available from the Surveys and Mapping Branch of Energy, Mines and Resources Canada, Ottawa. From these maps, you can discover details about the sites of several Canadian cities.
4. Slide sets (with accompanying notes) are available for some of Canada's major cities. Write to the Canadian Association of Geographers, McGill University, 805 Sherbrooke St. W., Montreal, H3A 2K6. After viewing several of these sets, make a slide set of your own city or town as a class project.
5. If your school is in a city, take a classroom survey to discover something about shopping habits in your part of the city. Make a list of types of groceries, clothing, furniture, appliances and other items purchased in a neighbourhood store, a suburban shopping plaza, the C.B.D. of your city, or another city.
6. For the major cities, the *Census of Canada* includes a great deal about the nature of city population. Compare some of this data for your city with another city of approximately the same size in another province. Attempt to explain the differences.
7. The following references contain excellent maps, diagrams, and other useful data:

    R.C. Bryfogle and R.R. Krueger (eds.), *Urban Problems* (Toronto: Holt, Rinehart and Winston, 1975).

    L.O. Gertler and R. Crowley, *Changing Canadian Cities: The Next 25 Years* (Toronto: McClelland and Stewart, 1977).

    J.N. Jackson, *The Canadian City* (Toronto: McGraw-Hill Ryerson, 1973).

    G.A. Nader, *Profiles of Fifteen Metropolitan Centres* (Toronto: Macmillan (Gage), 1975), Vol. 2: *Canadian Cities*.

    R. Preston and L.H. Russwurm (eds.), *Essays on Canadian Urban Problems and Form II* (Waterloo: Department of Geography, 1980), Publication No. 15.

    L.H. Russwurm, *The Surroundings of Our Cities* (Ottawa: Community Planning Press, 1977).

    J. Simmons and R. Simmons, *Urban Canada* (Toronto: Copp Clark, 1974).

    M. Yeates, *Main Street: Windsor to Quebec City* (Toronto: Macmillan (Gage), 1975).

9 AGRICULTURE

Climate, landforms, and soils have limited Canada's agricultural land to approximately 700 000 km², which is about seven percent of the total land area of Canada. Nevertheless, the Canadian agricultural industry is very important. It supplies the bulk of the food consumed in Canada and allows for the export of large volumes of food (mainly wheat and flour) to other countries.

Although farms employ about five percent of the Canadian work force, many other people have jobs that depend on agricultural production. Flour mills, food-processing and meat-packing plants, wineries, breweries, distilleries, cigarette factories, creameries and cheese factories, to mention only a few, all depend upon products from the farm. Farmers are also large purchasers of farm implements, trucks, tractors, fertilizer, and other farm supplies. Thus, in an indirect way, farming creates many jobs in the manufacturing industries. In addition, the transportation of equipment and supplies to farms, and of farm products to market, provides important business for the shipping, trucking, and railroad industries.

LOCAL STUDY

In this chapter, we are going to discuss farming trends and problems and describe different types of farming across the country. If you do not live on a farm, you will appreciate and understand this chapter on agriculture much better if you have an opportunity to visit one or two local farms. If it is not possible for all of the class to visit a farm, perhaps some of your class members who have friends or relatives on a farm can report to the class about that farm's operation.

The following are suggestions as to the kinds of questions that will help you gather useful information. You will be able to modify these questions so that they will be more suitable for the types of farms in your area.

1. How many hectares are on the farm? How much of it is cleared land? How many hectares are there in crops? in pasture?

2. How much would a farm of that size cost? How much would it cost to equip it? How much have these costs changed in the last ten years?

3. What type of soil does the farm have? Does the farmer plant different crops on different types of soils? Do any of the fields have to be drained by ditches or undertiling?

4. How often has the farmer suffered serious crop loss because of natural hazards such as drought, hail, windstorms, disease or insects? What field crops does the farmer grow?

5. When does the farmer plant and harvest different crops? Has the farmer changed the emphasis in crops in the last ten years? If so, why?

6. What special soil conservation measures does the farmer use?

7. What are the different implements used by the farmer? For what are they used?

8. What different kinds of livestock does the farmer keep? Is there a specialization in one kind? What breeds of cattle, hogs, or hens are kept? Has the farmer changed the livestock specialty within the last ten years? If so, why?

9. How does the farmer store fodder for the winter?

10. Is all the feed for the livestock grown by the farmer? If it is bought, from where does it come?

11. Where does the farmer market the crops, livestock, and other produce? What prices have been received in the past year? How are these prices determined? How much have they changed in the last ten years?

12. When does the farmer's day begin and end in summer? in winter? How does the length of the work day compare with that of a city labourer? How many holidays does the family take a year?

13. What are some advantages and disadvantages of living on a farm?

TRENDS IN THE AGRICULTURAL INDUSTRY

In Chapter 8 we said that Canada was a land of cities, and that most of the people in Canada were urban. This pattern has not always been true. At the turn of the century, Canada was a rural agricultural nation. Forty percent of the Canadian workers were engaged in agriculture—more than in any other single industry. The number employed in agriculture continued to grow until the 1930's when there were over 1 200 000 farm workers. In the early 1940's, the farm population began to decline and by

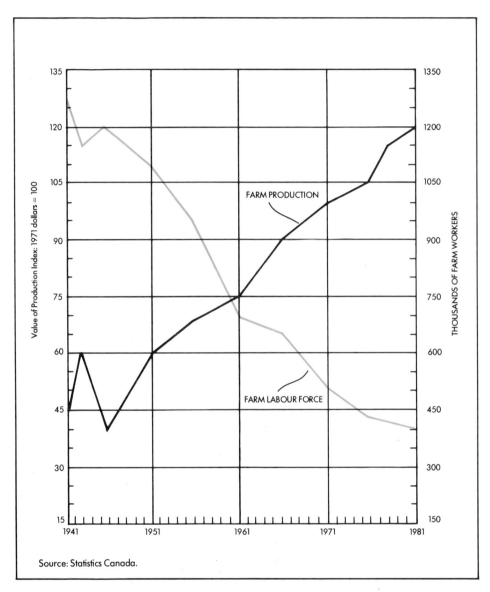

Figure 9.1 TRENDS IN FARM PRODUCTION AND LABOUR FORCE. Because inflation decreases the worth of a dollar, comparisons of farm production from one year to another require a common reference value. For this graph, the 1971 farm production value has been chosen. In a forty-year period, the amount of Canada's farm production more than doubled while the labour force was cut by more than half. How do you explain this situation?

1980 the number of farm workers had dropped to less than half of what the total was in the 1930's (Figure 9.1).

The decline in farm population has not resulted in a decline in agricultural production. In fact, since the 1930's, the volume of farm production has more than doubled. This increase has been made possible by:

(1) changing to larger farm units,
(2) mechanizing the farm operation,
(3) specializing in fewer products, and
(4) using scientific methods of farming.

Although farm population has decreased rapidly over the last forty years, the amount of agricultural land being farmed has declined only slightly. The result is that the average farmer today owns twice as much land as one did in the 1930's. The average farm size in the Prairies today is larger than one section (260 ha). In the drier areas it is common for a Prairie farmer to own as much as several thousand hectares. In most of the rest of Canada, the average size of farms is from 40 to 120 ha.

The switch from horse power to tractor power made it possible for one farmer to farm much larger areas. One farmer with a tractor can now plough in an hour as much as a farmer with a team of horses could plough in a day. Mechanization has speeded up and reduced the amount of labour needed for harvesting. Using old methods, the farmer cut the grain with a binder that tied the grain stalks into sheaves with binder twine, and then placed the sheaves in stooks. Later, the sheaves from the stooks were hauled by horse-drawn wagons to threshing machines. Today, a combine can cut and thresh the grain all in one operation, thus saving much time and work.

Farmers have also become more efficient by specializing in one particular type of farming. General mixed farming in which a farmer has a dozen milk cows, a few beef cattle, a dozen or so hogs, and a flock of 200 hens, is declining in importance. There is still much mixed crop and livestock farming, but most farmers now have one livestock specialty. If the farmers are specializing in dairying, they may have 40, 50, or even 100 or more dairy cows. If they are specializing in egg production, they may have 25 000 to 50 000 hens. To specialize in this way, farmers must mechanize their operations so that much of the labour is done by machinery. For instance, in a large "egg factory" the hens are fed and watered, and the eggs collected, completely automatically. By buying in large quantities, the farmers who specialize also save on the costs of feed and other supplies.

The agricultural industry in Canada has also become more efficient through the use of better farm methods, and improved crop varieties and livestock breeds. Canadian farmers are constantly using better methods of cultivation, new weed and insect sprays, more artificial fertilizers, and are planting new disease- and frost-resistant crops. All of these improvements result in higher production per hectare. By careful breeding, Canadian farmers are developing beef cattle and hogs that produce more high-grade meat on less feed. By feeding hens a well-balanced diet and by lengthening the day with artificial light, farmers can make their hens produce far more eggs. As a result of greater farming efficiency, the prices of farm products have risen less in the last twenty years than those of most other commodities.

To be successful, the modern farmer has to be a shrewd businessperson. Today's farm operation requires a capital investment of hundreds of thousands of dollars. Most farmers must carry a large mortgage on their farms. In addition, the farming operation usually requires a large investment before any income is received. Even with a cash crop, for example, there is much expense in planting, fertilizing, and harvesting before any money is received for it. In livestock farming, the delay in returns on the farmers' investment is even longer. For example, a steer must be fed for two to three years before it is ready for market. In order to arrange finances and ensure a profit, the farmers must use good business practices. They also have to watch the farm market trends to try to sell their products when prices are highest.

To be successful, the farmer also has to know a great deal about the growing of crops, and about looking after animals. Crops must be planted, cultivated, weeded, fertilized, and harvested at exactly the right times, and in the right ways. Animals must be given the right amounts of balanced feeds in order to increase in mass at the right rate. For instance, if hogs are fattened too quickly, they bring a lower price from the meat packer. Animal diseases have to be recognized and treated quickly before they spread.

Since the modern farm is so highly mechanized, modern farmers must also know something about mechanics. They cannot afford either the time or money to call a professional mechanic every time something goes wrong with their equipment. Farmers must do much of the servicing and repairing of their own machinery.

It was once thought that farming was just a matter of hard work. It is obvious that today's farmer also has to have a wide range of skills and abilities.

There was a time when most farm people had a lower standard of living than their city cousins. Prosperous farmers today live as well and have as many conveniences as the prosperous businesspeople who work in the city. The modern farm home has electricity and all the electrical appliances and gadgets of modern city homes. Rural mail delivery, the radio, and television keep the farmers as well informed as urban people about national and international news. Many farmers own a truck and at least one car, making it easy for them to get into the town or city whenever they wish. In most rural areas, buses pick up farm students at their front gate and take them to a school that is as modern and well-equipped as any city school.

In the past, horses were a slow but dependable means of pulling various pieces of farm machinery such as a binder (A). Later, powerful tractors were developed to do these jobs much faster.

By the 1940's, rubber-tired tractors had replaced horses on most commercial farms, but the binder was still commonly used. The binder cuts the grain, and binds it into bundles called sheaves that are gathered into stooks (B). These stooks are left out in the field for a week or so until the grain is dry enough to be threshed.

The sheaves were loaded onto wagons (C) and taken to be threshed by a threshing machine (D) that separated the grain from the stalks, called straw. This method of harvesting grain is still common in many parts of eastern Canada.

Today, combines (E) make harvesting a simple one- or two-step operation. In the photograph, a swather has cut and placed the grain into rows that the combine picks up and threshes. The grain is dumped into trucks and taken to storage bins or elevators nearby. Some combines cut and thresh the grain in one operation.

In very early times, after the hay was cut with a mower, it was heaped by pitchfork into piles called coils. In later times, the hay was raked into rows with a horse-drawn dump rake as shown in this photo. Still later, the dump rakes were replaced with side-delivery rakes that made much neater windrows.

After the hay is dry enough, a hay loader lifts the hay onto a hayrack mounted on a wagon. The wagon may be drawn by either horses or tractor. The hay is then hauled to the barn, where it is dumped into the haymow by use of either sling ropes or a hayfork. This method of "haying" was common until the late 1940's when most farmers turned to more modern methods.

Today, many farmers bale their hay as shown in this photo. These bales are drawn to the barn where they are lifted into the mow by means of an elevator belt.

An even more recent development is the forage harvester that cuts up the hay into fine pieces and blows it into a wagon. This finely cut hay is then blown into a haymow or a silo.

BETTER FARMING TECHNIQUES

Modern Cultivation Methods — Potatoes

Improved Crop Varieties — Sugar Beets

Spraying Crops with Herbicides — Grain

PROBLEMS OF THE AGRICULTURAL INDUSTRY

Natural Hazards

Many of the agricultural industry's problems are directly related to nature. As we pointed out in Chapter 5, the weather varies a great deal from year to year. While unusual weather changes are inconvenient to urban people, they often spell disaster to the farmers. Late spring frosts, early fall frosts, extreme low winter temperatures, extensive drought in summer, the lack of snow cover, hailstorms, windstorms, and rainstorms may cause crop losses. In addition, the farmer never knows when his crops may be destroyed by an onslaught of insects such as the army worm or the grasshopper. Diseases attack both crops and livestock.

Marketing Problems

Maintaining stable prices for their products has always been a major problem for farmers. In bumper-crop years, there is a greater supply of certain farm products than can be marketed; therefore, the price drops. If there is a high demand for a certain farm product for a number of years, more and more farmers specialize in that product until there is a surplus; again, the price drops. This pattern is particularly true for cash crops, eggs, poultry, and hogs, because farmers can quickly increase or decrease their production of these products. It takes a longer time to develop a herd of dairy or beef cattle, so that a farmer must decide several years in advance whether or not the market will be good. Prices of some products even fluctuate throughout the year. The price of fresh fruit and vegetables, for instance, may drop by fifty percent within several weeks of the start of a harvest.

Manufacturing firms usually base their price upon their costs of production plus a profit. Farmers cannot do this. They send their products to market and must accept whatever price the supply and demand dictates at that time. Farmers usually cannot hold their products from the market until the price is higher, since many of the items are perishable, or else decline in value with time. Also, farmers are usually forced to sell because either they need the money or they lack storage facilities.

While prices of farm products have gone up and down a great deal in the past twenty years, the average prices paid to the farmer for products have not kept pace with the costs of farmland, labour, machinery, and supplies. The result is what the farmers call "the cost-price squeeze." It has been one of the factors forcing farmers to increase the size of their farms, mechanize, and specialize in fewer products. Some farmers have had to take part-time jobs off the farm to supplement their incomes. Few farmers' children wish to remain on the farm because of the high risks, long hours of work, and low returns on investment of money and time.

Over the years, the federal government has given considerable help to the agricultural industry. It has established floor prices for some products. If the price drops below the floor price the government pays the farmers a subsidy. The government has passed legislation that has permitted farmers to form marketing boards to regulate the selling of farm products. The government gives grants to farmers in areas where there are major crop losses because of natural disasters. Both the federal and the provincial governments constantly conduct research to help farmers to grow better crops and to raise better livestock. Despite all of these efforts, many farmers find themselves in financial difficulties.

If this trend continues, enough farmland may be out of production that food shortages are created. Around some cities, the combination of low financial returns and urban demand for land has already put huge areas of some of Canada's best farmland out of food production. A food policy is needed to protect both the farmland and the agricultural industry. Consumers may have to pay higher food prices now in order to protect themselves against future food shortages.

Part-Time Farming

Within the last couple of decades, part-time farming has become increasingly important. In Ontario, for example, approximately one-third of the farmland is operated by part-time farmers. Part-time farmers are composed of two major groups: (1) farmers who supplement their farm income by taking "full-time" jobs off the farm, and (2) city people with full-time jobs who work part-time on the farm as a hobby. The number of hobby farmers is becoming very great around the heavily urbanized areas. These urban hobby farmers create a problem for full-time farmers because they bid up the price of land to the point where it can no longer be farmed economically.

The urban hobby farmer and other urbanites who build homes out in the country demand urban-type services. The increased municipal taxes to pay for these services impose an additional burden on the farmers who depend on the farm operation for their total income. When the price of land and the increase in taxes reach a certain point, the full-time farmers may sell part of their land, obtain jobs in the city, and join the ranks of the part-time farmers.

Farm Poverty

Although some farmers are very prosperous, and many are making a good living from their farms, there are many farmers who are unable to make a decent living. Farm poverty is difficult to define, but one good indication of the level of prosperity is the gross value of product sales. In 1981, well over 100 000 Canadian farms had a gross value of sales under $5000.

When you consider the high costs of production, these sales would provide meager incomes indeed. Many of the families represented by these statistics are living in severe poverty. They live in substandard houses, they often do not have electricity or running water, and they have inadequate health care. Their children are forced to leave school early in order to supplement the family income. However, with insufficient education and no job training, they find it difficult to obtain good jobs, and thus another generation of rural poor begins.

The distribution of farm poverty in Canada is widespread, but there are concentrations in certain areas (Figure 9.2). All of the areas shown on Figure 9.2 have poor physical conditions for agriculture: marginal climate, rough terrain, poor soils, and long distances from the major markets. Studies have indicated that much of this land should never have been cleared for agriculture. Those who do have good farmland do not have a large enough area and do not have the capital to buy additional land or to buy the modern equipment required. Nor do they have enough assets to borrow the needed money.

Federal and provincial government programs have been established to assist farmers to enlarge their areas, modernize their operations, and to utilize land for forestry and recreation purposes. Vocational training programs are also available to help people to obtain nonfarm jobs. Although some progress has been made, the farm poverty problem still remains a major problem (see discussion of regional disparities and regional development in Chapter 13).

Do you think the farmer shown in the photograph is able to make an adequate income from his farm? Give reasons for your answer. Consider the size of the field, the depth of the soil, and the steepness of the slopes. Why do you think the farmer is using a horse instead of a tractor?

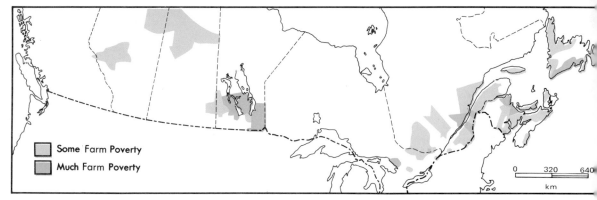

Figure 9.2 CONCENTRATIONS OF FARM POVERTY. In addition to the areas of poverty shown on this map, there are many small pockets of farm poverty scattered all across Canada.

AGRICULTURE REGIONS

Types of Farming

The different types of farming in Canada are shown in Figure 9.3. Compare this map with the landform, climate, natural vegetation, and the soils maps in the earlier chapters of the book. After studying these maps, answer the following questions.

QUESTIONS

1. What are the chief limiting factors for agriculture in the Atlantic Provinces and the Gaspé Peninsula of Quebec?

2. What landform region acts as the northern limit of extensive agriculture in Quebec, Ontario and Manitoba?

3. Compare the types of farming on the Prairies with the precipitation, natural vegetation, and soils patterns.

4. Explain why there is farming (a) around Lac Saint-Jean, (b) in several belts within the Canadian Shield south of James Bay, and (c) in the Peace River District. Which of these areas has the longest growing season? Which has the shortest growing season?

5. What are the major physical limitations to agriculture in British Columbia?

Because of the great variations in physical geography, farming is quite different from one region to another in Canada. The generalized map showing types of farming (Figure 9.3) shows that there is a great deal of mixed farming in the Prairies, in the Great Lakes-St. Lawrence Lowlands, and in the Atlantic Region. Although these regions all have areas of mixed crop and livestock farming, the kinds of crops and livestock are different, the sizes of the farms are different, and the farm operation itself is different. Because farming differs greatly from one part of the country to another, we will discuss agriculture region by region, followed by several detailed studies of specific types of farming.

The Atlantic Region

The climate of much of the Atlantic Region is moderately suitable for agriculture. There is plenty of precipitation and the frost-free season is long enough to let most grain and fodder crops mature. The major physical factors that hinder agriculture are the lack of level land and good soils. Most of the area of the Atlantic Region is composed of steep, rocky hills with soils too thin for farming. As a result, agriculture is confined to the lowlands bordering the sea and the river valleys. Prince Edward Island is the only province with most of its land under cultivation.

Despite the limited areas suitable for farming in the Atlantic Region, agricultural production could be doubled if markets were available. The total population of the whole Atlantic Region is less than that of either Metropolitan Montreal or Metropolitan Toronto! As a result, there is a very limited local market for farm products. The big markets in Quebec and Ontario are too far away, and tariffs discourage the sale of farm products to the United States. The only major agricultural exports from the Atlantic Region are apples and potatoes.

For Newfoundland, it is not the lack of markets, but the lack of agricultural land that is the major problem. Of the four Atlantic Provinces, Newfoundland has the shortest and coolest summers, the least level land, and the least fertile soils. Newfoundland produces only a small proportion of the province's food needs. The largest farming areas are in the Avalon Peninsula where vegetables, beef, and dairy products are produced for the St. John's market.

In recent years, some Newfoundland farmers have been successful in draining peat bogs and using the land for pasture and fodder crops for cattle. There are many large areas of bog that could be used in this way. Both blueberries and cranberries grow well in Newfoundland, but they are not grown commercially as they are in Nova Scotia and New Brunswick.

In Nova Scotia, the most productive agricultural land is in the Annapolis Valley. Apples are the most important crop in the Annapolis Valley, but the production of vegetables, poultry, and eggs is also important. The coastal plain of the north shore, along the Northumberland Strait, is an area of mixed farming with an emphasis on dairying. The same agricultural activity is found on this coastal plain as it continues on into New Brunswick.

In New Brunswick, the most important agricultural region is in the Saint John Valley. North of Fredericton, there are large areas of potatoes; south of Fredericton is an orchard district that supplies New Brunswick with its own apples. There are areas of mixed farming with an emphasis on dairying around the cities of Saint John and Moncton. Near the Bay of Fundy, tidal marshes that have been protected from flooding by dikes produce excellent hay crops.

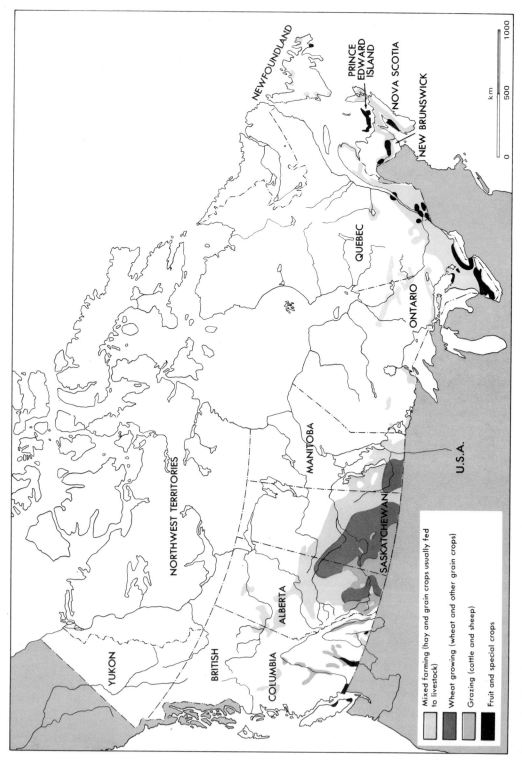

Figure 9.3 TYPES OF FARMING IN CANADA. Only the predominant types of farming can be shown on a map of this scale. For example, in the wheat growing region there are many other crops grown as well as some grazing and mixed farming.

Mixed farming is typical of the kind of agriculture that is best suited to the Atlantic Region. Most of the good farmland is located on gently rolling land, most of which is concentrated along narrow coastal plains or in river valleys farther inland.

Apart from the important farming areas mentioned above, most of the farms of Nova Scotia, New Brunswick and the Gaspé Peninsula are small, have poor soils, and do not provide an adequate living for the farm families. A small cleared area provides vegetables for the household, as well as hay and oats for a couple of horses, a few cows, and a small flock of hens. The farmers cannot afford modern implements. There are still some farms on which the hay is cut with a horse-drawn mower and is raked with a dump rake. Some farmers produce so little grain that it would not pay to buy harvesting equipment and threshing machines or even to share the cost of these machines with neighbours, and as a result, they cut their oats green and use it as fodder. They buy whatever grain is needed for their livestock. To make a decent living, these farmers must find work off the farm in the mining, lumbering, fishing, or manufacturing industries.

Farming along the rocky Newfoundland coastline is very poor. The soil is thin and a large number of rocks and boulders are scattered through the fields. At best, some of the land can be used for gardens or rough pastures.

Potatoes provide the most important cash crop in Prince Edward Island. P.E.I. potatoes are well known for their quality all across Canada.

Prince Edward Island presents a contrast to the rest of the Atlantic Region. Most of the rich, red soil of the rolling landscape is excellent farmland. Most of the farms grow oats, hay, and potatoes, and have either beef or dairy cattle, hogs, and hens. Potatoes are the most important cash crop. The sandy loam soils, the cool climate, and the freedom from disease produce very high-quality potatoes. P.E.I. potatoes are sold to different parts of Canada, and seed potatoes from the island are in demand from potato-growing districts all over North America. Because of the farmers' great dependence on potatoes as a cash crop, the farm economy fluctuates greatly with the rise and fall of potato prices. Another problem of the agricultural industry in Prince Edward Island is that many farms are too small to be economic operations.

At the head of the Bay of Fundy in New Brunswick, hay crops thrive on wet marshland protected from the sea by grass-covered dikes. These marshy meadows, known as the Tantramar Marsh, have been cropped for centuries.

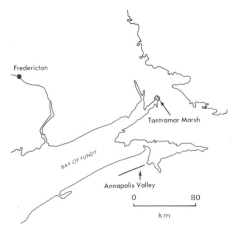

The Great Lakes-St. Lawrence Lowlands

The Great Lakes-St. Lawrence Lowlands is an excellent farming area. Its level land, good soils, long growing season, and adequate precipitation make it suitable for growing a wide variety of crops. The region produces all of Canada's grain corn and soybeans, about ninety percent of Canada's grapes and tobacco, and more than half of its potatoes, hay, milk, eggs, fruit and vegetables.

The large volume of agricultural production in the southern parts of Ontario and Quebec is partly the result of a favourable physical environment. In addition, the huge market provided by the many nearby cities explains the production of large quantities of many products such as milk, eggs, small fruits, and vegetables. Truck farming is usually carried on close to the cities because of the ease of transporting fresh fruit and vegetables to the market. In swamps that have been drained, muck soils are used for truck farming even though these areas may be some distance from a large city. The Holland Marsh just south of Lake Simcoe is a good example.

Most of the Great Lakes-St. Lawrence Lowlands can be considered a mixed farming area (Figure 9.4). Many of the farms closer to the large cities specialize in dairy cattle; those farther away specialize in beef cattle. Poultry, laying hens, and hogs are raised throughout the area. Although more and more farmers are specializing in only one livestock type, there is still a large number of farms on which cattle, hogs, and hens are kept.

Most of the mixed farms produce a number of crops including oats, barley, fall wheat, and hay. Corn is an important crop both for fodder stored in silos and for grain. Only the southwestern part of Ontario has a growing season long enough for grain corn, although with new hybrids that mature in shorter periods, the limit of the "corn belt" is gradually extending farther north and east.

In a number of areas, farmers depend upon cash crops for most of their income. The extreme southwestern tip of Ontario produces large amounts of corn, soybeans, and canning crops such as beans, peas, corn, and tomatoes. Irrigation is commonly used for these crops. Small areas around the towns of Leamington and Harrow produce orchard as well as vegetable crops. Because of the earlier springs and warmer summers, the Leamington and Harrow districts can produce fruit and vegetables earlier than any other part of the region.

Canada's most important tobacco-growing district is located midway along the shore of Lake Erie. Tobacco thrives on the very sandy soils found in this area. The growing season is long enough to prevent frost damage to the tobacco plants in most years.

The shore of Lake Ontario is important for its fruit production. The lake moderates the winter temperatures and protects the fruit blossoms from spring frosts. Between Niagara Falls and Hamilton is the famous Niagara Fruit Belt where grapes, peaches, pears, cherries, and plums are grown in abundance. The north shore of Lake Ontario has many large

The Norfolk Sand Plain is the largest of several small tobacco growing areas in southern Quebec and Ontario (see Figure 9.4 below). Here you see mature tobacco plants in full blossom. During late summer harvest, the largest leaves are picked from the ground up and tied in bundles to be placed on wooden laths in small barns, called *kilns*, for curing. Long hours of sunlight and the rich soils of the lowlands are suitable to make tobacco growing profitable. Peanuts require the same growing conditions and are being introduced to this area as a secondary cash crop.

apple orchards. Another apple orchard district is found near Collingwood, on the south shore of Georgian Bay.

In Quebec, there are major apple-growing districts on the slopes of hills within an 80 km radius of Montreal. Here, the winters are colder than in Ontario, but the trees survive because they are hardier varieties and grow on slopes that drain away the cold air.

Besides the Great Lakes-St. Lawrence Lowlands, the provinces of Ontario and Quebec have some isolated farming areas in the Canadian Shield. In the Lac Saint-Jean area, as well as in the Clay Belt of northern Quebec and Ontario, farming has developed on the clay soils of old lake bottoms. The growing season is very short and the land is poorly drained, but in most years, good crops of hay and oats are obtained. Both beef-raising and dairying are carried on in these areas, primarily to serve the local markets. The Clay Belt farms supply farm products for the mining

Figure 9.4 TYPES OF FARMING IN THE GREAT LAKES-ST. LAWRENCE LOWLANDS.

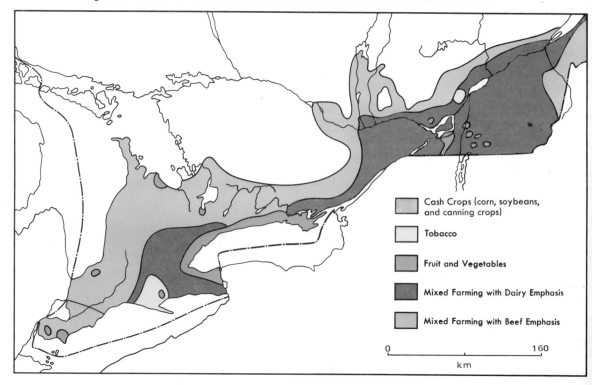

This photograph shows a farming landscape in Southern Ontario. Low, gently rolling farmland in the Great Lakes-St. Lawrence Lowlands supports an abundance of mixed farming. Most of the landscape is a patchwork of fields that are used to grow hay, grain, and other special crops. Herds of dairy and beef cattle graze on some of the fields.

In the extreme southwestern tip of Ontario, the land becomes very flat and the summers are long and warm. Here, a wide variety of cash crops are grown, such as tomatoes, fruits, vegetables, soybeans, sunflowers, corn, and squash. In the photographs you can see (left) a corn field and (right) a pumpkin crop.

Many mixed farms, especially in southern Ontario, grow winter wheat as an important cash crop. After fall planting, wheat grows until temperatures fall below 5°C and then remains green over the winter. A thick snow cover on the fields helps insulate young wheat seedlings from extremely cold winter temperatures and protects the crop from foraging wildlife, such as deer. Unprotected wheat near urban centres sometimes suffers damage from cross-country skiers or snowmobiles.

towns in northern Ontario and Quebec. The Lac Saint-Jean area supplies the rapidly growing industrial cities around Lac Saint-Jean and in the upper Saguenay Valley.

The Prairies

The Prairie Provinces have long been famous for wheat growing and cattle ranching. Wheat yields per hectare are not high, but the areas sown are so large that the Prairies are able to produce an average crop of over 14 000 000 t a year. Likewise, the pastures of the dry grasslands support very few cattle per hectare, but the vastness of the areas results in the fact that the Prairie Provinces have about six million beef cattle, or two-thirds of the Canadian total.

There are three main types of farming in the Prairie Provinces: livestock grazing, wheat farming, and mixed farming. The location of these types depends primarily upon the amount of moisture available.

The grazing of cattle and sheep is predominant in the dry belt where the very low precipitation supports a sparse, short-grass vegetation on a light brown, infertile soil. The foothills near the Rocky Mountains have more precipitation, but are used for grazing because the land is too steep to cultivate.

Ranching in the dry belt merges gradually into the wheat belt as the annual precipitation increases to about 325 to 375 mm. The wheat belt is located primarily on the dark brown and black soils that have resulted from the availability of more moisture and more abundant grass than in the dry belt.

The wheat belt, in turn, merges into the mixed farming belt found on the black soils, where the annual precipitation is closer to 500 mm. Beyond the mixed farming belt, there is an area where new farms have been carved out of the coniferous forest. In the Peace River District, there is mixed farming, with wheat providing an important cash crop on a large number of the farms.

These agricultural belts are named after the predominant type of farming found in each, but that does not mean that no other type of farming is carried out in each belt. For example, the grazing belt also has considerable wheat farming. In the wheat belt, there is considerable ranching, and towards the more humid edge of the wheat belt, other crops are grown.

In the mixed farming zone, the greater amount of moisture makes it possible to grow a greater variety of crops. Wheat is the most important grain crop, but large areas of oats, barley, and rye are also grown. Wheat is the most important cash crop; the other grains are usually fed to livestock. Other crops grown in the mixed farming area include hay, flaxseed, rapeseed, and sunflowers. Hogs, laying hens, and poultry are becoming more important in the mixed farming belt.

The northern edge of the mixed farming zone is called the *pioneer fringe* because farmers have recently cleared the forest to create new

THE REGIONS OF THE PRAIRIES

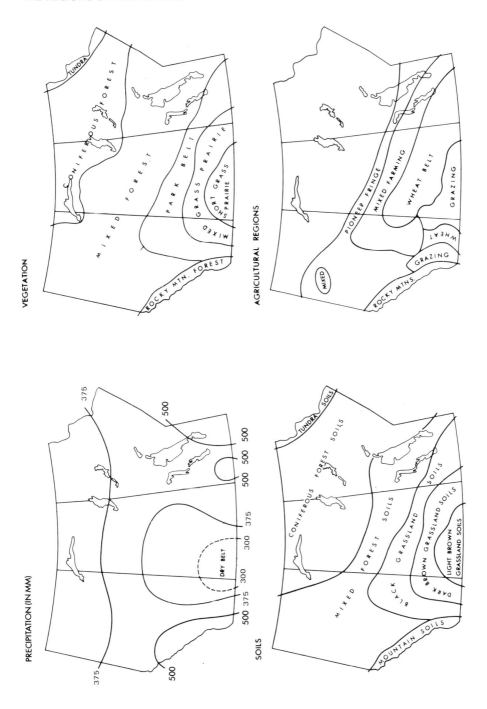

TYPES OF FARMING IN THE PRAIRIE PROVINCES.

Mixed farming in southern Manitoba.

Wheat farming in the Saskatchewan plain.

Ranching in the foothills of Alberta's Rocky Mountains.

farmland just as the early pioneers did years ago. The pioneer fringe farmers face many difficulties. Clearing the land is a difficult task. Transportation facilities are very poor. The winters are very cold, the growing season very short, and the soils range from fair to poor. Because of the many difficulties, many farmers are abandoning their farms in the pioneer fringe, and the land is reverting to forest.

The Peace River District is in many ways an extension of the mixed farming belt. Wheat growing, however, is more important than in the mixed farming area to the south. The growing season is very short, but the long summer days make up for this. Wheat yields in the Peace River District are good and the quality of the wheat is high except in years when the crop is damaged by frost. The Peace River District is about as large as Nova Scotia, but only a small portion of it has been cleared for agriculture. There are about 1 620 000 ha of land still open for settlement, but this land is not likely to be occupied unless there is a major improvement in the farm economy in Canada.

Irrigation is changing the farming pattern in some parts of the Prairies. Southern Alberta has large areas of irrigated hay, oats, and wheat. The additional moisture doubles the yields of these crops. Around Lethbridge, the irrigated lands have been planted with a variety of crops including sugar beets, potatoes, and other vegetables. The large artificial lake recently created on the South Saskatchewan River (the dam has been named the Gardiner Dam after a former federal minister of agriculture; the lake has been named Lake Diefenbaker after a former Canadian prime minister) can provide irrigation water for approximately 81 000 ha. This additional moisture could change the specialization of this area from wheat to mixed farming, with more emphasis on cash row crops such as sugar beets and soybeans, and on hay and beef cattle. However, many farmers around Lake Diefenbaker refuse to turn to irrigation farming because it would require them to learn new farming methods, and would require a great deal more labour than wheat farming.

Around each city in the Prairies are dairy farms that supply the urban market with fresh milk. The dry climate prevents the establishment of

Agricultural patterns in the drier parts of Alberta and Saskatchewan have been greatly altered through the use of irrigation. Dams built across some of the larger rivers such as the Gardiner Dam across the South Saskatchewan River have created huge reservoirs from which irrigation canals and ditches carry water to the fields. As a result, crop yields will increase and new kinds of crops can be grown.

truck farming immediately around the cities as it is in eastern Canada. Truck farming in the Prairies is limited to the irrigated lands around Lethbridge, Alberta, and to the area in Manitoba, south of Winnipeg, where there are over 500 mm of precipitation.

At one time the Prairie Provinces' economy was totally dependent upon agriculture. However with the development of forest, petroleum, and other mineral resources, and related secondary industries, the economy of the Prairie Region has diversified. This is particularly true for Alberta, and to a lesser extent, Manitoba. However, the economy of Saskatchewan is still primarily based on its agricultural industry.

British Columbia

The mountainous terrain of British Columbia has greatly restricted its agricultural industry. Nevertheless, there are some very important farming districts in the province.

The Fraser Delta, which extends some 128 km from Hope to Vancouver, and which is about 40 km at its widest point, provides British Columbia with excellent farmland. The deep, silty soils are rich; the land is level; precipitation is high; and the growing season is the longest in Canada. In addition, it is close to the large population of Metropolitan Vancouver.

Although the Fraser Delta, usually called the Lower Mainland, is labelled as "fruit and special crops" in Figure 9.3, it has the greatest variety of crops and farming types of any area in Canada. Because it is so close to a large urban market, there is a heavy emphasis on dairying. Beef cattle, hogs, laying hens and other poultry, are also raised. There are large areas of hay and pasture for the dairy and beef herds. Although the grass is frozen and brown in midwinter and therefore cannot be pastured, the winters are mild enough to permit the cattle to be out-of-doors all year round. The important field crops include potatoes and peas.

Market gardening provides the urban population with fresh fruit and vegetables. The long growing season permits two or more crops of some vegetables. Raspberries, strawberries, and boysenberries are grown in large quantities. The growing of flower bulbs is also important.

Despite the fact that the annual precipitation is high, and that dikes have to be built to protect some of the low-lying areas from flooding, summer drought is often a problem. In many parts of the Lower Mainland, the July rainfall is only 25 or 50 mm, which requires many farmers to irrigate their crops.

A major problem in the Lower Mainland is the amount of good farmland being taken up by urban development. If the urban sprawl continues, in the future the Fraser Delta will no longer be a major supplier of food for the population in the region. This event would be particularly serious because there is no other land nearby that could supply the

Garden produce thrives in the long, warm summers and on the deep, fertile soils near Victoria, Vancouver, and some southern interior valley sites, such as Oliver (above), in British Columbia. Besides fresh fruit and vegetables, some market garden farmers grow a few hectares of flowers to supply the local population with freshly cut flowers or with seedlings. Often, the bulbs and seeds are exported all across Canada.

Vancouver urban area with fresh milk and market garden crops. Thus, the loss of the farmland in the Lower Mainland would result in a decrease in quality (most small fruit and some vegetables deteriorate in quality soon after being harvested) and an increase in the cost of food products supplied to the city's people. Because of the seriousness of this problem, the British Columbia Government has established strict planning controls over the use of agricultural land. Insofar as possible, urban development is directed to nonagricultural land, and owners of good farmland are prohibited from selling their land for nonagricultural purposes.

Pockets of farmland are scattered across the mountainous interior of British Columbia. These are usually located in sheltered river valleys where soil and climate permit growing of crops. Mixed farming, with some cash crops and a dairy specialty, is common in the Fraser Delta south of Vancouver (below).

Vancouver Island's best farmland is located in the southeast corner of the island, near Victoria. The island produces most of its own milk and fresh vegetables.

In the dry interior valleys of British Columbia, there is a significant amount of cattle ranching. In some of the southern valleys, and where transportation access to markets is available, orchards and other fruit crops are produced with the help of irrigation. The Okanagan Valley is the most important of these, producing the bulk of the province's tree fruits and all of its grapes.

DETAILED STUDIES OF SOME TYPES OF FARMING

Wheat Growing

Over the years, Canada has gained an international reputation as a great wheat producer, because Canadian wheat has been shipped in large quantities to all parts of the world. Also, Canadian wheat is of such high quality that it is bought by many countries for seed.

Other countries, notably the United States, China, and the Soviet Union, produce more wheat than Canada, but most of their wheat is consumed at home. Because Canada has a relatively small population, most of its wheat is exported, and thus, Canada has become the leading exporter of wheat in the world.

World demand for Canadian wheat fluctuates greatly from year to year. In the 1950's Canada had large wheat surpluses. However, serious droughts in other parts of the world led to long-term sales to China and the Soviet Union, eliminating these surpluses by the mid-sixties. In fact, in 1966-67, Canada exported a record quantity of wheat and the Canadian Government encouraged farmers to plant more wheat. However, in the following export year, wheat exports fell to less than half those of the previous year, and by 1970 wheat surpluses were so great that the Government passed legislation to pay wheat farmers for not growing wheat. Farmers decreased their wheat plantings and increased their areas of summer fallow, pasture, oats, barley, flax, and rapeseed. Then in 1972 and 1973, export sales rocketed and again Prairie farmers were exhorted to produce more wheat. Wheat exports declined again later in the 1970's, but increased to even greater totals in the early 1980's (Figure 9.5).

Up to the early 1960's about half of all Canada's wheat and wheat flour exports went to the United Kingdom and the countries of the European Economic Community. By 1970 this trading pattern had changed. The United Kingdom and the European Economic Community received only a quarter of Canada's wheat and flour exports. The largest

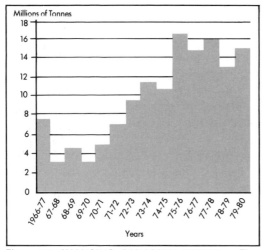 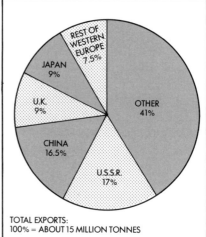

Figure 9.5 CHANGING CANADIAN WHEAT EXPORTS.

Figure 9.6 DESTINATION OF CANADIAN EXPORTS OF WHEAT AND WHEAT FLOUR.

gains were to the Soviet Union and China, which between 1970 and 1980 received over a third of Canada's wheat and flour exports. In 1980, the Soviet Union agreed to buy no less than 25 million tonnes of wheat and barley over a five-year period (Figure 9.6).

All Canadians are interested in giant wheat sales, because the production and export of wheat affects the entire economy and creates many jobs for people across Canada. When the wheat farmers are prosperous, they buy more farm machinery, automobiles, and trucks, as well as luxury items. Shipping wheat provides business for trucking companies, railroads, shipping companies, and flour mills.

A small amount of wheat (about 500 000 t) is grown outside the Prairie Provinces, most of it in the Great Lakes-St. Lawrence Lowlands. This wheat is sown in the fall and grows several centimetres before winter sets in. Snow protects it from the cold and in the spring it starts to grow as soon as the warm weather comes. This winter wheat (sometimes called fall wheat) fits in well with a mixed farm operation. It permits farmers to do some seeding in the fall, and it is the first grain crop to be harvested in the summer. If the price is good, farmers can sell it as a cash crop. If not, it can be mixed with other grain and fed to livestock.

The winter wheat that is grown in a humid climate is a soft wheat used for pastry flour. The hard wheat used for bread flour is grown in a dry climate.

PHYSICAL FACTORS

The winters of the Prairies are too severe for winter wheat to survive. However, the climate of the Prairies is quite suitable for spring wheat. The cool, wet weather and the long days in spring are good for the growth of the young wheat, and the dry and hot summers help in ripening and harvesting the wheat.

In the drier parts of the Prairies the total precipitation is not enough to permit wheat to be grown in the same field every year. In the dry belt, fields are left fallow every other year. During the summer fallow period, the land is cultivated at regular intervals to help the rain to soak in and to create a dry mulch at the surface that will prevent the soil moisture from evaporating. In this way, two years' moisture is accumulated in the soil, and is available for one wheat crop. With summer fallow in alternate years, wheat can be grown with as little as 275 to 300 mm of precipitation. In the more humid areas, the rotation is usually two years of wheat and one year of summer fallow. This method of alternating cropping and summer fallow to conserve moisture is called *dryland farming*.

The soils of the Prairies are also particularly suitable for wheat. The deep grassland soils permit the wheat plant's root system to grow deep where more moisture is available. The grassland soils are naturally fertile and permit cropping for many years without serious damage.

The vast extent of level land has made it possible to use large tractors and heavy equipment and has greatly reduced the cost of labour for wheat production. Thus, one man can farm a large area and make a good income even though the yield of wheat may be only 0.56 t or less to the hectare.

OVERCOMING PROBLEMS

Over the years, the wheat farmers of the Prairies have had to overcome many serious problems. The wheat that the pioneer settlers brought with them was suitable only for southern Manitoba. Farther north, the frost-free season was too short, and farther west, there was too little rainfall. Scientific research led to the development of a new variety of wheat, called Red Fife, which could be grown farther north and west. In 1904, an improved variety, called Marquis, helped to extend wheat growing to roughly its present limits.

In the early part of this century, wheat farmers were struck by a sudden outbreak of wheat rust, a fungus disease that attacks the stem of the plant. In 1916, Prairie farmers lost 2 800 000 t of wheat because of wheat rust. The losses continued to be severe until the late 1940's when a rust-resistant wheat variety was developed. In 1954, another kind of wheat rust reduced the wheat crop by 4 200 000 t. Canadian scientists are continuously working to develop new varieties of wheat that are resistant to as many diseases as possible. The Selkirk and Pembina are the most popular of the recently developed varieties.

The major insects that attack wheat are sawflies, wireworms, cutworms, and grasshoppers. In some years, the numbers of these insects reach epidemic proportions. Wheat varieties, with stems so solid that the sawfly cannot eat through them, are now available. Fortunately, summer fallow, which was initiated to conserve moisture, also controls insects by eliminating their natural habitat for a year. Today, most farmers spray their fields with insecticides whenever insects appear to be menacing the crops.

Weeds are a problem for all farmers. However, weeds on the Prairies are more damaging because they rob the crops of badly needed moisture. Proper cultivation is still the best form of weed control. The timing of the tilling operation, the depth of tillage, and the choice of implements are all important. Again, summer fallow is an excellent method of weed control, but herbicides are also used in vast amounts.

Wind erosion is another hazard that faces the wheat grower. In the early 1900's, the Prairies were in a weather cycle of high precipitation, and the pioneers pushed farming into the drier parts of the Prairies. Then, in the 1920's and 1930's, the precipitation decreased and exposed the dried-out soils to the winds. The topsoil of tens of thousands of hectares of land was blown away in great dust storms and many farmers had to abandon their farms completely. The Prairies were called the "dust bowl", with good reason.

A number of conservation practices have greatly eliminated wind erosion. Instead of ploughing, the farmers use a special kind of disk-packer that leaves the soil packed down instead of loose, and also leaves behind half-buried stubble that helps prevent wind erosion. Wheat, other crops, and fallow are laid out in long alternating strips so that the wind does not have a long sweep over bare, freshly cultivated soil. Windbreaks, composed of trees, have been planted around the fields in areas with enough moisture to grow trees. Summer fallow has reduced wind erosion by increasing the moisture content of soil.

In years of high demand for wheat, the farmers again plant more wheat in the dry belt and put fewer hectares into summer fallow. Fortunately, in recent years there has been sufficient precipitation to permit this greater production without the threat of another dust bowl.

THE FARM OPERATION

As soon as the land is ready to be worked, the wheat farmer and his family begin cultivating and seeding. The land cannot be worked until it is dry because the mass of a tractor would compress the wet soil, and later, it would bake into a bricklike surface. Also, the clay soil of much of the Prairies is appropriately named *gumbo* and tractors quickly mire in it when it is too wet.

There is a great rush to get the seed planted soon after the land is ready because every day gained is another day added to the growing season. Also, some spring seasons have heavy rainfall, and it is necessary to get the wheat sown during the period before the rains. Wheat should be sown in May or early June. If it is a late, wet spring the farmer often has to switch from wheat to crops that will mature in a shorter time such as oats, barley, rye or flax.

Even if it is an early spring, many farmers in the wheat belt plant other crops, some to sell for cash and others to feed to livestock. Because of

FROM WHEAT FIELD TO CARGO HOLD.

Golden fields of wheat are typical of the Prairie landscape in summer.

In early fall, combines in teams of six or more quickly harvest the grain and trucks take the wheat to local grain elevators scattered along railway branch lines.

From the local elevators trains carry most of the wheat to major ports both east and west.

At Thunder Bay, wheat is stored in huge elevators until it is loaded on lake freighters for the trip farther east. It requires about 20 250 ha of wheatland to produce enough cargo to fill some of the larger lake vessels.

the droughts and the marketing problems of the past, many farmers feel they have more security if they do not have to depend upon wheat entirely. The crops other than wheat are sown and harvested after the wheat, thus helping to spread the workload over a longer period of time. An example of a crop combination on a farm in the wheat belt of Saskatchewan is shown in Figure 9.7.

During the summer, farmers cultivate the fallow land as many as seven or eight times. They work it early in the spring to encourage the growing of weeds, and then, they till it at regular intervals to kill the weeds and create a loose mulch on top. Other summer jobs include spraying crops for weeds and insects, looking after livestock, checking fences around the pasture, and making repairs to farm buildings and implements.

The harvest begins in late August or early September. Before it can be combined, the grain has to stand until it is very ripe or it will not be dry enough to keep well. The recent trend is to cut the grain with a swather and to let it lie in the field before it is combined. A truck is driven beside the combine to pick up the grain and take it to the farm granary or local elevator.

All the wheat is sold as a cash crop. The oats and barley may be sold as cash crops, or else fed to livestock. Hay is baled and is used for winter fodder. The flaxseed (usually called linseed) and rapeseed are sold for the making of vegetable oils.

Since the farm in Figure 9.7 has livestock, the homestead is occupied all year round. Some large farms that grow cash crops only are run by farmers who live in the city. This trend is likely to increase as farming becomes more of a large-scale business and less a way of life.

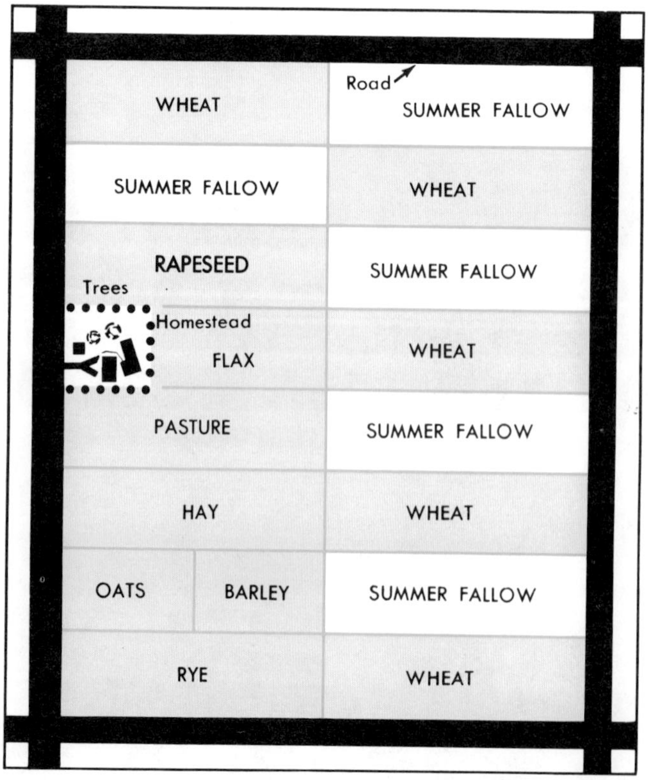

Figure 9.7 TYPICAL CROP COMBINATION IN THE WHEAT BELT. This map shows the crop combination on a 640-acre (one section) farm in Saskatchewan. The alternating of crops and summer fallow to conserve moisture, called dryland farming, is carried out only in the drier parts of the wheat belt. The trees around the farmstead provide protection from the wind.

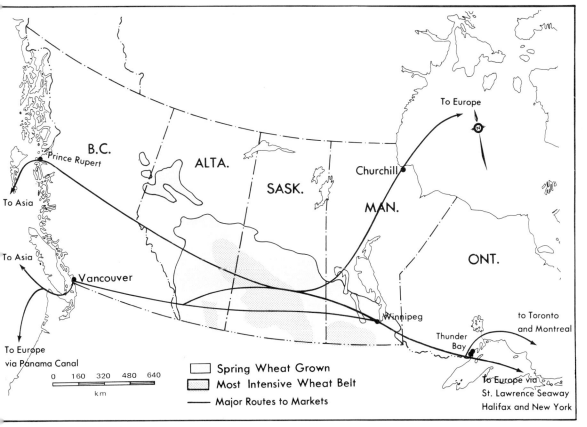

Figure 9.8 WESTERN WHEAT: ORIGIN TO DESTINATION.

TRANSPORTATION AND MARKETING

Very large farms have their own sheds or granaries where the wheat and other grain can be stored temporarily. The wheat from smaller farms must be trucked immediately to a local grain elevator before being transferred to railroad cars.

Most wheat going east is transferred to huge terminal elevators at Thunder Bay where it is stored until there is shipping space available. Although some ocean freighters transport the wheat directly to the overseas market, most of the wheat is transferred to ocean freighters at Montreal and lower river ports because lake freighters are not designed for ocean travel. Because movement of wheat on the Great Lakes is limited to about half the year, some wheat is shipped directly by rail to Toronto and Montreal, and some goes by rail to the ports of Halifax and New York for trans-shipment abroad (Figure 9.8).

With increasing sales to Japan, China, and other Asian countries, Vancouver has become a more important wheat-shipping port. In most years, Vancouver handles more wheat than Thunder Bay. Small amounts of wheat are exported through the port of Prince Rupert.

In peak years of export sales, there is a problem in transporting the huge volumes of wheat to market soon enough. If Canada cannot deliver on time, there is a danger of losing a particular market in future years. The Canadian Government has financed the building of wheat-carrying rail cars and has expanded the grain storage and port facilities at Prince Rupert. An extension of the length of the shipping season from Churchill would help in moving the grain destined for Europe.

Wheat crops grown in the Prairie Provinces are marketed through the Canadian Wheat Board which establishes prices for different grades of wheat. The wheat board pays farmers when they deliver their wheat to the elevator at a price somewhat below the prevailing world wheat prices. After the wheat crop has been sold, the wheat board distributes the surplus, if there is one, to the farmers.

Ranching in the West

RANCHING IN THE PRAIRIES

In the Prairie Provinces, ranching is the predominant type of agriculture in the short-grass country and in the foothills of the Rocky Mountains. Grazing is important in the dry belt because the drought makes cropping hazardous. The foothills country receives more precipitation, but the land is too rugged for cultivating crops. (See Figure 9.9.)

Since the late 1800's, grasslands in the dry belt on the Alberta-Saskatchewan border have been used for the grazing of cattle and sheep. In this region of light brown soils and short bunch grass, precipitation is not only light, but also very irregular. Since it takes 16 to 20 ha to support one head of cattle in this dry country, the ranches are very large. The average ranch has about 4050 ha, with some as large as 20 250 ha. Often, only a part of the ranch is owned by the rancher; the rest of the range is leased from the provincial government.

Sheep, as well as cattle, are grazed on the grasslands of the Prairie Provinces. Sheep are not usually grazed on the same range as cattle because they clip the grass too short. The most intensive sheep grazing is found in the foothills of western Alberta.

At one time, all livestock was allowed to roam over the range. Now, many ranchers realize that grasslands can quickly become overgrazed if the livestock remain too long on a plot of land. If the grass has been grazed too close to the ground, it may be so damaged that it will take years to return to its natural vigour. If new grass does not cover the ground thickly enough, wind and water erode the soil and cause gullies to form. To prevent the effects of overgrazing, some ranchers constantly move their herds of cattle and flocks of sheep from one pasture to another. By spacing salt blocks and watering troughs evenly over the range, ranchers can also keep livestock from trampling grass in any one spot.

In the past, livestock used to spend the winter in the open range, feeding on the dried grass that they could obtain by pawing through the

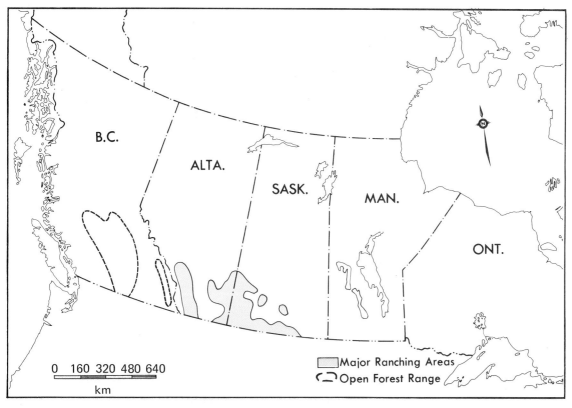

Figure 9.9 MAJOR RANCHING AREAS OF WESTERN CANADA.

Cattle herding no longer takes place only in the fall. During the summer when the cattle graze on the open range, ranchers will constantly move their herds from pasture to pasture to prevent overgrazing.

snow. If the snow was exceptionally deep, they would be in a half-starved condition by spring. Reduction in mass meant loss of income for the ranchers. For this reason, many ranchers are setting up feedlots for their young cattle in winter. They feed the cattle hay and grain that they buy or grow on irrigated land near the farmstead. Beef cattle that are being fattened for market are given commercial feed concentrates in addition to hay and grain. Sheds are built to protect the livestock from winter storms. By feeding and caring for cattle in this way during the winter, ranchers are able to "finish" the beef instead of selling young cattle to other farmers for fattening. Not only do the "finished" cattle have a greater mass, but the price per kilogram is much higher.

RANCHING IN INTERIOR BRITISH COLUMBIA

It may surprise you to learn that there are over 4 000 000 ha of grazing land in the Interior Plateau region of British Columbia. The valleys are as dry as the dry belt of the Prairies and have the same steppe vegetation, while the uplands have a mixture of grass and coniferous trees. The rancher usually owns his own grassland and leases the timber pasture from the province (Figure 9.9).

There is no typical size of ranch in British Columbia because the climate, the landforms, and the nature of the range differ so much from one area to another. However, all of the ranches are large. There are very few that are smaller than 1200 ha and some are several hundred thousand hectares.

In the spring, the cattle and sheep on the ranches of interior British Columbia are herded onto the lower slopes of the grassland range. They graze on this range, moving gradually toward the higher slopes until, by mid-June, they have reached the timber range. By this time, all the snow has melted from the timber range and the grass under the trees has matured. From mid-June until late September, the livestock graze in these high altitude forest pastures. Some livestock, particularly sheep, are driven into alpine meadows above the treeline when grass becomes dried out or depleted in the lower areas. (See Figure 9.10)

By early October, because of the danger from storms, all livestock are herded back down the mountain slopes to the grassland range. Pasture used in the spring may be grazed for a second time, or else part of the grassland range may be purposely set aside for autumn use. By late November, the stock is rounded up and shifted to the winter ranges and meadows in valley bottoms. Here, they are fed hay and grain in addition to whatever grass forage they can find. An average rancher may grow about 4 ha of alfalfa hay, as well as several hectares of oats and spring wheat, by irrigating the dry valley bottom with water from nearby rivers and streams. When warm temperatures return to the valleys in spring, the grazing cycle is repeated. Such seasonal migration is known as *transhumance* and is quite common in mountainous countries such as Norway and Switzerland.

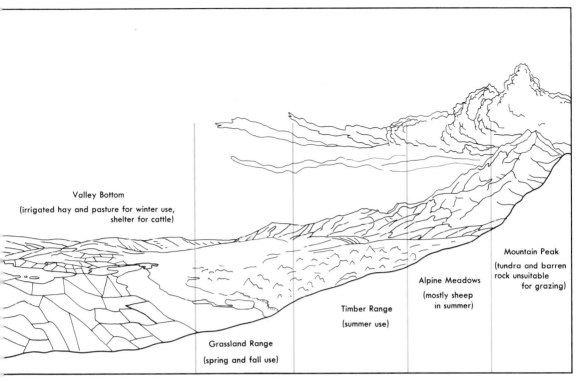

Figure 9.10 FORAGE AREAS ON A TYPICAL RANCH IN INTERIOR BRITISH COLUMBIA. By moving livestock from one elevation to another, ranchers in interior British Columbia are able to graze considerable numbers of sheep and cattle. The seasonal migration of livestock is called transhumance and is common in most mountainous countries.

Dairy Farming

Dairy products are an important part of the Canadian diet. Every year, each Canadian consumes about 450 kg of milk or milk products such as butter, cheese, and ice cream. The dairy farmers of Canada supply all of this Canadian demand and also provide enough milk for the production of large amounts of the cheese and concentrated milk products (milk powder and evaporated milk) that are exported to other countries.

In recent years, milk production has been increasing much more rapidly than the number of dairy cows in Canada. This increase has been possible because farmers have been constantly introducing better milk cows and better methods of feeding. In addition to feed grown on their own farms, dairy farmers give their cows well-balanced feed concentrates prepared by commercial feed companies.

Although there are many farmers producing milk, they are not considered to be dairy farmers unless most of the farm income is from milk. A dairy farmer usually grows other crops including cash crops, and may raise some other livestock. However, the chief interest and most of the work is related to the dairy herd. On this basis, only about ten percent of Canadian farms producing milk could be classified as true dairy farms.

As with other farm operations, dairying is becoming more specialized. Some of the larger dairy producers have one hundred or more milk cows. A great deal of money is required to have such a herd because good milk cows are expensive, and the farmer must have a large barn with one or more large silos, as well as a refrigerated room for keeping the milk. Milking machines, automatic feeding equipment, manure loaders, and other expensive equipment are also required.

Most of the large dairy farms are found in the areas surrounding the city. They provide the fresh whole milk consumed by the city dwellers. The areas that produce fresh milk around urban centres are called *milksheds*.

Farther away from the cities, dairy farms usually supply milk for cheese, other processed milk products, or cream for butter. A farmer who sells cream uses the remaining skim milk to feed the calves and hogs. Lower prices are paid for milk sold to processing plants, creameries, and cheese factories than for milk sold for fresh whole milk consumption. Although there is a good demand for cheese and concentrated milk products, the demand for butter has fallen because many Canadians are now using oleomargarine instead.

Many farmers with small dairy herds are finding that the dairying business is not profitable. Large numbers are turning to other types of farming or are leaving the farm for jobs in the city. A few, however, are enlarging their herds and are buying modern equipment in order to make the dairy enterprise more efficient and profitable.

A DAIRY FARM IN QUEBEC

The dairy farm in Figure 9.11 is located in the St. Lawrence Lowlands of Quebec. This farm supplies fresh whole milk to the city of Montreal, which is only about 24 km away.

Although the flat land of the St. Lawrence Lowlands with its silt and clay soils makes good farmland, it must be drained by ditches before it can be cropped. In Figure 9.11, you will notice main drainage ditches down the centres of the fields, and a number of smaller tributary ditches. The shape of the farm, resulting from the original French strip-farm settlement pattern, makes it easy to drain the land in this way. This farmer owns another 8 ha across the road that has not yet been drained, which he uses only for hay and pasture. He has also purchased, for his son, a 40 ha farm 400 m down the road.

The mixed farm operation is closely integrated with the dairying. Oats and barley are grown to be fed to the cows in the winter. Alfalfa, red clover, and timothy are sown with oats and barley in some of the

Large-scale modern dairy farms now supply much of the milk for Canada's large cities. The farmstead shown above is about 48 km from Montreal. These modern dairy farms use expensive equipment such as milking machines which take the milk from the cows through pipes directly to a stainless steel storage tank and automatic feeders which move the ensilage from the silo to the cows' feeding trough.

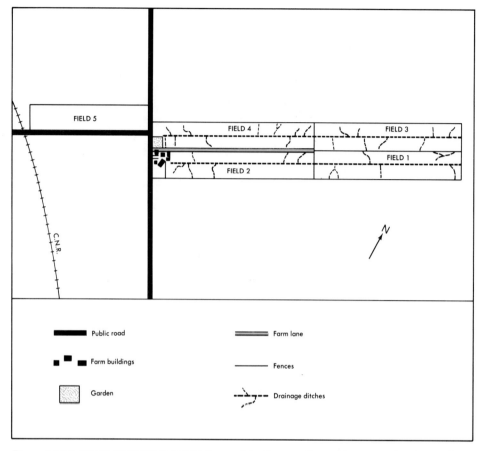

Figure 9.11 A DAIRY FARM IN QUEBEC. Parts of the Eastern Townships are so flat that drainage ditches must be dug across farmland to carry excess water away from the fields.

fields. The grain grows quickly and protects the young alfalfa, clover, and timothy plants. When the grain is cut, in August, the hay crop begins to grow. In a good season, it will grow sufficiently to provide pasture that fall, and by the next summer, there will be an excellent hay crop. The hay is first cut in early July, and a second crop is either pastured or cured for hay later in the summer. The hay is cut with a mower, dried in the field, baled, and is then stored in the barn.

The grain from the oats and barley is ground up and fed to the cattle in the winter, and the oat and barley straw is spread under the cows in winter for bedding. This straw, when mixed with cow dung, forms a manure that makes excellent fertilizer for the fields. The farmer spreads manure on at least two fields every year. He also uses commercial fertilizer to increase the yield of his crops.

One field of corn is grown each year. This crop is not grain corn, but it is chopped up and blown with a corn blower into a large cylindrical concrete silo to create *ensilage*. In addition to the hay, grain and corn grown on the farm, large amounts of commercial feed concentrates must be purchased to feed all of the farmer's cattle.

This farmer also plants small amounts of sweet corn, peas, and beans that he sells as cash crops to canneries in the Montreal area to supplement the income from his dairy herd.

The dairy herd on this farm has only fifty cows. This is not considered a large dairy herd, but it is about average for the Montreal milkshed. Thirty of these cows are registered purebred Holsteins. The rest are not purebred cattle, although they all have some Holstein in their background. The farmer is breeding his registered Holstein cows to a purebred Holstein bull and is keeping the heifer calves for cows in order to build up the quality of his herd. The young bull calves are sold for veal when they reach a few weeks of age.

There is much work associated with dairy farming. The land must be worked, and the crops must be sown and harvested in the spring and summer season. In winter, the cattle have to be fed, and the manure has to be cleaned from the stables every day. In addition, the cows must be milked regularly, twice a day, all year round. The stables must be swept regularly and must be whitewashed occasionally to keep down dust that would contaminate the milk. The farm buildings must be painted and kept in good repair, fences must be kept up, and farm machinery looked after. Since he has no hired man, the farmer and his family, in this study, do all these jobs themselves. They have not been away from the farm for more than a day in the last ten years.

The Orchard Industry

The orchard industry is important in five Canadian provinces (Figure 9.12). Ontario leads in tree fruit production by a large margin. Ontario also produces more of the soft fruits such as peaches, pears, cherries, plums, and prunes. In the last twenty-five years, Quebec has become an important apple producer. The Province of Quebec has a much higher apple production than the famous Annapolis Valley, and in some years, produces almost as many apples as Ontario or British Columbia.

As with most farming operations, fruit farming is becoming much more specialized and efficient. Orchardists are planting improved varieties, are using fertilizers, insecticides and fungicides in a scientific manner, and are increasingly mechanizing their operations. As a result, a smaller number of orchardists are growing more fruit on less land. For example, between 1931 and 1981, the number of apple trees in Canada decreased by about two-thirds, while the size of the annual apple crop doubled in the same period.

Everywhere, except in the Prairie Provinces, one of the first farm activities of the early Canadian settler was usually the planting of a few apple trees to provide his family with much-needed fresh fruit in the fall

Right: An orchard district in the Monteregian Hills region east of Montreal. The orchards are located on the terraces that surround the rocky hills. The photograph shows a roadside stand that sells apples, vegetables, flowers, and other items grown or made in the local area.

and winter. Where the climate was favourable, orchards were enlarged year by year until they also provided a welcome cash crop. Thus, the growing of apples became a part of the mixed farm operation.

In eastern Canada, the present orchard regions developed from the original farmstead orchards. When Canada began exporting apples to Europe, only those who had the good varieties of apples and a high-quality product could market their apples. Wormy and scabby apples could not be sold. It did not pay farmers to invest in expensive spraying equipment unless they had large orchards. Farmers in areas with the mildest winters and longest growing seasons increased their orchard areas. Those in less favourable areas ceased to look after their orchards and often did not even pick the fruit. Several very cold winters in the 1930's killed many of the fruit trees in areas with a less favourable orchard climate. Orchard specialization occurred in those areas where the winters were mildest and where there was the least chance of spring frost damage to blossoms.

Below, left and right: Orchards in the Deux Montagnes region of Quebec. The orchard districts are among the most prosperous farming areas of Quebec. The province of Quebec produces about one-quarter of Canada's apple production.

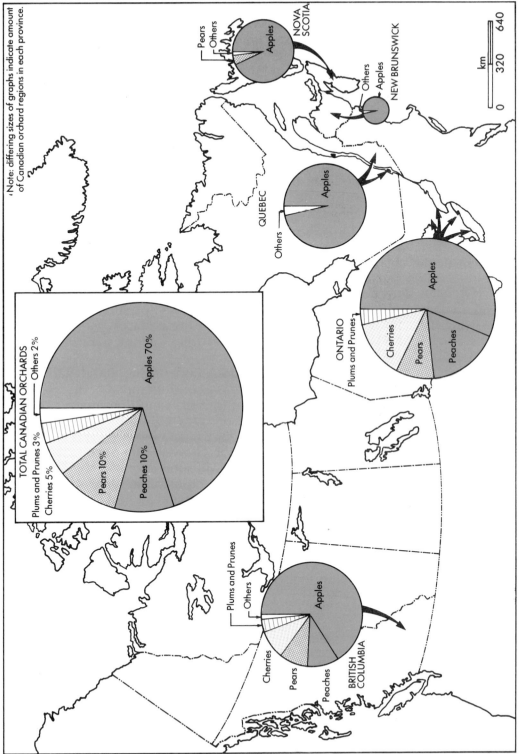

Figure 9.12 ORCHARD AREAS IN CANADA. Ontario has by far the largest orchard area. However in the last few decades, Quebec has shown the most rapid increase in orchard plantings. This is illustrated by the fact that in 1930 Quebec produced only 5% of Canada's apples, while today it is producing over 20% of the Canadian total. Explain why most of the soft fruits are grown in British Columbia and Ontario.

In the Okanagan Valley of British Columbia, the change from cattle ranches to orchards was an abrupt one. Just before the turn of this century, the steppe land was purchased by real estate promoters for $2.50 per hectare. After it was irrigated and planted to orchards, the land was sold at $2500 per hectare. Much of this irrigated orchard land was sold to British immigrants. By 1920, the Okanagan had almost two million fruit trees.

Orchard crops have very specific physical requirements. Apple trees are the most hardy and will survive temperatures as low as $-34°C$. A temperature of $-29°C$ will kill peach and apricot trees and a temperature of $-24°C$ will kill the dormant peach and apricot buds. In the spring, a temperature of three or four degrees below freezing ($0°C$) will kill all blossoms. If the blossoming period is cloudy and rainy enough to prevent bees from working, the blossoms will not be pollinated and there will be no crop. The balance of the spring must have sufficient rain to assure good-size fruit, but the harvest season must be dry and sunny or the skin of the fruit will be damaged, or rot will set in.

Orchards must be planted on sloping sites so that cold air will drain away. Orchards do not survive long in depressions, valley bottoms, or any low-lying flat land because cold air collects in these places. Most tree crops also require well-drained soil, for fruit trees cannot stand "wet feet."

Orcharding is a profitable but hazardous undertaking. Besides the problems related to climate and soils, there is a constant threat of disease and insects. To combat these, the different fruit trees must be sprayed at exactly the right times throughout the season. When the soft fruits, such as peaches, cherries, and plums, are ripe, they must be picked and sold immediately, whether the price is good or not. Growers cannot hold back fruit if prices are too low because most fruit will spoil in only a few days time. Nor can growers determine the size of this crop; this is done for them by the weather. A grower must guess about the demand for fruit five or ten years ahead, because it takes that long for a newly planted orchard to come into production.

The following section is a brief description of four major Canadian orchard areas: the Annapolis Valley of Nova Scotia, the Orchard Regions of Quebec, the Niagara Fruit Belt of Ontario, and the Okanagan Valley of British Columbia.

THE ANNAPOLIS VALLEY

Using the map in Figure 9.13, your atlas, and other parts of this book, answer the following questions:

QUESTIONS

1. What accounts for Nova Scotia's moderate winters, considering its northern latitude?
2. In what way is the Annapolis Valley protected against winter winds from the north, northeast, and northwest?
3. What is the average annual precipitation for the Annapolis Valley?
4. How long and how wide is the valley? What two main rivers are found in it?
5. What is the nearest city with a population of more than 100 000?
6. What advantage does the Annapolis Valley have over other Canadian orchard regions for exporting apples to Europe?

The winters in the Annapolis Valley never become cold enough to injure apple trees, but winters are too cold for the tender peach, apricot, and sweet cherry trees. Small plots of peaches planted near Kentville have suffered winter injury on the average of every second year. Consequently, most of the Annapolis orchard area is apples; pears are the only other significant tree crop.

The apple orchards are concentrated in the eastern end of the Valley. The orchards are planted mostly on loam soils, as the sandy soils have been found to be too dry and the clays too poorly drained.

If you drive along the country roads, you may gain an exaggerated impression of orchard density. The reason is that the orchards are all located along the roads, while the back parts of the farms generally grow other farm crops. Although there is a trend towards specializing completely in orchards, most farmers have anywhere from 2 to 12 ha of orchard. They combine this area with a mixed farm operation that may include livestock and general farm crops, as well as a cash crop such as potatoes.

Until 1939, most of the Annapolis apples grown were exported to Great Britain. The Annapolis growers specialized in hard-textured apples that could be shipped long distances in barrels. The British market was lost during World War II. After the war, when the British could afford to import apples again, the demand had changed to bright red dessert apples such as the Delicious and McIntosh apples, neatly packed in boxes instead of in barrels. The result was that British Columbia, instead of the Annapolis Valley, supplied the British market with apples.

With the loss of the British market, the Annapolis orchardists decided that they should reduce their orchard areas, change their apple varieties, and use modern methods of packaging. In a twenty-year period, more than a million apple trees were removed. New varieties were planted and were grafted onto healthy young trees. Apples were packed in boxes instead of barrels and were kept in cold storage. As a result, part of the British market was regained, and other, new markets were found in

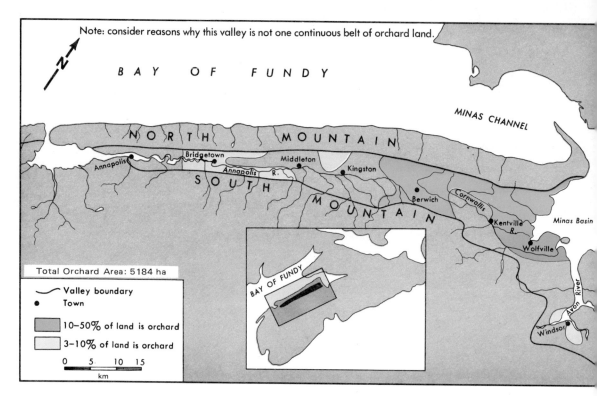

Figure 9.13 ORCHARD AREA IN THE ANNAPOLIS VALLEY OF NOVA SCOTIA.

Europe. These exports, along with the production of large amounts of processed apple products, have made it possible for the Annapolis Valley to remain one of Canada's major orchard regions.

THE ORCHARD REGIONS OF QUEBEC

By consulting Figures 9.12 and 9.14, earlier chapters in this book, and other sources, answer the following questions about the orchard industry of Quebec.

QUESTIONS

1. Compare the apple orchard area of Quebec with those of British Columbia, Ontario, and Nova Scotia.
2. Compare the average January temperatures of the Quebec orchard regions with those of the Okanagan Valley, Southern Ontario, and the Annapolis Valley (see Figure 5.6).
3. Why are the orchards in Quebec predominantly apple?
4. What are the landform and soil conditions that make the Monteregian Hills region so favourable for apple orchards? Why are there no apple trees on the flat plain between the mountains?
5. What marketing advantages have the Deux Montagnes and Monteregian Hills regions?

Only a small portion of the Annapolis Valley is covered with orchards. The orchard area has decreased considerably since the 1940's. In this photograph you can pick out the rows of trees that suggest the scattered location of the apple orchards.

A farmstead and adjacent orchard in the Annapolis Valley. Most of the orchards are located around the farmsteads while the rest of the farm area is used to grow hay, grain, and vegetables. How does this contrast with fruit farming in the Niagara Fruit Belt and in the Okanagan Valley?

Apple orchards are located throughout the southern portion of Quebec from the Ontario border to a point some 100 km northeast of Quebec City. Approximately 95 percent of the commercial orchards are found in the southwestern corner of the province, within 80 km of Montreal (Figure 9.14).

The Monteregian Hills region is the most important orchard region in Quebec, annually producing more than one-third of the total provincial apple crop. The highest yields are obtained in the Deux Montagnes and Missisquoi regions. The Missisquoi and Huntingdon regions are the most recently developed, while the Châteauguay-Napierville is the only region in which the orchard area has been decreasing since the 1930's. Because the land is too flat and the soil too poorly drained, there has been severe winter tree injury and few damaged orchards have been replanted. As a result, this region is gradually disappearing as an orchard district.

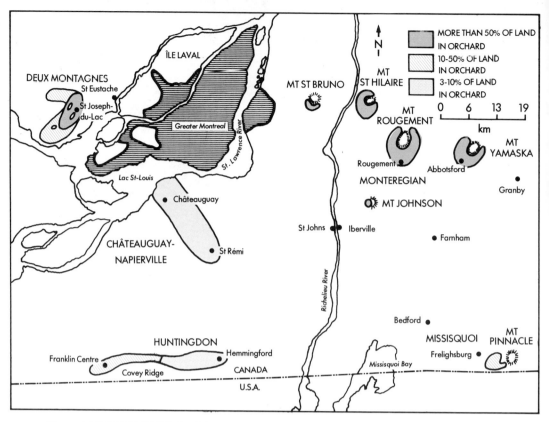

Figure 9.14 THE ORCHARD REGIONS OF QUEBEC. What factors make the Monteregian area the most important apple-growing district in Quebec?

All the major orchard regions of Quebec have similar terrain and soil characteristics. The orchards are planted on sloping sites that provide good air drainage. Light-textured soils ranging from sandy to clay loams encourage good root development and provide drainage of excess moisture. Through trial and error, growers have discovered that orchards do not produce well nor survive long on low, flat areas with heavy, poorly draining clay soils.

The winters of Quebec orchard regions are more severe than in any other orchard region in Canada. Quebec orchardists face the hazards of tree and crop losses because of extremely low winter temperatures. In the winter of 1917-1918, Quebec lost about half of its bearing apple trees. Another cold winter in 1933-1934 again resulted in serious tree losses. In the following decades, most of the less hardy apple varieties, such as Fameuse (Snow), Spy, and Ben Davis, were replaced with hardier varieties like McIntosh. Nevertheless, in the winter of 1980-1981, extremely low temperatures killed about one-third of all apple trees and reduced the 1981 apple crop by about eighty percent.

Because of their proximity to Montreal and its sprawling suburbs, some of the orchard lands are also threatened by urban development. The same terrain and soil characteristics that make these sites desirable for orchards also make them desirable for residential subdivisions. This problem has been especially serious in the Monteregian district. Fortunately, in recent years, the Quebec Government has passed legislation protecting good farmland from urbanization.

As in most other orchard regions in Canada, Quebec orchards have become increasingly productive. More trees are being planted to each hectare. Pruning, fertilizing, and control of insects and diseases have all been improved. As a result, although the area in orchard has been declining, the average annual production of apples has been increasing constantly. Although the cost-price squeeze has pushed many small, inefficient growers out of business, large efficient specialists have managed to increase their orchard size and to produce better quality apples.

Although Quebec has found some export markets in the United Kingdom and Europe for its apples and apple products, the bulk of the annual production is sold in Quebec. Local growers can sell their apples in Montreal at prices that cannot be met by competition from other provinces. Quebec does import some apples from the United States and British Columbia—mainly the Delicious variety that does not grow well in Quebec. About three-quarters of the Quebec apple crop is sold as fresh fruit; the balance is processed as applesauce, apple juice, and apple cider.

THE NIAGARA FRUIT BELT

The Niagara Fruit Belt lies along the south shore of Lake Ontario, between Hamilton and the Niagara River. From Figures 9.15, 9.16, and other maps and references, answer the following questions:

QUESTIONS

1. What are the climatic, landform, and soil factors that permit the growing of grapes and tender tree fruits in the Niagara Fruit Belt?
2. What are the limits to growing peaches in the Niagara Fruit Belt?
3. Apples grow very well in the Niagara Fruit Belt. Why are very few grown?
4. Name the major highway and the railway that connect Hamilton, St. Catharines, and Niagara Falls.
5. What other location advantages can Hamilton and St. Catharines offer to industry?
6. What attractions have caused housing to spread out over the post-glacial Lake Iroquois plain (north of the Niagara Escarpment) from Hamilton to St. Catharines?
7. How far is St. Catharines from Toronto? How long would it take to truck fruit that far?
8. If you were an urban planner, where would you direct future urban development in the Niagara Peninsula so that it would not use up valuable fruitland?

The Niagara Fruit Belt has an ideal combination of climate and soils to grow a wide range of fruit crops, including peaches and grapes. Although most of the fruit production is contracted to canneries and wineries, some is sold as fresh fruit both at roadside stands and in stores. Vineyards are on sloping land in the Niagara Fruit Belt. More than ninety percent of Canada's grapes are grown in the Niagara Peninsula.

The Niagara Fruit Belt has the best climate in Canada for growing grapes and soft fruits. Lake Ontario protects the trees from cold air masses in winter. In the spring, the cooling effect of the lake delays the opening of blossoms until the risk of frost is over. The lake plain dips gently towards the north, thus helping the cold air to drain away to the lake. In the fall, the lake delays the first frost so that fruit is seldom damaged before harvest. In fact, outside of California, there is no other fruit district in North America where there is as little risk of frost damage to a tender fruit crop like peaches.

The soils in the Niagara Fruit Belt range from very sandy to heavy clay. Peaches and sweet cherries, as well as small fruits and vegetables, are grown on the well-drained sandy or gravelly soils. The clay loam soils are planted with apples, grapes, pears, plums, and sour cherries. Low-lying, poorly drained, heavy clay soils are not good for any of the fruit crops.

As you can see from Figure 9.15, the most intensive orchard and vineyard areas are located north of the escarpment, where a very large percentage of land is taken up by orchards, vineyards, and small fruits and vegetables. Most of the farms are smaller than 8 ha, although a few prosperous farmers own as much as 40 or 80 ha. Many of the small farm owners are part-time farmers who also have jobs in the city.

South of the Niagara Escarpment, most of the orchards and vineyards are part of a mixed farm operation. The intensive area of fruit growing around Fonthill is located on a huge gravelly hill known as a *kame*. The hill provides excellent air and soil drainage, thus compensating for its distance from the lake. This location is the only place south of the escarpment where peaches are grown.

About ninety percent of Canada's grapes are grown in the Niagara Fruit Belt. The recent introduction of European hybrid and vinifera grapes has helped this area to produce high quality wines. Because of a rapidly growing demand for wine, the Niagara grape growing and wine industries are thriving.

The greatest threat facing the orchard industry of the Niagara Fruit Belt is urban development. Urban land uses have been rapidly spreading into prime fruitlands, particularly along the Queen Elizabeth Way where the best fruit climate and soils are located. The low density urban sprawl pattern of development has ruined much more agricultural land than is actually needed for urban purposes. Because of urbanization, the price of orchard land has risen to the point where it is uneconomic to farm it, and increasing land taxes to provide urban-type services have created an additional burden for the fruit growers.

If the trends of the 1950's to the 1970's had continued, the orchard industry in Niagara would have collapsed well before the year 2000. Hardest hit by urban encroachment have been the peach orchards that can grow only on a limited amount of deep, well-drained, light-textured soil. Peach production declined by about fifty percent between 1951 and 1981.

Because grapes can be produced on a wider variety of soils, including clay loams, they do not face the same threat of extinction as do peach orchards. In fact, grape areas are continuing to increase. This reflects improved profits to grape growers in recent years, as well as the availability of large areas of clay loam land suitable for grapes in areas beyond the influence of urban development.

In 1970, the counties of Welland and Lincoln were reorganized into the Regional Municipality of Niagara (Figure 8.4). The new Regional Government, unfortunately, was more committed to urban growth than it was

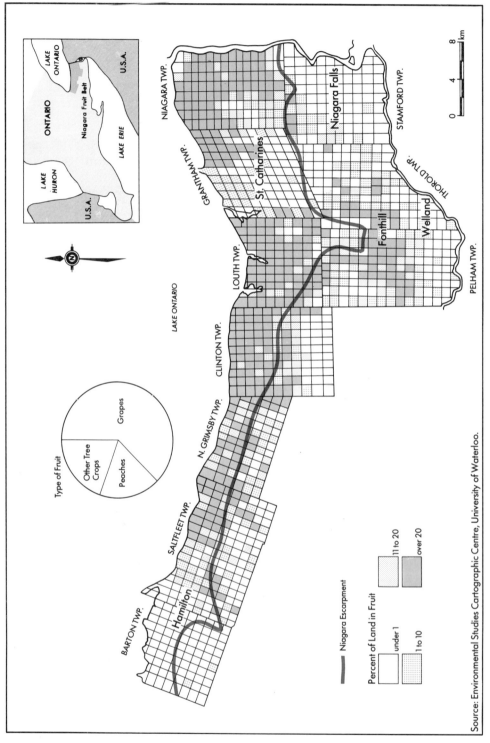

Figure 9.15 THE NIAGARA FRUIT BELT. The most intensive orchard and vineyard areas are located north of the Niagara Escarpment, particularly peaches, sweet cherries, and the more tender vinifera and hybrid grapes. The more hardy apples, pears, plums, and labrusca grapes are grown south of the Escarpment. The blocks shown on this map (and on Figure 9.16) are concession blocks bounded by roads. The municipalities named on these maps are those that existed before the Regional Government was established for the Niagara Peninsula. For an explanation of the township survey system, see Figure 8.3.

Source: Environmental Studies Cartographic Centre, University of Waterloo.

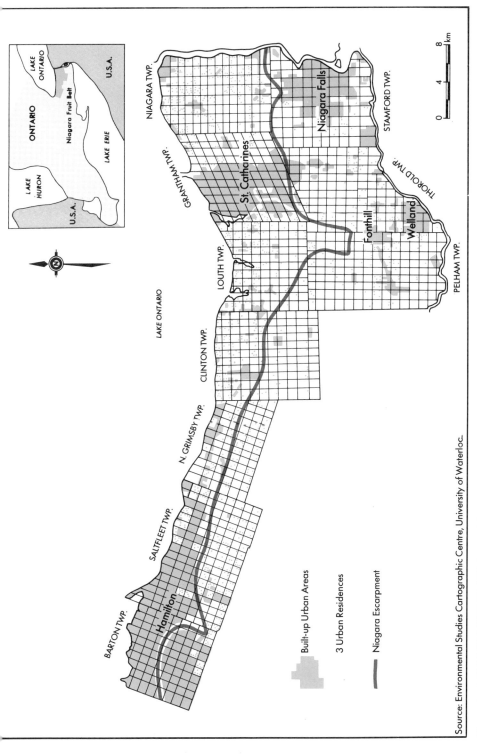

Figure 9.16 NIAGARA URBAN SPRAWL. This map shows the way in which urban land uses had spread across the Niagara Fruit Belt by the late 1970's. Compare the urban development pattern here with the fruit-growing pattern in Figure 9.15. Why has urban development occurred in the most intensive fruit-growing areas? What alternatives are available for urban growth in the Niagara Peninsula? What planning tools are available to direct urban growth? How successful has the Ontario Municipal Board's 1981 decision been to make urban boundaries in the Niagara Fruit Belt permanent?

to preserving the fruitlands. Its proposed land-use plan provided more area for urban development than was required for the projected population growth. A group of conservation-minded citizens protested that the unique Niagara fruitland was being needlessly destroyed. The Provincial Government referred the controversy to the Ontario Municipal Board (O.M.B.), which held hearings that lasted for two years. Finally, in 1981, the O.M.B. made a decision that may save the remaining fruitland. It ruled that the urban boundaries be rolled back by some 2100 ha and that these boundaries should be considered permanent. It further recommended that future urban development should be directed south of the Niagara Fruit Belt, along the Welland Canal. Only time will tell whether this O.M.B. decision will be upheld so that the Niagara Fruit Belt can be saved for future generations.

THE OKANAGAN VALLEY

The Okanagan Valley accounts for over 90 percent of British Columbia's tree-fruit and all of its grape production. Using Figure 9.17 and other references, answer the following questions about the Okanagan Valley.

QUESTIONS

1. In what landform subdivision of the Western Cordillera is the Okanagan Valley located?
2. How do January and July average temperatures compare with those of the Niagara Peninsula and the Annapolis Valley? What is the annual precipitation in the Okanagan Valley?
3. What is the natural vegetation of the Okanagan Valley?
4. Low temperatures are a major hazard for Okanagan orchards. What measures can orchardists take to protect their crops from freezing?
5. Why are few peaches grown in the North Okanagan?
6. The Okanagan orchard lands are also being threatened by urban development. Why do retired people and tourists find this valley an attractive place?

On flat land on the floor of the Okanagan Valley, some towns have been built where orchards cannot grow. The orchards in the photograph were planted on a terrace about 15 m above the level of Lake Okanagan. Orchards are protected from frost damage here because of good air drainage. Recently urban development has been spreading into the orchard terraces.

Top, left: Orchards in the southern part of the Okanagan Valley. In the foreground you can see the natural steppe vegetation with short grass, sagebrush, and other shrubs. In the middleground are the irrigated orchards on terraces along Lake Osoyoos. In the background is the valley slope.

Top, right: Okanagan orchard land for sale. Urban land uses have spread over much of the good orchard land in this valley. As a result, orchard land sells at a very high price. British Columbia has established an Agricultural Land Reserves system to protect the orchard lands from further urban encroachment.

Bottom, left: A pond used to store irrigation water. A dam on a small stream coming down from the plateau has created the pond. The source of the stream is usually a lake on the plateau. Extra water is also stored in the plateau lakes by using control dams at their outlets. The pond in the photograph is about halfway up the valley slope. Note the mixture of grass and coniferous tree vegetation. A heavier stand of coniferous forest is on top of the plateau because of the greater precipitation there.

Bottom, middle: A concrete irrigation flume. This flume west of the Summerland area takes water from the storage pond down to the orchard areas where smaller flumes carry the water to individual farms.

Bottom, right: Sprinkler irrigation. From the flume, the orchardists divert water into aluminum irrigation pipes to which sprinklers are attached. It is common to sprinkle an orchard for eight hours at a time. After one part of the orchard has been watered, the pipes are moved to another place.

The Okanagan Valley can be divided into two distinct regions. In the South Okanagan, pears, peaches, apricots, and cherries are all important crops. The climate in the North Okanagan is too cold for peaches and apricots, and so more apples are grown. The chief apple varieties grown are the Delicious and McIntosh. Because they are hardy trees, more McIntosh apples than other varieties are grown in the northern part of the Valley. (See Figure 9.17.)

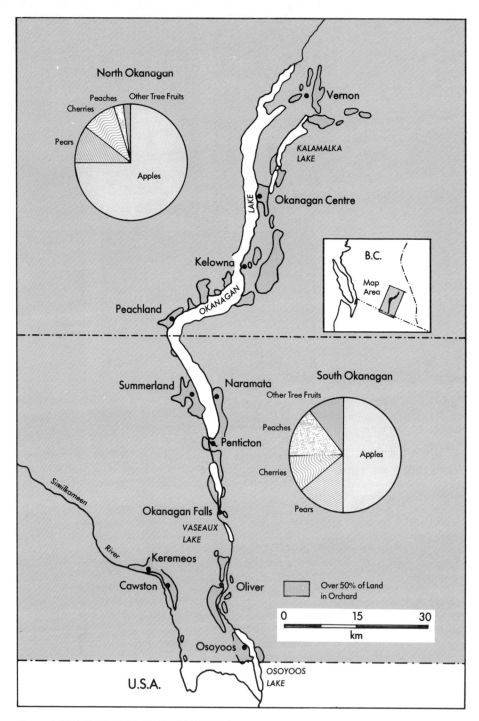

Figure 9.17 THE OKANAGAN ORCHARD REGION. The Okanagan is well known for its bright red apples. The orchards are planted on the terraces along the sides of the valley. Winter and spring frosts are the greatest threat to Okanagan orchardists.

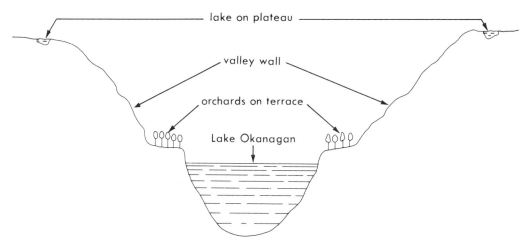

All the orchards are located on terraces at a considerable height above the Valley floor. This location permits excellent air drainage that helps to protect the orchards from low temperatures.

Because of the dry climate (300 mm annual precipitation at Penticton) trees will not grow without irrigation. Most of the irrigation water is obtained from streams that come from the surrounding plateau surface, several thousand metres above the Valley floor. From streams, the water is carried by troughs called *flumes* to the orchards. At one time, the Okanagan used the furrow system of irrigation that allows water to run in little ditches between the rows of trees. Today, most fruit growers use sprinklers. With this method, the water is carried through portable aluminum pipes, and specially designed sprinklers spray water under the trees. The cost of irrigation is very high, but, by controlling the moisture, Okanagan orchardists are able to obtain yields twice as high as in eastern Canada.

Dead peach trees at Peachland. The tender fruits of peaches and apricots can be grown successfully only in the southern part of the Okanagan Valley. This photo shows peach trees that were killed in under −18°C temperatures of the previous winter. On the opposite side of the valley in the background, there are no orchards because the steep slopes come right down to the edge of the lake, leaving no terrace for orchards.

As on the lake plain of the Niagara Fruit Belt, the Okanagan fruit growers specialize completely in the growing of fruit and some vegetables. There are no general farm crops or livestock associated with the orchard operation, except north of Vernon where there is a transition area into mixed farming and grazing. The farms of more than half of the growers are smaller than four hectares. These farms are too small to provide a living for a family, and so the farm income must be supplemented with off-the-farm jobs.

Although the climate of the Okanagan is generally considered favourable for fruit growing, frost damage is the most serious threat facing the Okanagan orchardists. There is a chance of spring frost damage to some tree-fruit crops about every other year. In addition, once in about every seven years, there is a winter with temperatures cold enough to kill thousands of fruit trees.

All of the fruit grown in the Okanagan Valley is sold through one selling agency known as B.C. Tree Fruits Limited. This agency grades, packs, and sells the fruit for the growers. By having one selling agency, the growers are assured of the best prices because they are able to prevent growers from competing with one another by cutting prices. B.C. Tree Fruits grades fruit very carefully so that only the best quality product reaches the consumer. In this way, a good reputation has been established for B.C. apples across Canada and in the United Kingdom and Europe as well.

Vegetable Farming

Vegetables are one of the specialty farm crops like fruit, tobacco, field beans, and potatoes that thrive in only a few favourable locations across Canada. (See Figures 9.3 and 9.4.) Many different kinds of vegetables are grown to be sold as fresh produce or processed for canning or freezing.

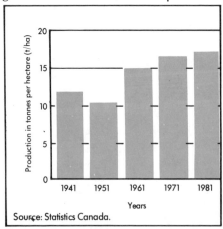

Figure 9.18 CHANGING VEGETABLE CROP PRODUCTION. Explain why these changes in amount of production for farm vegetables have occurred.

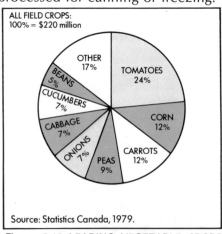

Figure 9.19 LEADING VEGETABLE CROPS OF CANADA. Only some of the major commercial vegetable crops grown in Canada are shown here.

Long, narrow plots of green vegetables stretch across the flat Holland Marsh as far as the line of hills in the distance. Drainage ditches between the fields control the water level in the soil and empty into canals dredged around this former swamp. When harvested, some vegetables are packed in wooden crates that can be seen piled beside the canal road and in the background of these photographs.

Small garden plots are cultivated every summer by many homeowners across Canada. Most towns and cities have pockets of suitable farmland nearby on which farmers can produce field vegetables as cash crops. This type of farming is known as *market gardening*. Regions specializing in field vegetables, as well as greenhouse and hothouse produce, are found in the Fraser Delta in British Columbia and in parts of southern Quebec and Ontario.

With rapid annual population growth, the demand for more Canadian vegetables has been rising constantly. Growers have been increasing their output significantly by using improved farming methods and by planting crop varieties with better yields. Since the 1950's, production has risen from about 10 t of vegetables grown on each hectare of cropland to an average of over 15 t/ha today (see Figure 9.18). Vegetables, like other specialty crops, earn a high dollar value for growers. With annual farm sales of nearly a quarter of a billion dollars, field vegetables are one of Canada's most valuable cash crops.

Cropland suitable for vegetables is comparatively limited because very few areas in Canada provide enough favourable soil and climate conditions to make such farming profitable. Experiments have shown that most vegetables thrive on light, moist clay or sandy loam soils with a high content of humus from decayed organic matter. Because these dark brown to black soils retain much moisture, the soils warm up slowly in the spring. Therefore, planting is delayed until well after the last threat of frost in May. During the summer growing season, vegetables require many hours of sunshine, high average daily temperatures, and up to 100 mm of rainfall per month. Vegetable regions must have a long enough frost-free period to allow the different crops to be harvested undamaged by October. Such climate and soil requirements help explain why 90 percent of Canada's vegetable production comes from parts of Ontario, Quebec, and British Columbia. Nearly 60 percent of Canada's market produce is grown in southern Ontario, where the large population creates a demand for vegetables and where appropriate environmental conditions are found in many districts (see sketch map on page 335). Also of importance for vegetable production is the Fraser Delta region in British Columbia; this area has excellent climate and soil conditions and provides fresh produce for Metropolitan Vancouver.

In total value of production, tomatoes are the leading crop. (See Figure 9.19.) The value of corn is only half that for tomatoes, but it is grown on almost one-third of all land suitable for growing vegetables. Although carrots too rank second to tomatoes in total dollar value, they are produced at a much higher value per hectare than most other crops. Much of the produce is sold by farmers fresh at roadside stands, in farmers' markets, or through grocery stores. Some vegetables, like cooking onions, are kept fresh for later use by storage in temperature and humidity controlled warehouses. Many growers are under contract to food companies to grow vegetables for freezing and canning. A few surplus vegetables are exported to the United States.

Despite the limited amount of land available for market gardening, more vegetables could be produced in Canada. A major problem for the industry is competition from United States canning companies. Most of the Canadian canneries have been bought out or closed down by U.S. corporations. This situation reduces the number of Canadian canning operations and the Canadian markets are supplied largely by canneries in the United States. As a result, vegetable growers under contract to Canadian canneries find that Canadian markets are constantly shrinking.

Vegetables are not generally grown over the winter months and Canadians must rely on expensive imported produce from the southern United States and Mexico. However, during late spring and early summer, before field crops ripen, tomatoes and cucumbers grown in greenhouses help to reduce Canada's dependence on imported produce. Hothouse mushrooms are the only major crop grown all year long in Canada. The total annual value of all greenhouse and hothouse vegetables is only about half that of field vegetable crops.

THE HOLLAND MARSH

Of the many growing areas in Ontario, the Holland Marsh is one of the richest and largest. It is located astride Highway 400, about 50 km north of Toronto, near the southern tip of Lake Simcoe. The extremely flat land surface lies in a narrow, oval depression about 3 km wide and 12 km long,

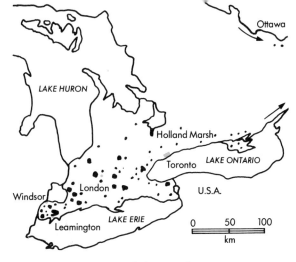

bordered by moraine ridges. Using the sketch map to the right, other maps in this book, and additional references, answer the following questions about vegetable gardening in the Holland Marsh.

QUESTIONS

1. Compare the average annual precipitation, frost-free season, and degree-days for the Holland Marsh with other vegetable regions in southwestern Ontario, in the Montreal region, and in British Columbia. (See Figures 5.8, 5.14, 5.15, and 9.3.)

2. What production and marketing advantages does the Holland Marsh have over other market gardening areas in southern Ontario?

3. Are there any pockets of market gardening in your area? If so, how are they similar to and different from the Holland Marsh?

4. If you have an agricultural research station, canning company, food wholesaler, or farm supply plant (for fertilizers, boxes and crates, machinery, etc.) in your local area, arrange to visit one of them to discover how important they are to market gardening.

More than fifty years ago, the Holland Marsh was a swamp left behind by an older, expanded glacial Lake Simcoe. Through a series of dikes, ditches, and field tiles, over 3000 ha of marshland were drained and converted to valuable farmland. Most farms were laid out in long, narrow fields about 25 ha in size. Just one hectare of this black soil called *muck* can produce over 10 t of carrots or 10 000 heads of lettuce. The loose spongy muck is easily cultivated and, although carrots and onions are the major crops, other root crops planted closely together—like potatoes, parsnips, and beets—grow quickly to full size. The black soil holds the heat of summer days to hasten their growth, while leafy vegetables—such as lettuce, celery, cabbage, cauliflower, and brussel sprouts—prefer the colder night air that moves down into the marsh from the surrounding ridges. While the marsh generally receives enough rainfall, during prolonged dry periods, crops are irrigated using sprinklers.

To grow and harvest vegetables, Holland Marsh farmers are constantly busy. In early spring, they plant seeds of some leafy vegetables in unheated greenhouses before ploughing their fields. In late May, after all frost danger has passed, they transplant the sturdiest seedlings into parallel rows. Carrot and onion seeds are usually sown directly into the newly ploughed soil, using precision seeders that select the proper soil depth and row spacing. At harvest time later in the summer, most leafy vegetables are hand picked for best quality fresh produce. All root crops and celery are picked by harvesting machines; however, if the fields are too wet, the harvesters bog down and farmers have to hire temporary workers to pick these crops before water or early frost damages them. Most harvested crops are sold by farmers to packing houses in the town of Bradford at the north end of the marsh. Here, they are graded, cleaned, packaged, and shipped across Canada. Smaller amounts of vegetables are sold locally for the Toronto markets, or are exported.

The most serious problem facing Holland Marsh farmers is the loss of precious muck slowly washing or blowing off the fields into drainage ditches. To reduce this erosion, they try to keep the topsoil as compact as possible by ploughing in the spring, by removing weeds chemically, and by rolling the fields between ploughing and seeding. Each year, ditches must be dredged and some of this topsoil is spread back over the fields. If this thin layer of muck erodes totally, the field vegetable industry here could be saved by draining more of the Keswick Marsh northeast of the Holland Marsh to begin market gardening there.

PROBLEMS AND PROJECTS

1. In a newspaper, or from a radio farm broadcast, check the market prices for beef, hogs, and eggs. Compare these prices with those you pay in a store. How can you account for the differences?

2. In your grocery supermarket, check where the following items have come from: fresh apples, potatoes, canned peaches and pears, and apple juice.

3. Explain why most of Canada's urban development is located on the best agricultural land.

4. For additional interesting reading on types of agriculture in Canada, consult the following sources:
    W.B. Braund and W.C. Blake, *Studies in Canadian Economic Geography* (Toronto: McGraw-Hill Ryerson, 1982).
    Canadian Geography Resources, *Agriculture in Southern Ontario* and *Wheat Farming* (Toronto: Dent, 1972).
    Ginn Sample Studies of Canada series, *Fruit Farming in the Okanagan*, *Market Gardening on the Fraser Delta*, *Mixed Farming Near Carmen, Manitoba* and *Wheat Farming Near Regina* (Toronto: Ginn, 1968).
    People and Places in Canada series, *Alberta Foothills*, *Okanagan Valley: Life on an Orchard Farm* and *Manitoba Lowlands: A Mixed Farm* (Toronto: Holt, Rinehart and Winston, 1968).

5. From the following statistics of livestock and crops, list the provinces for each item from highest to lowest. Try to explain the reason for the leading province in each case. In how many cases is the natural environment the explanation and in how many cases is it the closeness of the market? Excluding Newfoundland, which Atlantic Province ranks lowest in most of the livestock and crop production items? Why does it rank so low? Why is P.E.I. so well known for potato production when Ontario, Quebec, and New Brunswick also produce large amounts of potatoes?

Agricultural Statistics

Livestock (in thousands)

	Nfld.	P.E.I.	N.S.	N.B.	Que.	Ont.	Man.	Sask.	Alta.	B.C.	Can.
Dairy Cattle[1]	–	30	50	38	960	785	111	83	183	110	2350
Beef Cattle[2]	–	39	43	33	292	1275	618	1353	2275	322	6250
Hogs (all pigs)	–	99	100	50	2900	3150	850	660	1150	137	9096
Sheep	–	5	25	8	54	155	14	60	127	33	481
Poultry[3]	386	646	12 746	8789	98 702	124 585	19 331	11 927	35 715	37 876	350 703

Note: [1] cows and heifers, one year old or over, milking or about to be milked; [2] cows, heifers, and steers, one year old or over, kept for meat; [3] chickens and turkeys laying eggs or raised for eating; – means data unavailable.

Value of Farm Crops[1] ($ millions)

	Nfld.	P.E.I.	N.S.	N.B.	Que.	Ont.	Man.	Sask.	Alta.	B.C.	Can.
Wheat	–	0.4	0.7	0.4	4.3	99.2	381.6	2017.9	619.8	11.1	3135.6
Potatoes	0.6	45.0	4.2	31.5	32.3	42.9	20.7	2.1	14.9	9.0	203.2
Corn[2]	–	–	–	–	6.0	357.5	8.5	–	–	–	372.0
Fruit	0.8	1.2	13.4	4.3	31.7	95.9	2.1	–	3.0	91.6	244.0
Vegetables	2.1	2.3	4.2	4.5	59.0	194.4	7.9	–	13.0	44.7	332.1

Note: [1] only crops sold off farms are included here – other crops not listed are grown and used on farms mainly for livestock feed; [2] field corn harvested for grain or seed; – means value of production too low.

(Source: Statistics Canada, 1980.)

10 MINING, FORESTRY, AND FISHING

In Chapter 9, we discussed how the combination of climate and soils in various parts of Canada has enabled farmers to produce different agricultural products. We noted that only a small portion of Canada was suitable for agriculture; but this does not mean that the rest of Canada is unproductive. There is a wealth of minerals in the bedrock as well as in the glacial deposits in both the settled and unsettled regions of Canada. The vast forest areas of Canada yield valuable lumber, while fishing, in both the ocean and freshwater lakes, provides a living for thousands of Canadians.

Mining, forestry, and fishing are known as *extractive industries*. In an extractive industry, some substance is taken from the land or water for human use. Minerals can be extracted only until the supply is exhausted. With the proper management of forests and fisheries, these resources can always be harvested because both trees and fish will renew themselves under proper conditions.

The extractive industries of mining, forestry, and fishing have always played an important part in Canada's growth and development. All three are very old industries that date from the days of the early explorers. Soon after Cabot had sighted seas "swarming with fish" in 1497, vessels from many European nations began to fish in the Atlantic Ocean off the coast of Newfoundland. Most of their catch was shipped back to Europe, or else was sent to the West Indies as food for slaves. When the first settlers in Canada cleared the forests to cultivate their crops, they used some of the timber to build houses, barns, and other farm buildings. Later, in the 1700's, companies were formed to cut forests in order to supply English and French shipyards with lumber. During this period in the eighteenth century, wood was in high demand for the building of sailing ships and for many other purposes. Because of the large quantities of heavily forested land that were easily accessible along the coasts and inland waterways, Canada became a major supplier of lumber to Europe.

The first mining in Canada, apart from the use of some copper by the Indians, was the digging of coal along the Atlantic coast and the excavation of bog iron ore in the region of Trois Rivières. The coal was used for heating homes, and later, as fuel for steam engines. The iron was used to make the nails, pots, pans, and ploughshares needed by the early settlers.

At one time, Canada was known primarily as a producer and exporter of raw materials. Agriculture and the extractive industries of mining, forestry, and fishing were the mainstay of the economy. Today, the manufacturing and service industries have become more important. Nevertheless, the extractive industries are needed to provide raw materials for the manufacturing industry. The products of mines, forests, and fisheries also contribute greatly to Canada's exports.

THE MINING INDUSTRY

Canada is one of the leading mineral producers in the world. In the early 1980's Canada was the world's largest producer of asbestos, nickel, zinc, and potash; it was second in the production of gold, molybdenum, gypsum, uranium, titanium, and silver; and was among the leaders in the production of coal, platinum, copper, and iron ore.

Although Canada produces some sixty different minerals, the eleven leading minerals constitute eighty percent of total production (see accompanying table). Can you name one important mining place in Canada for each of the leading minerals?

The great size of Canada's mining industry is based to a great degree on export sales. In the early 1980's, ninety percent of the total Canadian production of minerals was exported. Because so much of Canada's mineral production is exported, the success of the industry depends, to a large extent, on world prices for minerals. In times of high world demand and resulting high prices, mining companies step up their production and new mineral deposits are opened up. However, if world prices for certain minerals suddenly drop, mining companies are forced to cut back their production and to lay off workers. If low prices continue for an indefinite period, some companies are forced to close their mines permanently so that the miners must seek work elsewhere. When towns in which mining is the only industry become deserted, they are said to be "ghost towns."

LEADING CANADIAN MINERALS
(excluding crude oil and natural gas)

	Percent of Total Value
Iron Ore	13
Copper	12
Nickel	9
Zinc	8
Coal	7
Potash	6
Gold	6
Cement	5
Uranium	5
Asbestos	5
Silver	4
Others	20

Source: Statistics Canada.

Note: the percentages given are averages for 1978 to 1980 (about $22 billion) to offset slight yearly shifts that result from changing world markets and prices. Crude oil and natural gas are excluded because of their rapidly increasing values.

Mining companies must anticipate the demand for minerals many years in advance, for it takes a long time to bring a mine into production. First, explorations must be carried out to discover where the chances of mining certain minerals are best. Then, test holes must be bored. Next, roads have to be built, and a mining camp has to be set up before the mining operation can begin. If it is deep mining, much drilling and excavating must be done before the first ore is brought to the surface.

In earlier days, prospecting for minerals was done by men who went into the wilderness on foot and brought out samples of rock. Today, mineral prospecting has been modernized. Geologists study geological maps that show the nature of the bedrock in an area. When they find an area that has the right kind of rock for a certain mineral, they use large-scale air photos for more detailed study. Some minerals can be identified

Right: Dawson City in the Yukon was once a booming gold mining town. When the gold ran out and the mines were forced to shut down, Dawson City became a ghost town. As people moved away, store owners had to close their businesses.

Below, left: A survey team pushing into Canada's northland to search for mineral deposits or to stake out valuable mineral properties. In the Coniferous Forest region the muskeg conditions of summer make exploration difficult.

Below, right: A geologist breaks chips of rock from large rock formations. For further detailed study rock chips from the most promising mineral deposits will be carried back to the laboratory.

by special ore-detecting devices attached to an airplane. It is only after the company is sure that there is a good chance of locating a large mineral deposit that geologists are sent into the area to obtain rock samples and to drill test holes.

If the tests show a mineral ore that can be mined and sold at a profit in view of the existing world prices, a new mine is opened up. The opening of a new mine creates thousands of jobs. For every person employed in the mine, four others are needed in those industries that supply the mines with equipment or in industries that smelt and refine the extracted minerals.

Mining has also been a key factor in developing Canada's northlands. In many areas, roads and railway lines, built to serve new mining operations, have encouraged the development of forest and recreational resources. In some areas, farming has been established to provide food products for the mining towns. Unfortunately, along with development has come damage to the natural environment. A major problem facing the mining industry in the 1980's is how to exploit minerals with a minimum of environmental damage.

LOCAL STUDY

If you live in a mining town, you probably will know all about the local mining operations, what the minerals look like, and what they are used for. You will also appreciate how important mining is to your community. If you do not live in a mining town, you may be surprised to learn that there are minerals extracted from your region, and that the use of minerals is very important to you and your community. The answers to the following questions should help to prove this point.

1. List all the items you can think of that contain nickel (most steel products contain nickel), as well as items that contain copper, iron, gold, silver, lead, and zinc. These metallic minerals are known as *metals*. Where are these metal ores mined in Canada?

2. The majority of homes in Canada are heated by fuel oil, natural gas, or coal. These are known as mineral *fuels*. From where do most of the fuels used in your community come? How are they transported?

3. The siding on many houses often contains asbestos. The wallboard on the inside of houses contains gypsum. Commercial fertilizer used by farmers contains phosphate. Salt is used in most foods to give them flavour. Asbestos, gypsum, phosphate, and salt form another group of minerals described as *nonmetallic*. Are there any industries in your region that use these raw materials? Where are these minerals mined in Canada?

4. What minerals are mined in your local area? If your answer is "none," you may be mistaken. Have you ever noticed any sand-and-gravel pits, cement plants, stone quarries, or brick factories (using clay) in the surrounding countryside? Minerals such as sand, cement, stone, and clay are also nonmetallic minerals. They are usually called *structural materials* because they are used mainly in the building and construction industry.

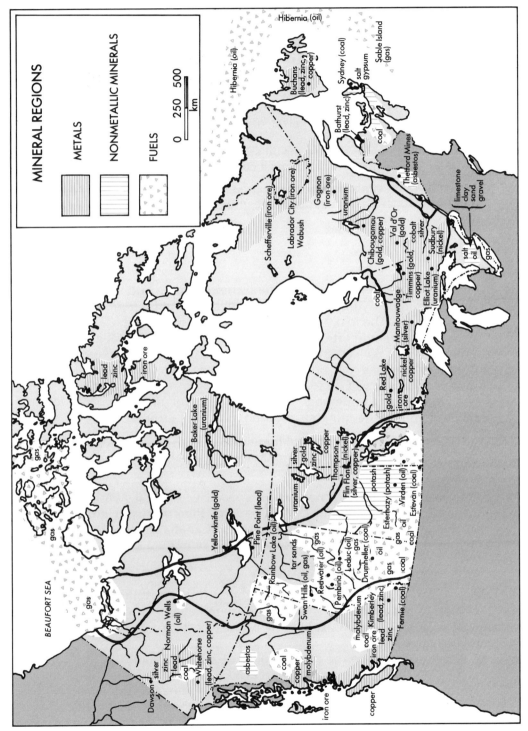

Figure 10.1 PRINCIPAL MINERAL REGIONS AND SOME MAJOR MINING AREAS IN CANADA. This map shows only some of the important mining places. For a detailed map of mining operations in Canada, see Map 900A from Energy, Mines and Resources Canada, Ottawa. For the most recent "Canadian Petroleum Highlights", write to the Canadian Imperial Bank of Commerce, P.O. Box 2595, Calgary, Alberta, T2P 2P2.

5. Few mineral deposits are found in a pure state, that is, without any waste rock around them. For this reason, minerals need to be refined or smelted to extract the minerals from the ore. Are there any smelters, refineries, or blast furnaces in your area or in nearby centres? If so, what kind of minerals do they process? Where have the raw materials come from?

6. Every transportation system has its own special equipment for carrying minerals or mineral products. Railway companies own hopper cars for coal, iron ore, and so on, and tank cars for crude oil. On the highways, you may see trucks hauling sand and gravel, or tank trucks carrying gasoline. Shipping companies have special carriers for bulky cargoes such as coal, iron ore, and cement, and special tankers for oil and gas products. Which of these types of mineral transportation is the most common sight in your area?

7. What different kinds of minerals are used in making, running, and lubricating an automobile? Can you group these into types of minerals: metals, nonmetallics, and fuels?

Where Minerals Are Found

Because of the vast number of minerals being mined and the wide distribution of mines, trying to remember *what* is mined *where* is an almost impossible task. The locations of most of Canada's mining operations can be found in a recent edition of a Canadian atlas. The federal government produces a map, revised annually, that shows the names and locations of all producing mines and principal oil and gas fields in Canada (Energy, Mines and Resources Canada, Map 900A).

It is easier to remember general patterns of mining operations in Canada if you divide minerals into the following groups:

(1) metallic minerals called metals,
(2) nonmetallic minerals, including structural materials,
(3) mineral fuels.

Each group has been formed under special geological conditions, and thus, each is found only in certain parts of the country (Figure 10.1). It may be useful at this point to re-read "Canada's Rock Foundation" (Chapter 4) in which the origin of a number of minerals is described.

METALS

Most metals are found in igneous rock. Thus, metallic minerals are common in mountainous areas where there have been many intrusions of igneous rock, which explains why the Western Cordillera is rich in metallic minerals.

Lead and zinc are the most important minerals found in the Cordillera. The lead and zinc smelters at Trail, B.C., are the largest in the country. Important lead, zinc, and copper mines are found near Whitehorse, in the Yukon; the Dawson area is also famous for its production of silver and other metals.

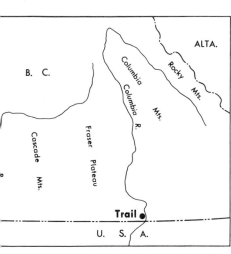

Lead and zinc smelters at Trail, B.C. This smelter complex has been located on a terrace overlooking the Columbia River, from which the water needed in the smelting process can be obtained. At one time the smoke from the huge chimneys you see in the photograph so polluted the air that all the vegetation on the valley sides was killed. The harmful chemicals are now filtered out of the smoke. The minerals processed at Trail are obtained from mines in different parts of British Columbia.

On the east coast, the folded mountains of the Appalachians also contain metallic minerals. There are important copper, lead, and zinc mines at Buchans, in Newfoundland, and lead and zinc mines near Bathurst, in New Brunswick.

Because the Canadian Shield is composed primarily of igneous rocks, it contains vast amounts of metallic minerals (Figure 10.1). There are far too many important mining places in the Canadian Shield even to mention them all in this book. (There were 125 operating mines in the Shield in the early 1970's.) However, a few places are so important for certain minerals that their names are well known even beyond the boundaries of Canada. For example, the Sudbury area produces more nickel than any other place in the world.

International Nickel Company (INCO) mine and plant at Thompson, Manitoba. Nickel ore is brought to the surface by an enclosed conveyor (on the right) from the mine deep beneath the surface of the Canadian Shield. The ore is smelted down (in the building with the tall chimney) to obtain a purer form of nickel before it is shipped to many parts of Canada and the world. The town of Thompson shows up as the white patch on the dark, forest-covered landscape. Thompson is typical of the many mining centres that have sprung up across the Canadian Shield as a result of discoveries of rich metallic minerals.

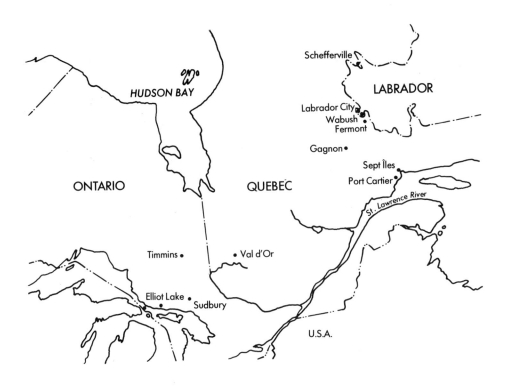

Elliot Lake is Canada's major source of uranium. The region of Timmins and Val d'Or and the Red Lake area are major gold producers. In a physiographic subregion called the Labrador Trough are rich deposits of iron ore. Mining activities here have led to the development of towns along the Quebec-Labrador boundary such as Schefferville, Labrador City, Wabush, and Gagnon. (The mines at Schefferville were closed in 1982.)

NONMETALLIC MINERALS

Nonmetallic minerals are formed in a variety of ways. Asbestos is found in bands of igneous rocks that have been intruded into the sedimentary rock layers during a period of mountain building. Thus, it is not surprising that Canada's large supply of asbestos is found in folded mountains of the Appalachian region. The mines at Thetford, Quebec, have made Canada the leading producer of asbestos in the world.

Accumulations of salt, potash, and gypsum were deposited in sedimentary rock formations when the salty water of ancient seas slowly evaporated. Large deposits of gypsum are found in New Brunswick and Nova Scotia. Most of Canada's salt is mined in Southern Ontario. The largest known deposits of potash in the world are located in Saskatchewan. Potash is a major ingredient in most commercial fertilizers.

Structural materials, such as limestone and granite, are quarried from the bedrock wherever the right kind of rock is found, and preferably, close to where it is going to be used. Some limestone is good for cement and lime, and other limestone is cut up into large blocks for building purposes. Sands, gravels, and the clays are found in abundance in glacial deposits. The meltwater from the glacier sorted the sands and gravels and deposited them in the form of kames, outwash plains, and deltas. These sands and gravels are in great demand for the construction of roads, buildings, dams, and any other construction that requires concrete. Concrete is a mixture of cement (made from limestone) and sand or gravel. Clay deposits that have the right texture for bricks and under-tiles are found all across the country. Structural materials are used most extensively in Southern Ontario because of the large amount of construction there related to industrial development and urban growth.

NONMETALLIC MINERALS.

Open-pit asbestos mine at Asbestos, Quebec.

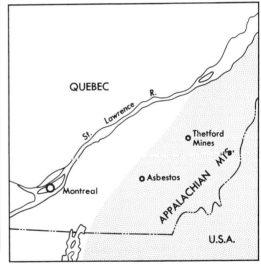

Gypsum mine north of Halifax, Nova Scotia.

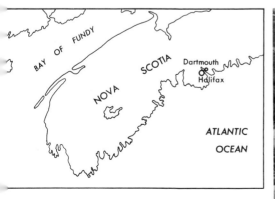

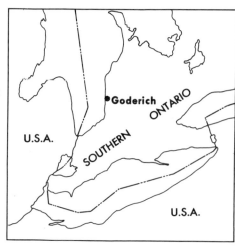

Salt mine at Goderich, Ontario.

Potash mine at Esterhazy, Saskatchewan.

Sand and gravel pit near Guelph, Ontario.

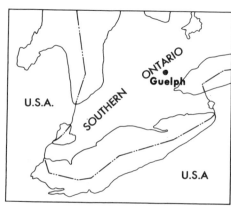

MINERAL FUELS

The minerals that are used as fuel are also found in areas with sedimentary bedrock. Oil deposits were formed from the millions of tiny marine plants and animals that lived in the Devonian and Cretaceous Seas. You can check Figures 4.8 and 4.9 again to see the location of these seas, and compare their extent with the oil deposits shown in Figure 10.1. Oil deposits were also formed in several other geological periods when seas covered major portions of land in North America. Coal was formed when masses of fernlike trees growing in swamps along the edge of the sea were submerged and buried under sedimentary rock. Woody parts of the vegetation did not decay, but were pressed, first into peat and then into coal, by the mass of the overlying rock layers. The proper conditions for the formation of coal occurred late in the Paleozoic era in three regions of Canada: the Appalachian Region, the Interior Plains, and parts of the Western Cordillera.

Where Minerals Are Mined

The location of minerals is determined by geological conditions, but the location of operating mines is determined by economic factors. Production of a mineral in any given location does not begin until a mining company believes it will be a profitable venture.

Following are some of the factors that must be considered in making a decision to begin exploiting a given mineral deposit:

(1) Accessibility. Are transportation facilities available to move mining equipment in and mineral products out?
(2) Extent and richness of deposit. How rich is the deposit, what is its quality, and how long will the mining operation last?
(3) Depth of deposit. How deep is the deposit? What kind of mining operation will be necessary? How expensive will it be?
(4) The natural environment. The nature of the terrain, natural vegetation, and climate can affect the costs of establishing both the mine and the living quarters for the miners. In the Arctic, for example, additional expenses are involved in overcoming the problems of constructing roads on permafrost and protecting workers and machinery from the severe winter temperatures.
(5) The cost of transporting minerals to market. A low-cost mining operation may be offset by additional costs of transportation. Thus, on occasions it is more profitable to mine lower-grade deposits close to the market, than high-grade deposits in remote areas.

(6) The world supply and demand situation. Before investing millions of dollars in a mining operation, a company must assess what the future supply and demand is going to be.
(7) Government subsidy and tax policies.

In practice, mining companies consider a combination of all of these factors and decide to start new operations, continue existing operations, or close down existing mines, depending on whether a profit can be made. There is a great concentration of mining operations in certain areas of Canada because the districts close to existing mines can be most easily explored, and because of the economic advantage of being close to an existing transportation facility.

The Oil And Gas Industry

The correct name for oil is *petroleum,* which comes from the Latin word meaning "rock oil." The raw petroleum that first comes out of the earth

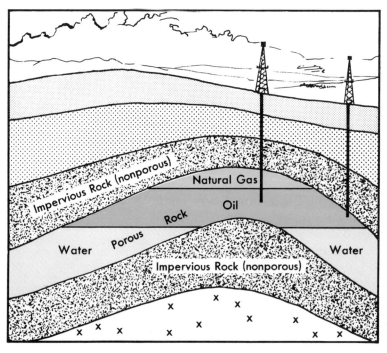

Figure 10.2 TYPICAL OIL FORMATION. The impervious (nonporous) rock layers trap the oil and gas which float on the water in the porous rock. Not all rock formations like this contain oil and gas. Scientists can determine where oil might be found, but the final test in the search for oil is the drilling of a well. In the Canadian Prairies, oil is often found in the porous coral reefs that were deposited in the time of the Cretaceous Sea (see Figure 4.2).

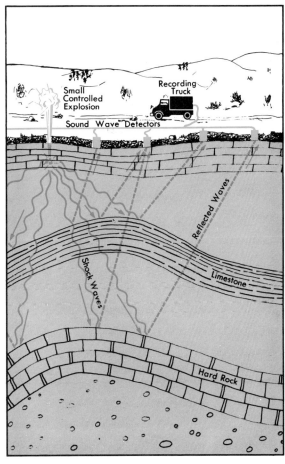

Figure 10.3 SEISMOGRAPHIC TESTING. Shock waves from small controlled explosions near the surface travel downward, strike underlying rock layers and are reflected back to the surface where they are recorded by the seismograph. The depth of the different layers of rock is measured by the length of time it takes for the shock waves to return to the surface.

is called *crude oil*. This crude oil is refined into many products including gasoline, fuel oil, and kerosene, as well as grease and oil used for lubrication, and a number of *petrochemicals* (chemicals made from petroleum). Petrochemicals are the raw material for a wide range of products such as wax, cleaning fluids, detergents, plastics, antiseptics, insecticides, ointments, shampoos, and many others. *Natural gas* is not a by-product of petroleum, but is usually found in the same areas. It is a very efficient, clean, and relatively inexpensive fuel that has a wide variety of industrial and domestic uses.

The oil and gas industry has been very important to Canada's economy. Production of oil and gas adds wealth to the country in a number of ways. The oil companies pay a tax (royalty) to the government of the province in which the oil is found. Oil royalties have greatly assisted the finances of the governments of the Prairie Provinces, particularly Alberta. The drilling for oil creates many new jobs, and the new people in the region require housing, schooling, and other services. The availability of oil and gas attracts industries that are dependent on those products. The piping of crude oil and natural gas across great distances makes it possible to have refineries and petrochemical industries (industries that make chem-

Drilling for oil and gas offshore in the Arctic is more expensive than on land and the chances of finding valuable deposits are no better. Exploration holes must be drilled deep into the sea-bottom rock, using specially constructed *drilling rigs*, or *derricks*. Some are mounted on huge floating platforms or on ships, when the ocean floor is hundreds of metres below. For shallower water, as in parts of the Beaufort Sea (above), a drilling tower can be erected on an artificial island made from material dredged up from the sea floor.

ical and other products from petroleum) in places as far from the oil fields as Sarnia and Toronto, in Ontario. The petrochemical products are used in many other factories to produce a wide variety of manufactured products. The equipment used by oil drillers and the thousands of kilometres of oil and gas pipelines require many specialized steel products made in the steel mills of eastern Canada. Thus, the petroleum industry creates jobs not only in western Canada, but all across the country.

LOCATING PRODUCTIVE FIELDS

Petroleum was formed from billions of tiny marine plants and animals that became trapped in sediments in the bottoms of ancient seas. After millions of years, the particles of these tiny creatures changed into oil and gas that seeped into the pores of the bedrock. When oil-producing rock was sandwiched between two layers of nonporous or impervious rock, the oil and gas could not escape and it lay there for millions of years until some miner tapped it with a drill (Figure 10.2).

The search for oil is a costly venture that involves many skilled scientists and technicians and a great deal of equipment. Geologists study maps and air photographs to learn where oil might be discovered, and then check their findings in the field by using a seismograph which provides information about the rock formation. The seismograph records the time that it takes for the sound waves from dynamite explosions to bounce back from rock layers. In this way, geologists can tell when the rock formation is the right kind for oil, but they cannot tell if there really is oil there until a well is drilled. In drilling for oil, there are more failures than successes. The oil companies consider that drilling and finding a dry well is a part of their normal expenses. (See Figure 10.3.)

A number of oil wells in one area is known as an *oil field*. The only signs that indicate the location of an oil field are the scattered oil pumps. In the oil and gas fields of western Canada, the land in between the wells is being farmed in the same way it was before oil and gas were found. Many natural gas wells are found indirectly in the search for oil; a group of such gas wells in one area is known as a *gas field*. (See Figure 10.4.)

The Athabasca tar sands in northern Alberta are regarded as the world's largest reserve of crude oil. Until recently, this resource had gone undeveloped because of the difficulty in extracting the oil from the sands. In most oil fields, the oil is trapped in liquid form in porous rock. In the Athabasca deposit, the oil has collected around individual sand particles as a tarlike substance that does not flow. After numerous attempts, a way was discovered to separate the tar from the sand and a recovery plant began oil production in 1967. There are now several plants in operation and more are planned for the future. The amount of oil taken from the sands will depend on further technological developments and on the demand and price for oil.

To mine the oil in the Athabasca tar sands, giant earthmovers must scrape away up to 20 m of muskeg, soil, and rock that cover the sand. Excavators then mine the tar-covered sand from the pits. After being transported by conveyor belt to the separation plant, the sand is mixed with water and is heated. When the molten tar has risen to the water surface, it can be skimmed and processed further to produce high-quality crude oil. A refinery at Edmonton receives the crude oil by pipeline for processing. (See Figure 10.4.)

There is also heavy oil along the Alberta-Saskatchewan border near Cold Lake and in an area west of Saskatoon. This oil is extracted by drilling wells but, because it is very thick and has a high sulphur and asphalt content, it is difficult to pump and expensive to refine.

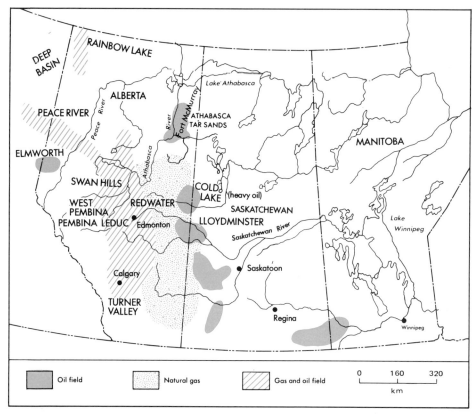

Figure 10.4 OIL AND GAS FIELDS IN WESTERN CANADA. The first oil was found at Turner Valley in 1913. The real petroleum boom of Alberta began in the late 1940's with the discovery of huge oil and gas fields at Leduc and Redwater.

Piles of Athabasca tar sands are dug from surface pits by huge bucket-wheel and drag-line cranes near Fort McMurray, Alberta. Nearby recovery plants separate the embedded sand grains from the tar-like bitumen before upgrading it into crude oil. In winter's freezing temperatures, the tar sands are rock-hard and must be blasted from the ground. During the heat of summer, they become very soft and sticky.

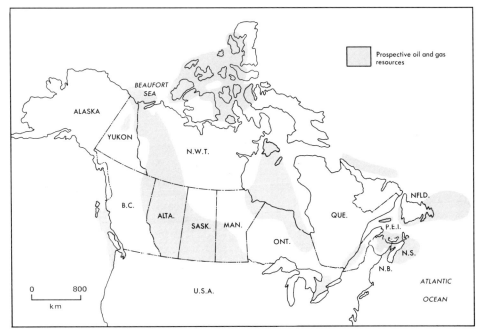

Figure 10.5 AREAS WITH POTENTIAL OIL AND GAS RESOURCES.

DEVELOPING OUR RESOURCES

In the Geological Survey of Canada, 1851, bitumen gum beds were reported near Sarnia, Ontario. The first free-flowing crude oil well was drilled at Oil Springs (near Sarnia) in 1857, two years before the first oil well developed in the United States. However, Canada did not retain its lead in oil production very long. The Oil Springs oil deposit was not very large compared to the huge oil discoveries made in Texas and Oklahoma.

The first major oil field discovered in Canada was at Turner Valley, Alberta. By 1924, Turner Valley was the largest oil-producing field in the British Empire, but it still did not come near meeting Canada's oil needs as the gasoline-burning automobile became increasingly popular.

The next major oil discovery was in 1946 at Leduc, just southwest of Edmonton. The oil deposit at Leduc proved to be a large one and marked the beginning of the oil boom in western Canada. By the early 1970's, there were some seventy oil fields and even more gas fields in Alberta, and additional oil and gas fields were being developed in British Columbia, Saskatchewan, and Manitoba. Very small amounts of oil and gas continued to be produced in Southern Ontario. (See Figure 10.4.)

After the discovery of oil and gas at Prudhoe Bay on the north slope of Alaska, oil and gas exploration increased greatly in the Canadian Arctic and off the east coast. Large deposits of gas and some oil have been discovered in the Mackenzie delta and the Arctic islands, as well as in the

Petro-Canada Tank Truck

Railway Tank Car

Oil Tanker

Oil Pipeline

METHODS OF TRANSPORTING OIL

A modern refinery beside the St. Clair River at Sarnia. In the tower-like structures, crude oil is processed into a variety of petroleum products such as gasoline, motor oil, and grease. Many of the tanks (white) you see store crude oil while others hold the finished petroleum products until they are ready for distribution. Petroleum products are shipped from the refinery by truck, railway car, or boat.

Beaufort Sea. The Hibernia oil field was also discovered off the coast of Newfoundland and gas was discovered on Sable Island off the coast of Nova Scotia. However, most of these recent discoveries will not actually produce oil and gas until well into the 1980's. (See Figures 10.1 and 10.5.)

Canada's peak production of oil was reached in 1973. Since that time there has been a constant decline in production. What will happen in the future depends on how much oil is produced from frontier sources and from tar sands and heavy oil deposits. (See Figure 10.6.)

Because Canada's petroleum resources are located so far away from the largest markets, transportation has always been a difficult problem. The current and proposed major pipelines are shown in Figure 10.7. In addition, there is a proposal to ship liquefied natural gas from the Arctic by ocean tankers to the east coast for distribution inland by pipeline.

Between 1961 and 1973, Canada's national oil policy was to reserve all of Canada west of the Ottawa Valley for Canadian crude oil. This policy guaranteed the oil-producing provinces a large portion of the Canadian market at higher-than-world prices. During the same period, areas east of the Ottawa Valley chose to import lower-priced oil from Venezuela and the Middle East. However, after the Arab-Israeli War of 1973, the Arab countries rapidly raised oil prices far above Canadian oil prices. At that time the Canadian Government began subsidizing those areas dependent on high-priced foreign oil and authorized the extension of the oil pipeline from Toronto as far east as Montreal. This policy resulted in all parts of Canada paying the same price for oil, a price that was considerably lower than world prices.

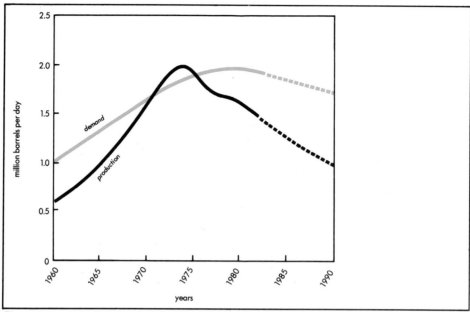

Figure 10.6 CHANGING OIL DEMAND AND PRODUCTION. The projected demand curve (broken line) is based on proved oil reserves and does not include oil from the tar sands or from the frontier such as the Arctic or off the Atlantic coast. Do you think Canada will become self-sufficient in oil by decreasing demand, by increasing production, or both?

A low oil price policy pleased all parts of the country except the oil-producing provinces that preferred high oil prices in order to encourage oil exploration and to increase their petroleum royalty income. To continue the development of oil in the Arctic, the Athabasca tar sands, and the heavy oil areas, the Canadian Government has been forced to increase oil prices gradually up to the world level.

Most of the major oil companies in Canada are subsidiaries of American or European based multinational corporations. Therefore, they do not always act in Canada's best interests. For this reason, the Canadian Government has established a policy to make the petroleum industry at least fifty percent Canadian. To this end, Canadian-owned companies have been given certain tax and exploration advantages. Also a national oil company, Petro-Canada, was created that has bought out nonCanadian companies such as Pacific Petroleum and Petrofina. Petro-Canada is an active partner in many of the frontier explorations.

Canada's goal is to become self-sufficient in oil by 1990. To achieve this goal, it will be necessary not only to bring many more oil fields into production, but also to require conservation of energy by all Canadians and much greater use of alternate energy resources.

Throughout the 1960's, both production and demand for crude oil increased rapidly. For only a few years in the early 1970's did Canada produce more oil than it used; the surplus was exported to the United States. Since about 1974, oil production has declined and Canada has been forced to import increasing amounts of oil at high prices. In the future, it is hoped that the demand for oil will decline with conservation measures and increased use of natural gas, electricity, and renewable energy (e.g. wind, solar, tidal power). It is also hoped that more oil will be produced from the tar sands and from frontier sources such as those in the Arctic and off the east coast.

In the early 1980's, a major world-wide economic recession plus energy conservation measures led to a lowering of the world price for oil. This made oil imports more attractive and the production of high-cost oil from the tar sands and the frontier areas less attractive. As a result, several of the plans for large tar sands projects were postponed.

Figure 10.7 MAJOR OIL AND GAS PIPELINES AND REFINERY CENTRES IN CANADA. Marketing the oil and gas produced in western Canada requires a complex network of pipelines to eastern Canada, the west coast, and various points in the United States. A system of pipelines for petroleum products also runs between Sarnia and Montreal. When the technology is completed, a shipping route from the Arctic is proposed to carry liquefied natural gas (LNG).

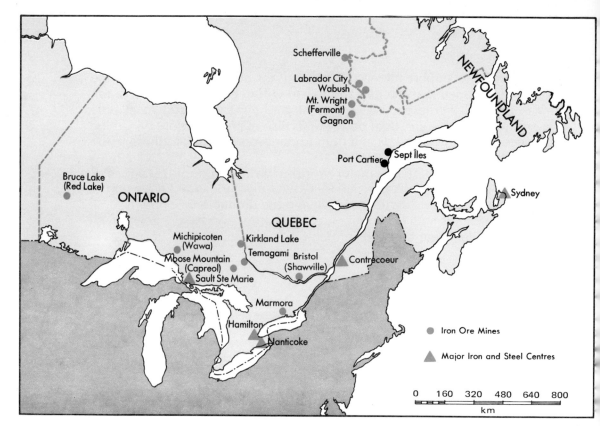

Figure 10.8 IRON ORE MINES AND MAJOR IRON AND STEEL CENTRES IN EASTERN CANADA. Most of Canada's iron ore is found in the Canadian Shield. Much of the iron ore must be upgraded by concentrating it into pellets before it is shipped to Canadian steel mills or to United States or overseas steel centres. Names in brackets indicate larger settlements near the actual mine sites.

Mining Iron Ore

No important mining of iron ore was done in Canada until World War I when the American steel companies started developing Canadian iron ore mines to help supply the steel needed for guns, tanks, ships, and other arms. Because most of the mines are American-owned, a large proportion of the iron ore mined in the Canadian Shield is exported to steel mills in the United States. Most of the iron ore mined in British Columbia is exported to Japan. Because Canada uses so little of the iron ore mined in the country, it has become the leading iron ore exporter in the world. The export of iron ore is an important factor in Canada's attempts to balance its trade with other countries.

Although Canada is a large exporter of iron ore, it also imports considerable iron ore from the United States. For example, the steel mills at Hamilton, which produce about three-quarters of the total Canadian steel production, obtain large amounts of their iron ore from mines in Minnesota near Lake Superior.

The largest volume of Canadian iron-ore production comes from the

Sept Îles on the north shore of the St. Lawrence. At this port, iron ore from Schefferville is loaded aboard large freighters. The freighters shown in the photograph can carry about 50 000 t of ore. At Sept Îles inbound ships also unload food, mining equipment, oil, gasoline, and general cargo for the mining communities. On the left is part of the new town built at the port. The stockpiles of iron ore show up as a reddish-brown colour in the middle-ground of this photo.

Figure 10.9 CHANGING IRON ORE PRODUCTION AND USE IN CANADA.

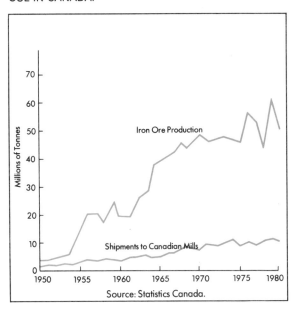

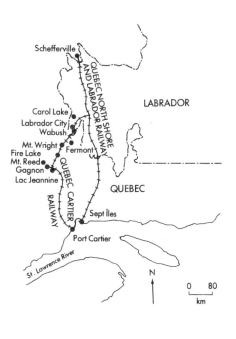

vicinity of the Labrador Trough, a part of the Canadian Shield that runs from Ungava Bay to a point about 320 km north of the St. Lawrence River. Estimates have placed the reserves of iron ore in the Schefferville area alone at close to 364 million tonnes of high-grade iron ore. Since it is found near the surface, the ore can be obtained by *open-pit* or *strip mining*. In this process, no shafts are necessary, for the ore is scooped out in the same way that gravel is excavated from a gravel pit.

The Quebec-Labrador iron ore was discovered late in the nineteenth century but it was not economic to develop it until many years later. When it became known in the late 1930's that high-grade iron ore resources were running out in the Mesabi and other ranges in the Lake Superior region, development of the Labrador Trough iron ore became more attractive. The decision to build the St. Lawrence Seaway meant that transportation costs to steel centres in the Lower Great Lakes would be reduced substantially. In 1954, a 580 km railroad (The Quebec North Shore and Labrador) was completed over the rugged terrain of the Canadian Shield from what is now the town of Schefferville to Sept Îles, and iron ore mining began.

The Quebec-Labrador iron ore was discovered late in the nineteenth century but it was not economic to develop it until many years later. When it became known in the late 1930's that high-grade iron ore resources were running out in the Mesabi and other ranges in the Lake Superior region, development of the Labrador Trough iron ore became more attractive. The decision to build the St. Lawrence Seaway meant that transportation costs to steel centres in the Lower Great Lakes would be reduced substantially. In 1954, a 580 km railroad (The Quebec North Shore and Labrador) was completed over the rugged terrain of the Canadian Shield from what is now the town of Schefferville to Sept Îles, and iron ore mining began. The mine produced until 1982.

The location of Sept Îles and Port Cartier on the St. Lawrence River makes the Quebec-Labrador iron ore readily accessible to the steel centres of the Lower Great Lakes, the United States east coast, and western Europe. Most of the ore currently goes to United States steel centres on Lake Erie and along the Atlantic Seaboard.

Another large deposit of high-grade iron ore is located at Steep Rock Lake, near Atikokan, west of Lake Superior. The ore was discovered in the bottom of Steep Rock Lake, so the lake had to be drained before mining could take place. The mining operation began with an open-pit mine on the lake floor. Soon, however, the mining company had to tunnel underground, which increased production costs until the mine closed in 1979.

In summary, it may be said that Canada has vast reserves of iron ore. The development of these ores is primarily dependent upon the demand from steel mills in the United States. As the supply of iron ore in the United States dwindles, more Canadian iron ore will be used. It would appear that there is a bright future for the iron-ore mining industry (Figure 10.9).

Open-pit mining of iron ore at Schefferville. The conveyor on the right elevates the iron ore to the surface where it is loaded on the Quebec North Shore and Labrador Railway to be taken to Sept Îles.

Figure 10.10 COAL RESOURCES AND MINES OF CANADA. Note that the coal resources are all far removed from the major manufacturing belt of Canada.

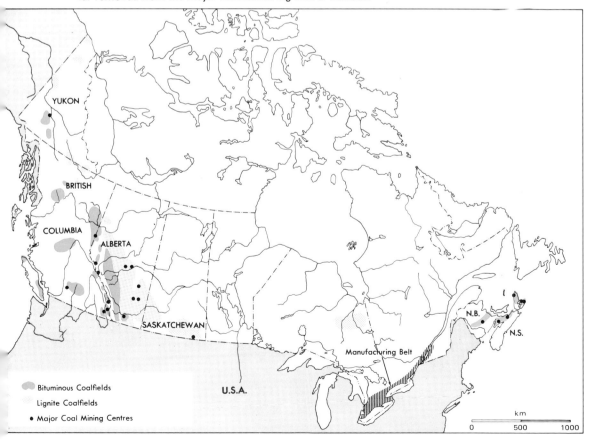

Modern open-pit coal mining in Saskatchewan, using the drag-line method.

A miner takes a load of coal out of one of three modern mines being operated by a Crown Corporation on Cape Breton Island.

Coal is loaded by conveyor belts aboard ocean freighters for export overseas at Roberts Bank (southwest of Vancouver), an offshore coal storage depot constructed from landfill. Much of the coal mined in Alberta and British Columbia is transported to the coast by long unit trains that haul only coal cars. Roberts Bank handles much of this coal.

Mining Coal

Canada has huge amounts of coal resources, but unfortunately they are located far from Canada's manufacturing belt, where there is the greatest demand (Figure 10.10). Because of this, vast amounts of coal are imported from the United States to the Great Lakes-St. Lawrence Lowlands to supply the needs of the iron and steel industry and to feed the increasing number of thermal electric plants.

By the middle of this century, most railways and shiplines that once used coal for power had converted to diesel engines that use fuel oil. Likewise most Canadian homes turned from coal for heating their homes to fuel oil, natural gas, and electricity. Consequently the 1950's and 1960's saw the coal mining industry in serious decline. By 1970 the trend had changed. Huge sales of coal to Japan led to the rapid development of the coal industry in British Columbia. As a result of these sales, plus increased use of coal in Canada for thermal electric plants, by 1981 coal production in Canada rose to a record 36 million tonnes a year. It is predicted that this amount will be doubled before 1990.

Coal has been mined in New Brunswick and Nova Scotia for many years. However, Maritime coal mining has never been able to compete successfully with that of the United States for the Great Lakes-St. Lawrence Lowlands' market. The coal is mined by the underground shaft and tunnel method. The seams of coal are thin and quite deep so that production costs to extract a tonne of coal are very high. Moreover, the quality of the coal for steel-making purposes is inferior to that of the Appalachian region in the United States. Thus, for many years only a small proportion of local coal was used even in the Sydney steel mills.

The Canadian Government paid a subsidy to the Maritime coal industry for many years, as well as a transportation subsidy to help defray the cost of moving coal to markets in central Canada. However, even with these subsidies, the industry declined, and finally in 1966 the Canadian and Nova Scotia Governments decided to buy out the coal mining industry on Cape Breton Island. They planned gradually to phase the mines out of operation. However, under government ownership and new management, the Cape Breton coal mines have been revitalized. With new technology and higher coal prices, these mines are now able to provide 80 percent of the coal required by the nearby steel mills at Sydney.

The increased use of coal has a number of environmental consequences. Strip mining destroys the soil and natural vegetation over large areas and causes serious stream silting. The additional use of coal in thermal electric plants has increased air pollution and resulting acid rain. Therefore, while Canada has enough coal to last for centuries, changing from petroleum products to coal for energy may be costly in environmental terms.

Uranium Mining: Boom and Bust

The history of uranium mining in Canada illustrates the way in which geographical and external factors influence the growth or decline of a mining industry and of particular mining operations. In the early part of this century, radium, which is found in a uranium-bearing ore called

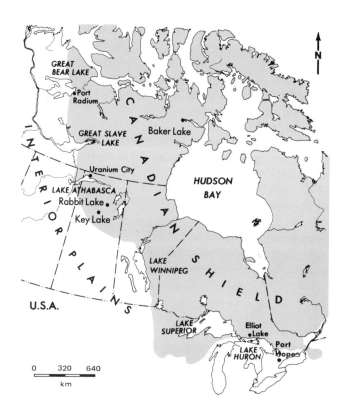

pitchblende, was discovered and became an important treatment for cancer. At this time only a few deposits of pitchblende ore were known in the world, and the price of radium was about $880 per gram. Because of the high price, the pitchblende ore at Port Radium, on the east side of Great Bear Lake, was brought into production in 1933, despite the fact that the place was very inaccessible.

Canadian uranium was used to make the atomic bombs that were first dropped on Japan near the end of World War II (1939-45). After the war, the United States continued to develop more powerful atomic bombs and built up a stockpile of these bombs. All of this activity created a high demand for uranium and resulted in a boom in the Canadian uranium mining industry. In the six years between 1953 and 1959, Canada became the world's leading uranium producer.

After the "boom" came the "bust." By the late 1950's, the United States had invented the hydrogen bomb which did not require uranium, and at the same time other large quantities of uranium ore were discovered in the United States and elsewhere. Consequently, the uranium ores from the Canadian Shield were no longer in such great demand and the production of uranium fell drastically.

During the boom years, uranium ore (pitchblende) was first mined at Port Radium, later at Uranium City on Lake Athabasca, and still later at Elliot Lake. The Port Radium ore was depleted in 1961. Uranium mines at Uranium City became the leading producers until larger uranium ore deposits were found north of Lake Huron, which was closer to markets. Within months, a new town, called Elliot Lake, sprang up out of the forest, and for several years became a boom town to which people moved from all across Canada. Mining at Uranium City ceased in 1982.

With the sag in demand and prices for uranium in the early 1960's, many of the mines at Elliot Lake were closed. As a result, the miners were forced to find employment elsewhere. When the miners left, schools, banks, stores and other businesses also had to shut down. Homes were abandoned, their windows were boarded up, and Elliot Lake began to look like a ghost town.

By 1965, a number of countries, including Canada, were turning to nuclear reactors for the production of electricity. Things were looking up for the uranium mining industry, and some mines were re-opened. However, a real boom did not materialize, as the development of nuclear-based electricity did not proceed as quickly as had been anticipated. Moreover, the United States placed an embargo on imports of uranium from Canada. By 1971, one of the largest mining companies at Elliot Lake announced that it would have to close down for lack of sales. At this point the federal government made a financial arrangement with the company which enabled it to stay in production and to add to its uranium stockpile for several years, at which time it was hoped that demand would pick up again. This government action saved Elliot Lake from another major decline.

Fortunately, in 1973, Japan signed a contract to buy large amounts of uranium for nuclear power purposes extending over a twenty-year period. This, along with other long-term contracts with Ontario Hydro and the United Kingdom, guaranteed a high level of uranium production, and a high level of employment in Elliot Lake until 1994. By that time it is expected that there will be a high world demand for uranium, so that Elliot Lake's future now seems assured. However, the town is still looking for secondary industry to expand its economic base so that it will not be totally dependent on the mining industry.

In the early 1980's about 80 percent of the Canadian production of uranium ore was mined at Elliot Lake, with the balance coming from other centres in Saskatchewan and the Northwest Territories (see map, page 366). The only uranium refinery in Canada is in Port Hope, Ontario.

THE FOREST INDUSTRY

> *LOCAL STUDY*
> You may be surprised to know how important the forest industry is to your community. Which of the following are found in your town or city? From where do the raw materials come?
> — a television and radio cabinet factory
> — a cardboard box factory
> — a wax-paper factory
> — a baseball bat and hockey stick factory
> — a specialty paper producer
> — a book publishing company
> — a newspaper publishing company
> — a handle factory
> — a lumber dealer
> — a furniture factory
> — a planing mill
> — a pulp mill
> — a paper mill
> — a sawmill

The *Canada Year Book* states: "Perhaps in no other country is the national wealth so dependent upon its forest resources and the success of its forest industries as in Canada" (1972 edition, p. 648). The forest industry is actually a complex of industries, including logging and a multitude of wood, paper, and allied industries, which employ many people across the country. Forests are the source of approximately one-fifth of Canada's total exports. Canadian forest products are well known in many parts of the world. In fact, almost half of all the newspaper pages printed in the non-Communist world are printed on Canadian newsprint.

The frontier of the logging industry moved across Canada from east to west, helping to open up the country to other development as it progressed. In pre-Confederation times, forestry was one of the most important industries in what are now the Maritime Provinces. The great demand for wood for ships, the British trade preferences, and the location on the Atlantic Seaboard, were advantages which led to a very prosperous forestry industry in the Maritimes.

By the 1850's the logging frontier had moved to the St. Lawrence Lowlands; by the 1870's lumbermen were cutting the white pine forests east of Georgian Bay; by the turn of the century lumbering was being carried on in Ontario west of Lake Superior (the north shore of Lake Superior was skipped at this time because of a lack of white pine). By the early part of this century most of the good white pine had been removed from the southern edge of the Canadian Shield, and the lumber industry in eastern Canada began to decline. At about the same time increasing amounts of high quality lumber from British Columbia were beginning to reach both Canadian and world markets.

Fortunately for the economy of Ontario and Quebec, the pulp and paper industry began to grow rapidly at about the same time that the lumber industry was beginning to decline. The pulp and paper industry began using the previously untouched spruce forests farther to the north, and so the logging frontier began moving northward instead of westward.

Tree nurseries contribute to the forest resources.

When logging camps and wood-processing industries invade virgin forest areas, changes are bound to follow. Many new towns such as Kapuskasing and Fort Frances in Ontario are established and are almost wholly dependent upon forestry. New and permanent homes are built and new businesses are set up to serve the needs of the people living in these lumber towns. Often, roads and railway lines are constructed to link new communities with the populated south and to help in marketing the wood products. Small pockets of suitable agricultural land may be farmed in the area to supply settlements with food. In this way, Canada's settlement frontier has been pushed forward.

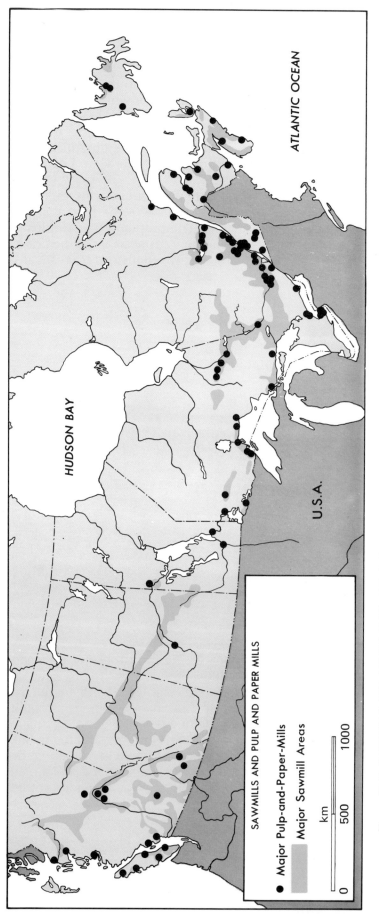

Figure 10.11 SAWMILLS AND MAJOR PULP AND PAPER MILLS. The sawmills are always located close to the supply of logs, and near a transportation route that leads to the market for the lumber. Pulp and paper mills must always be located where the raw materials can be transported cheaply to them. However, for the pulp and paper mills, the presence of a large body of water into which chemical waste can be dumped, and closeness to the market are also important.

Canada's Forest Resources

A glance at the natural vegetation map of Canada would lead one to believe that Canada has almost unlimited forest resources (see Figure 6.1).

Although the forest resources are vast, they are not as great as the map implies. Although trees do grow as far north as the treeline, those in the Subarctic region do not grow large enough for commercial lumbering. The northern part of what is mapped as the Coniferous Forest region is a zone of stunted and slow-growing trees interspersed with large areas of tundra vegetation and numerous muskegs and other land too poorly drained for forest growth. Even south of this zone, there are some tracts of land where there are insufficient soils or the land is too poorly drained to support timber growth on a commercial scale. Moreover, much of the Coniferous Forest region is too remote from the markets to make it economic to carry out logging at the present time. In British Columbia the rugged mountainous terrain makes some forest stands inaccessible for commercial production. Much of the potentially productive forest land in the Mixed Forest region has been cleared for farming. Thus, only a small part of Canada's forested land is available for the commercial production of lumber and wood products. Fire, disease, insects, and poor conservation practices are diminishing Canada's accessible forest resources each year. Nevertheless, Canada does have substantial accessible forest resources—more, in fact, than any other industrial nation except the Soviet Union.

In recent years, government agencies and forestry companies have shown great interest in managing Canada's forest resources so that they can continue to produce valuable timber crops. Millions of trees are being planted in areas that have been denuded by fire or that have failed to seed themselves after timber cutting. Reforestation is also common on abandoned farmland, and many farmers are beginning to manage their farm woodlots properly. In addition, companies are finding ways to eliminate waste in their logging operations. Instead of cutting the entire forest plot, they select only the mature trees, leaving the young ones for a future crop. If almost all the trees are ready for harvest, a few mature trees are left as seed trees. Lumber companies that have very large areas of forest divide their holdings into huge plots and rotate their lumber harvesting from plot to plot over a long period of time. By the time they have covered the entire holding, the first plot is again ready for selective harvesting. In the Coniferous Forest region, this cycle may be as long as 80 to 100 years.

New methods are also being developed in the wood-processing industries. Today, a greater quantity of pulp and paper can be produced from one cord of wood than ever before. New products are being made from what used to be the waste materials from the pulp and paper industry. New processes are also making it possible to use inferior varieties of trees for pulp and paper, and for products such as chipboard, fibreboard, and plywood.

This sawmill is located along the Vancouver Island shoreline near Victoria, B.C. in the West Coast Forest region. Logs from the surrounding mountain slopes are floated to the mill in booms to be sawn into lumber. The lumber that you can see stored in piles is ready to be trucked to nearby markets. Sawdust from the mill is either burned in the small black furnace (middleground) or shipped away on barges.

A pulp and paper mill on the edge of the Canadian Shield east of Winnipeg. Pulp logs, cut from the softwood forests of the Shield, are floated down rivers and stored in booms on the river next to the mill. The finished products like newsprint are shipped to market by rail.

Top, left: Using a power saw, a logger cuts felled trees into even lengths to obtain the maximum amount of lumber.

Top, right: A portable spar is often used to collect cut timbers into piles that can then be loaded onto trucks.

Bottom, left: A powerful truck hauls the large timbers toward the Pacific coast.

Bottom, right: Groups of logs that are lashed together into large booms have been towed to a sawmill in Vancouver. Here the booms are anchored in the water until the logs in them can be sorted for size and grade. A more modern method is to use automatic self-loading and unloading barges instead of booms.

The most recent trend in forest management recognizes that forest resources have multiple uses. They are an important aspect of any water conservation program, they provide protection against soil erosion on steep slopes, they maintain a habitat for wildlife, and provide a wide range of recreational activities for the public.

The Location of Canada's Forest Industries

The hardwood trees in the Mixed Forest region of eastern Canada are important for the furniture industry. There are very few cherry and walnut trees left, but birch and maple are still commonly used for furniture as well as for hardwood floors. Oak, ash, and hickory are used for handles as well as some sports equipment such as hockey sticks and baseball bats.

In the Coniferous Forest region, the spruce, balsam, fir, hemlock, and poplar provide the soft woods required for the pulp and paper industry. The West Coast Forest region has dense stands of Douglas fir, western hemlock, and red cedar trees that grow to enormous sizes because of the long growing season and the abundance of rainfall. These trees make excellent lumber that is in great demand in many parts of the world. Because of the large diameter of the tree trunks, the British Columbia trees are well suited to the making of plywood, a product that is in very high demand in the construction industry.

Figure 10.11 shows the location of the majority of Canada's sawmills and pulp and paper mills. The sawmills are always located close to the supply of logs and near a transportation route that leads to the market for lumber. The precise location of the sawmill is strongly influenced by the availability of hydroelectric power.

Pulp and paper mills are usually located on the market side of the forest resource, or at least on a major transportation route which facilitates getting the product to market. They are not always located close to the forest resources either. For example, the huge pulp and paper mills at Thorold, Ontario (in the Niagara Peninsula), obtain their pulp logs from

considerable distances. These mills have several locational advantages that offset the distance from raw materials. Pulp logs can be transported cheaply on the Great Lakes; plenty of hydroelectricity is available; and the mills have access to large bodies of water into which the chemical wastes can be dumped. In addition, the finished paper can be transported cheaply to major centres in the manufacturing belt of the United States and Canada.

In recent years laws regulating the pollution of bodies of water have had an effect on where pulp and paper mills can be located. New mills are not permitted to be established unless they have sufficient pollution controls so that the water body can assimilate the treated wastes adequately. In some cases established mills have been ordered to reduce their pollution or suspend operations.

THE FOREST INDUSTRY IN BRITISH COLUMBIA

In terms of number of people employed and net value of production, the forest industry is British Columbia's most important resource-based industry. Over half of British Columbia's lumber production comes from the coastal forests of hemlock, cedar, Douglas fir and Sitka spruce. British

Figure 10.12 BRITISH COLUMBIA'S FOREST INDUSTRY. The coastal districts of Vancouver and Prince Rupert still produce more total timber than the combined, larger interior districts of Kamloops, Prince George, and Nelson. Why?

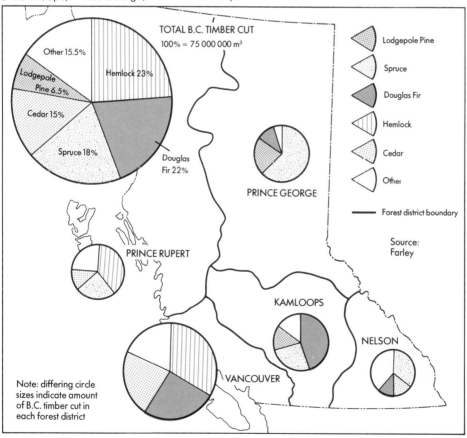

Columbia also accounts for more than one-quarter of the national production of wood pulp.

The potential value of British Columbia's timber resources was recognized at an early date. However, until external markets and transportation to the markets were developed, logging was confined to the areas around the major settlements, mainly in the southern coastal area. By the turn of this century, logging was widespread on the coast, but it was not until about 1950 that it had spread over much of the southern interior. During the 1960's logging was extended into a number of areas in the northern interior forest (Figure 10.12).

For many years the forestry industry of British Columbia has been more dominant along the coast because of certain environmental and locational advantages:

(1) a moist mild climate favours rapid dense tree growth, with large trees, and species of high value;
(2) low cost water transportation is available all year round;
(3) large amounts of water are available for pulp production; and
(4) the ocean provides a vast body of water to assimilate wastes.

Although the forest industry was still more important along the coast during the 1960's and 1970's, there has been a rapid expansion in the interior. The increase in pulp mills in the interior is partially a result of special location inducements provided by the provincial government. However, it also reflects the availability of low-cost wood chips that are waste material from the sawmills of the interior. With the establishment of new pulp mills, many interior sawmills have been rejuvenated and thus the production of lumber in the interior has also been increasing.

In the early years, there was intense competition for forest resources, with companies adopting a "cut out and get out" policy. Recently, the British Columbia Government has taken action to set logging standards, to exercise stronger controls on allocation of cutting rights on public lands, and to implement what is known as a sustained-yield program. Under these conditions the forests will, over a period of time, renew themselves and the forest industry will have a permanent supply of raw materials.

Besides supplying lumber to other parts of Canada, British Columbia exports huge quantities of lumber, particularly to the United States, Japan, and the European Economic Community. High quality, and low costs resulting from large scale operations, give British Columbia a competitive advantage in world lumber and pulp and paper markets. The seaboard ports, open all year, are also an advantage. Based upon the quantity and quality of the resource, the locational advantages, and world trends in consumption of wood products, the future prospects of the British Columbia forest industry look very good.

THE FISHING INDUSTRY

Commercial fisheries, dating back hundreds of years, were the country's first industry. Fishing is still an important industry today, yielding more than one million tonnes of fish annually. About 65 000 Canadians make their living by fishing and many others are employed in fish processing plants and in industries that supply the fishing industry with boats, nets, and other fishing gear.

Of all the fish products produced in Canada, very few are sold on the home market. Seventy percent of the output is exported, largely to markets in the United States and Europe. Canada's small population and the fact that Canadians generally eat little fish account for the large volume of fish that must be exported to maintain the present status of the fishing industry.

Because so little of Canada's fishing catch is consumed at home, the success of the industry depends on successful competition with other fishing countries such as the United States, Japan, the Soviet Union, and countries from northwestern Europe. In international competition, the Canadian industry has some natural advantages. Canada is closer to the great fishing grounds off the east coast of North America, and has within its borders greater areas of fresh water than any other country. It is also next door to the United States, the world's greatest fish importer. However, in competition for existing fish markets, Canada has not fared so well. Fishermen of other nations with modern equipment have been working the fishing banks off the coasts of Canada to the point of reducing

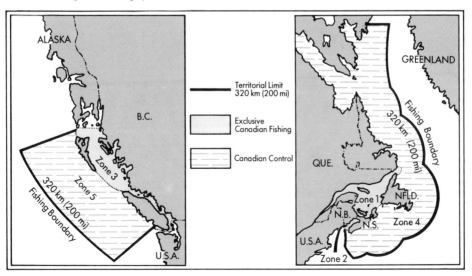

Figure 10.13 CANADA'S NEW FISHING ZONES. In Zones 1, 2, and 3 Canadian fishermen have exclusive fishing rights. In Zones 4 and 5, Canada controls all fishing, but certain foreign countries are given fishing quotas.

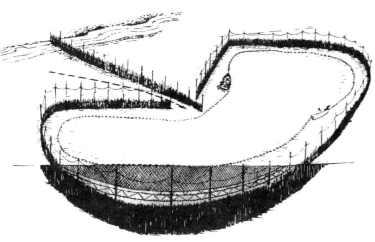

Trolling.

The troller used for salmon catching is marked by long poles that extend over the sides of the vessel while fishing. To these poles are attached weighted lines bearing lures that are dragged through the water at a slow speed. The salmon are caught on the lures.

Weir Fishing.

Weirs are large structures resembling corrals. Weirs are used near shore to catch small herring (commonly called sardines). The fish are diverted into the weir by a fence that stretches from the shore to the weir. The fishermen then run a seine net around the inside of the weir and gather the catch inside. In certain places, at low tide, the fishermen can walk out to the weir to take in the catch.

Gill Netting.

From a drum on the stern of the boat, the net is set out. It hangs like a curtain in the water. The fish are caught when they become entangled as they swim into the net. This method is the most popular one for salmon fishing on the Pacific coast.

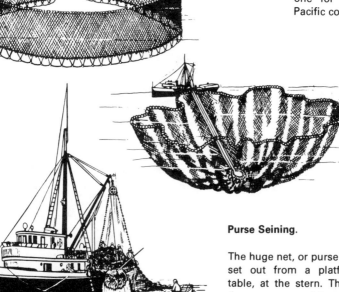

Purse Seining.

The huge net, or purse seine, is set out from a platform, or table, at the stern. The net is then manoeuvred to encircle a school of fish. The seine is then drawn together, and the fish are scooped into the boat with a large dip net on a long pole.

Otter Trawling.

Pacific groundfish (fish that feed on the ocean floor) include ling, grey and black cod, ocean perch, sole, and flounder. These fish are all caught by otter trawls.

The otter trawl is a large bag-shaped net that is dragged along the ocean floor. The net is held open by boards called "doors". The fish become trapped in the closed end of the net.

Longlining.

The long line is used to catch halibut and Pacific Coast cod. The gear consists of a strong rope made up in lengths to which are attached shorter lines that have baited hooks. The long line is set out on the bottom of the sea and is later hauled in with its catch, using a powered winch.

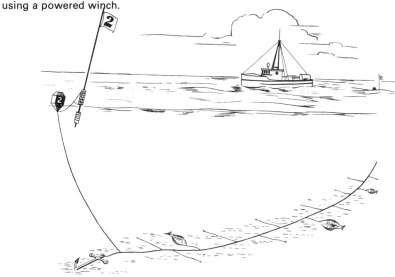

Cod Trapping.

Cod traps are used by thousands of Newfoundland inshore fishermen. The net, which resembles a huge room with a net floor and walls, is set at varying depths in the water. The net hangs down from the surface of the water. Cod swim into the net and become trapped. The net is then hauled up with the catch of fish in it.

Newfoundland fishermen hauling in their catch.

the Canadian catch. In 1977, Canada unilaterally extended its fisheries management zones from 19.2 km (12 mi) to 320 km (200 mi). The Soviet Union, Norway, Spain, and Portugal have been given fishing quotas within these zones (Figure 10.13).

In addition to food for humans, fish provide a number of other important by-products. Oils extracted from the livers of cod and halibut are used as a vitamin supplement for both humans and animals. Fish meal and fertilizer can be processed from the less profitable species.

LOCAL STUDY

Even if you do not live in an area that is important for fishing, you will be able to learn something about the fishing industry through your own investigation.

1. What kinds of fish are caught locally by sports fishermen? Is there any commercial fishing in your area?

2. Visit a local meat market, or better still, a special fish market. Find out from where the store obtains the different kinds of fish. How much of the fish is sold fresh, frozen, smoked, or salted?

3. Check cans of salmon, tuna, and lobster to see where the product was processed.

4. Find out how many times a month each family in your classroom eats fish. (Do not forget to include canned salmon or tuna.) From this information, and assuming each person eats a 0.1 kg serving at a meal, calculate how many kilograms of fish the average person in your classroom eats in a year. You may be interested in comparing your answer with the Canadian average annual consumption of 6.4 kg per person.

The Atlantic Fisheries

Fish gather in large numbers off the coasts of Newfoundland and Nova Scotia because of the great amount of available food. All fish basically depend on plankton for food. Some feed on it directly, while others eat the fish that eat the plankton. Plankton consists of tiny one-celled plants and animals that require minerals from the water, and multiply where the water is shallow enough for sunlight to penetrate to the ocean floor.

Off the east coast, part of the continent projects under the sea as a low, flat platform called a *continental shelf*, over which the water is very shallow compared to the rest of the ocean. The parts of the continental shelf where the sea is shallowest are known as *fishing banks*. The largest and most famous of the fishing banks are the *Grand Banks*, southeast of Newfoundland.

The banks off the east coast are excellent fishing grounds because of the abundance of plankton that thrives there because the shallow water permits needed sunlight to penetrate to the sea floor. Also, the meeting of the cold Labrador Current and the warm Gulf Stream keeps the water constantly stirred up to keep supplying the different levels of water with the minerals on which plankton feeds. (See map, page 382.)

Canada's Atlantic fisheries can be divided into two classes: inshore fishing, in which small boats are used, and offshore or banks fishing, in which large trawlers are employed. The majority of fishermen in the Atlantic Provinces earn a living in the inshore fisheries. Every morning they journey to coastal waters to harvest fish and other seafoods with a variety of handlines or traps. At night they return to the small villages that

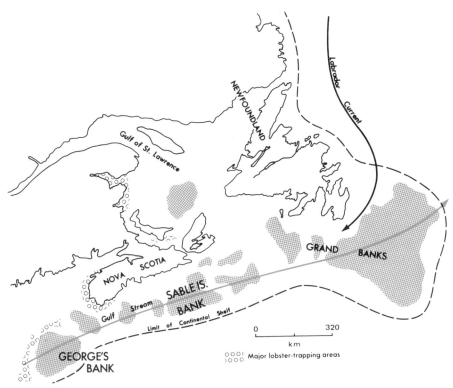

dot the coves and bays of the Atlantic coastline. Here the fish are either dried on rough platforms or taken to processing plants in the larger centres.

The inshore fishing is carried on by individual fishermen. Each family has its own boat and each cures the catch of fish in the way that it has been done for centuries. These fishermen cannot afford large boats, modern fishing gear, or modern processing equipment. Their catch is small, the price for fish is low, and consequently, many of them live on the edge of poverty. However, they like being independent, and prefer being able to return home each night, instead of having to stay out at sea for weeks at a time as do the offshore fishermen.

Some inshore fishermen have formed co-operative companies to help them to buy modern fishing equipment that they can share. Co-operatives also permit them to purchase supplies in large quantities, at a lower cost. Some co-operative canning companies have been formed, so that the fish can be processed in a way that will bring higher prices. If any of the co-operative ventures shows a profit, it is shared with the co-operating members. The federal and provincial governments have provided financial assistance to help the fishermen procure more efficient boats and fishing gear and to establish fish processing plants.

The offshore fisheries are controlled by companies, not individual fishermen, and are operated from a few larger ports, instead of from many

In the offshore fishing industry, large companies can hire many fishermen and can purchase the necessary equipment to load, operate, and unload modern fishing boats. This photograph shows a large fish processing plant at Lunenburg, south of Halifax, N.S. The trawlers you see can unload their catches directly at the plants where the fish are quickly processed into a variety of fish products. On the right you can see some of the refrigerated railway boxcars that keep the fish fresh until they reach distant markets.

small villages. However, many fishermen are not willing to move their families from the small villages where they have lived all their lives, nor learn the new methods required for offshore fishing.

Most larger companies rely on the steam or diesel-powered trawlers that visit the banks about every two weeks. They catch the fish by hauling a large "trawl" net through the water, trapping all species of fish. On board the ships, fish that are of commercial value are sorted from the rest, cleaned, and packed in ice. Trawlers can operate in stormy weather and during the winter, when it is impossible for the small boats to go out. Thus, trawlers help provide a regular supply of fish to canneries and fish-freezing plants throughout the year and, of course, a steady income for the fishermen.

The offshore fisheries, which employ only about fifteen percent of the Atlantic fishermen, produce almost two-thirds of all the fish landed in the region. Only in Newfoundland is the inshore fish production greater than offshore. The offshore fisherman's average value of fish production is ten times as great as that of the inshore fisherman. However, the capital cost of the offshore fishing operation is much greater.

Until 1977, when Canada imposed the 320 km (200 mi) fishing limit, fishermen in the Atlantic Provinces faced severe competition from fishermen from other countries. In the mid 1970's, four times as many foreign fishing vessels as Canadian could be found off Canada's east coast. As a result, the Canadian fishing and fish processing industries suffered from unemployment and low incomes. Moreover, because of overfishing, there was an overall decline in the fish catch. Deep-sea fishing substantially reduced the salmon take; the Canadian Government authorized a ban on salmon fishing in the Atlantic Provinces for a period of time and paid the fishermen subsidies to help cover their losses.

Left: A motorboat has been loaded with baited traps before heading toward the lobster grounds. Other fishing boats surround the wharf ready to take on their load of lobster traps.

Right: Lobsters like the one in the photograph will be kept alive in a tank until they reach the processing plant or the restaurant.

In the past, cod was considered the most important fish in the Atlantic region, both in terms of the quantity caught and the value of production. By the 1970's, cod was no longer king. The leading species ranked by mass of catch were as follows: herring, cod, plaice, redfish, haddock, lobster, scallops. However, the ranking was different when done by value of production. Because of the very high prices received for lobsters, they led all the rest in value of production, followed by cod and scallops. The size of the catch of these different fish differs from place to place in the Atlantic region and also changes over time.

LOBSTER FISHING

Lobsters are one of the Atlantic Region's most valuable ocean harvests. The most prolific lobster trapping area extends along the south and southwest shore of Nova Scotia from Shelburne to Digby. Other important lobster trapping areas are found along the north shore of Nova Scotia and New Brunswick and the east end of Prince Edward Island. (See page 382.)

Each day, from hundreds of harbours in these areas, punts, skiffs, dories, or other motorboats set out to harvest lobsters. Most of the open motorboats, up to 15 m in length, are now fitted with power-driven haulers to lighten the backbreaking task of raising hundreds of traps out of the water. Most boats operate with a crew of from one to three people, often a family team.

The most popular lobster trap used is in the form of a half cylinder with two compartments. The bottom, sides, and top of the trap are closed

An Atlantic fisherman proudly displays a lobster trap he has just made. The netting is so constructed that the lobster can enter the trap but cannot work its way out.

in with wooden laths. One end is closed in with hand-knitted cotton or nylon netting. The other end has a net ramp that permits the lobster to climb into the trap but prevents it from getting out. A door at the top, for baiting the trap and removing the lobsters, runs the length of the trap. There is a wide variety in the size and mass of these traps because most fishermen build their own.

When the lobster grounds are reached, traps are weighted with stones and are lowered to the ocean floor by rope. Groups of two or three traps are fastened by their ropes to a brightly-painted cedar buoy to help the fishermen find their traps. Each year, east coast fishermen set a total of two million traps.

As each trap is hauled up, the lobsters are removed. Those obviously below legal size, as well as those carrying eggs, are released. By doing this, the fishermen are assuring themselves of a good harvest of lobsters in years to come. Lobsters that are large enough to harvest are placed in wooden crates and are protected as much as possible from the sun and wind. In severe winter weather, they are often put in a tub of salt water to keep them from freezing. Such careful steps are taken to keep the lobsters alive until they are to be processed because the quality of the lobster meat deteriorates quickly once the lobster is dead.

Most of the larger-sized lobsters are shipped alive to markets in eastern Canada and the United States. The smaller lobsters are prepared as chilled fresh lobster meat or else they are canned in canneries located in the major lobster trapping areas.

Pacific Coast Fisheries

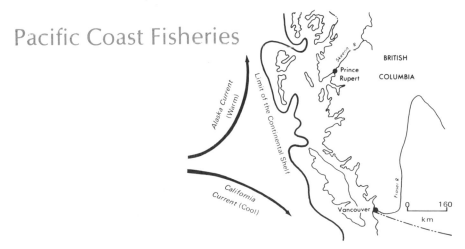

A number of factors have combined to make the west coast a good fishing area. There are numerous mountain streams suitable for salmon to use for spawning. Off the coast there is a constant mixing of warm and cold currents that favours the production of plankton which provides food for the fish. The many sheltered passages created by the island coast and the fiords provide protection for fishing boats in time of storm.

The Pacific coast fishery accounts for approximately one-third of Canada's total fish production (by value). The largest catch is salmon, the second largest herring; these two compose about ninety percent of the total production. The fishery also produces small amounts of halibut, cod, flounder, clams, crabs, and oysters. Until about 1970, halibut was second in importance; since then, halibut stocks in the North Pacific have declined rapidly because of overfishing by foreign offshore trawling fleets. As halibut stock declined, more fishermen turned to salmon and herring, thus putting more pressure on the stock of those fish.

Canada has special bilateral agreements with the United States to help preserve a sustained yield of both salmon and halibut. However, for many years foreign fishing fleets (from outside North America) fished heavily beyond Canada's fishing limits. The new 320 km (200 mi) fishing boundaries are designed to protect both the fish stocks and the British Columbia fishing industry. The new fishing limits protect the halibut stock because halibut are caught offshore in the area now included in Canada's fishing boundary. However, how much the British Columbia fishing industry will benefit is not clear, because the fishing vessels of its inshore fleet are too small to take advantage of the new limit. Nor are their methods geared to the factory-ship industry, as are the fishing fleets of the Soviet Union and Japan.

The British Columbia Government has a number of programs to control and regulate the fishing industry in order to protect fish stocks and to increase the earnings of fishermen. For example, in recent years the numbers of vessels given licences to fish for salmon was reduced substantially. Regulations about when and where fishing can be done have been introduced and quotas are established for specific species.

A Pacific coast fisherman catching salmon by trolling.

SALMON FISHING

Many kinds of salmon—sockeye, cohoe, pink, chum, and spring—thrive on the west coast. Of these, the sockeye, with its red-coloured flesh, is the most valuable.

Salmon are migratory fish. Every fall, they swim from the sea up the numerous long fiords of the coast looking for fresh-water pools and lakes in which to spawn. Only the mature fish make this trip after spending several years in the Pacific Ocean. The fish seek out sheltered spawning grounds in mountain rivers and lakes, and there they lay millions of eggs. Shortly after laying the eggs, the adult female salmon dies. After the eggs hatch, the young salmon, called *fry*, remain in these inland waters for a year, and then they swim out to sea. After several years in the ocean, the fry become full-grown and return to the river or lake where they were hatched to lay their eggs. In this way, another catch of fish is born.

Most fishermen try to catch the large, mature salmon on their way to spawning grounds, when the fish have reached their maximum size. Most fishermen put the fish on ice until it can be delivered to a cannery. Some fishermen, however, transfer their catch of fresh fish to ships called "packers" which act as a shuttle service taking the catch to onshore canneries at Prince Rupert or the Vancouver area.

Depletion of the salmon resources is of vital concern to fishermen and government officials alike. In some years, too many full-grown salmon are harvested and very few new fry are born. Restrictions have

A salmon seiner in operation off the British Columbia coast. Working with a winch and a dory, fishermen can lift the fish caught in the long seine net into the boat.

A fish ladder allows salmon and other species of fish to by-pass high dams that are built across the rivers they travel up to spawn.

been placed on the amount Canadian and American fishermen can catch. Foreign fishermen cannot be stopped, however, from taking salmon beyond Canada's 320 km (200 mi) fishing control limit. Only some kind of international agreement will prevent serious overfishing of salmon by foreign fishermen.

There are also other threats to the salmon industry. For example, dams have been built along the Columbia River for flood control and for the generation of hydroelectricity. These dams are barriers that prevent the salmon from reaching their spawning grounds. Many salmon die while attempting to jump over the dams. In some places, fish ladders have been erected to help the salmon climb over the dams. These ladders have been only partially successful. Many salmon die in climbing the ladders and even more young salmon perish in the turbines on their trip downstream to the ocean. Because of the damage that it would do to the salmon industry, British Columbia has not developed hydroelectric power on two of the largest salmon-producing rivers, the Skeena and the Fraser. Pollution of the spawning streams is also a threat to the salmon.

The Inland Fisheries

Besides supplying sports fishing, Canada's inland waters also support some commercial fisheries. The bulk of the catch comes from the Great Lakes, Lake Winnipeg, and Great Slave Lake. More than 600 smaller lakes are also fished commercially. The total fish catch from the inland lakes is less than five percent of the fish production of the two coastal fisheries.

In summer, the inland fishermen use boats up to 12 to 15 m in length, as well as skiffs and canoes. Gill nets and pound nets are the chief gear. The fish catch is taken to permanent shore stations that have docking, icing, cooling, grading, and warehousing facilities. Winter fishing in large and small lakes, through holes in the ice, is carried on by teams, many of whom are only part-time fishermen with their main occupation in farming, lumbering, or the fur-trapping industries. Huts or mobile cabooses shelter these fishermen. Horses, trucks, and snowmobiles are used to haul fish and equipment.

Most of the catch of perch, whitefish, and yellow pickerel is marketed fresh or frozen, mainly in the large urban centres of eastern Canada and the northeastern United States. Ontario leads all provinces in fresh-water fish production because of its access to the Great Lakes. Manitoba is second in the value of its catch.

Great Lakes fishing has declined greatly in recent years for a number of reasons. Some years ago, the sea lamprey invaded these inland waters and killed off large numbers of whitefish and lake trout. Another threat to the fishing industry in the Great Lakes has been pollution of the water. Lake Erie, the shallowest lake with the greatest number of cities dumping sewage into it, has suffered most. The United States and Canada have co-operated on lamprey and pollution control programs that will ultimately help make Erie a good fishing lake again.

PROBLEMS AND PROJECTS

1. The following minerals are listed in the order of their value within each group. The values given are for 1978-1980. You will be able to bring these statistics up to date by consulting the most recent *Canada Year Book* or *Canada Handbook*. Draw a bar graph for each group of minerals.

   Canadian Mineral Production (1978-80 average) (Source: Statistics Canada.)

Metallic Minerals	Value in Millions of Dollars
1. Iron Ore	1584
2. Copper	1484
3. Nickel	1048
4. Zinc	913
5. Gold	665
6. Uranium	624
7. Silver	516
8. Lead	323
9. Molybdenum	276

Nonmetallic Minerals	Value in Millions of Dollars
1. Potash	742
2. Asbestos	594
3. Sulphur	245
4. Salt	111
5. Titanium	92
6. Gypsum	41

Structural Materials	Value in Millions of Dollars
1. Cement	628
2. Sand and gravel	462
3. Building stone	334
4. Clay products (brick, tile, etc.)	115

Mineral Fuels	Value in Millions of Dollars
1. Crude oil	7454
2. Natural gas	5157
3. Natural gas byproducts	1418
4. Coal	862

2. Select one of the above minerals that was not discussed in this chapter and write a report about it. You will be able to obtain information from sources such as geography textbooks, encyclopedias, the *Canada Year Book*, the department of mines in the provinces where the mineral is important, and from mining companies. The following is a suggested outline for your report:
   - Where the mineral is found and why it is found there (use maps)
   - The mining processes (use sketches or photos)
   - Processing or refining the mineral (use flow-arrow diagrams)
   - Transportation routes (use maps)
   - Uses of the mineral (use photos)
   - Trends in production (use graphs)
   - Problems of the industry (use sketches or photos)

3. On a page in your notes, draw a divided circle graph for each production item listed in the following tables. The percentage values are 1978-1980 averages in order to balance out minor yearly changes in the rank position of provinces or regions. You can update the statistics by referring to a recent *Canada Year Book* or *Canada Handbook*. Why does each leading province produce the most of each of these items?

Value of Mineral Production by Province (Source: Statistics Canada.)

Province	Share of Production (%)
Alberta	51%
Ontario	14%
British Columbia	9%
Quebec	9%
Saskatchewan	7%
Newfoundland	4%
Manitoba	3%
Others	3%

Value of Paper Production by Province

Province	Share of Production (%)
Quebec	44%
Ontario	27%
British Columbia	15%
Others	14%

Value of Lumber Production by Province

Province	Share of Production (%)
British Columbia	72%
Quebec	15%
Ontario	7%
Atlantic Provinces	3%
Others	3%

Value of Fish Catch by Region

Region	Share of Production (%)
Atlantic Coast	63%
Pacific Coast	34%
Arctic and Inland Fisheries	3%

4. Energy of all types (electricity, fuels, etc.) differs greatly from region to region across Canada. Describe the demand for and supply of various kinds of energy in each region. What is each region's potential for developing alternate energy sources (i.e. solar, wind, and tidal power)? For useful energy picture summaries, look at these two sound-filmstrip series: *Energy in Canada* (Ontario Hydro, 1981) and *Energy Management for the Future* (National Film Board, 1979).

11 MANUFACTURING

Although Canada has become famous around the world for the products from its mines, forests, and farms, Canada's most important industry is manufacturing. The manufacturing industry employs more people than forestry, fishing, mining, and agriculture put together.

Manufacturing may be divided into two types, *primary* and *secondary*. The primary manufacturing industries process raw materials. The secondary manufacturing industries make new products from materials that have already been processed by primary manufacturing industries. For example, the making of steel from iron ore is considered a primary industry. The making of automobiles from the steel supplied by the steel industry is considered a secondary industry. Likewise, flour milling is a primary industry, but baking is a secondary industry.

Figure 11.1 shows the value of production of some of the leading manufacturing industries in Canada. Although the processing of raw materials such as metals, coal, petroleum, timber, and farm products is still very important to Canada, the total value of secondary manufacturing is almost double that of primary manufacturing. There are literally thousands of secondary manufacturing industries in Canada. A few not included in Figure 11.1 are the rubber, leather construction, and animal feed industries.

Secondary manufacturing in Canada is faced with stiff competition from other countries. Many countries in Europe and Asia have lower costs of production because factory workers receive lower wages. Factories in the United States often produce more cheaply because they have a larger volume of production. Large factories can obtain raw materials at a lower cost because they buy larger amounts and they can also afford to invest more money in large-scale modern equipment. To protect Canadian manufacturing from cheaper imports, the Canadian Government charges a tariff on most manufactured goods entering Canada. The tariff is usually high enough so that a Canadian buyer can purchase a Canadian-made product at almost the same price as the imported one.

Many Canadian secondary manufacturing industries are branch plants of United States firms. Because of this fact, many Canadian-made products are identical to those made in the United States. For example, it is often difficult to tell whether an automobile was made in Canada or in the United States. This similarity in products, along with higher costs of production, has made it difficult for many Canadian firms to export their products. Increasingly Canadian manufacturers are developing distinctive products of high quality that they can sell in other countries.

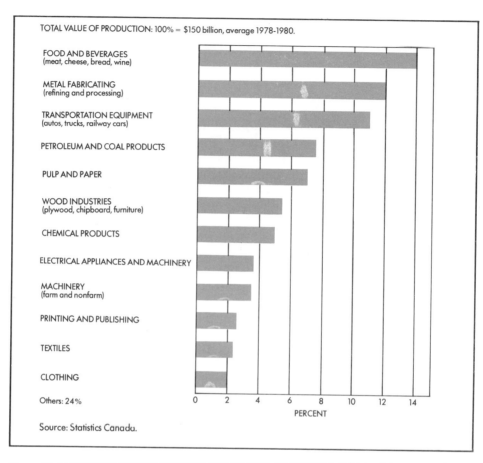

Figure 11.1 LEADING MANUFACTURING INDUSTRIES OF CANADA. Which of these are primary and which are secondary manufacturing industries? Which of these groups include both primary and secondary manufacturing industries?

LOCAL STUDY

If you live in a town or city, the following questions and suggested activities will give you an appreciation of the local manufacturing industry.

1. Find out how many of the parents of students in your classroom work in a primary and how many work in a secondary manufacturing industry.

2. From your local industrial commissioner, or from your chamber of commerce, obtain a list of all the manufacturing firms in your town or city. Ask also for the number of employees in each firm. Divide the list into primary and secondary manufacturing.

3. Plot the location of the manufacturing firms on a city map. What proportion of the firms are located in or around the C.B.D.? in the suburbs? What kind of manufacturing industries tend to locate close to railroads? close to highways or expressways?

4. Arrange to visit several manufacturing firms. From these firms, try to obtain answers to the following questions:
   (a) From where do the raw materials come?
   (b) To where are the finished products shipped?
   (c) What kind of transportation is used for the movement of raw materials and finished products?
   (d) What is the major source of power?
   (e) How much water is used per day? In what way are the industrial wastes treated before being released into the sewers?
   (f) How many people does the firm employ? How many of these employees are office workers? unskilled labourers? highly skilled labourers?
   (g) What other Canadian and foreign companies provide competition?
   (h) Are new methods and new machinery putting workers out of jobs?
   (i) If the plant has been established recently, why did the company choose this particular town or city?
   (j) Why did the company choose its particular location within the city?

LOCATION FACTORS

There are six major factors that determine where a manufacturing industry will be located. These six factors are the availability of the following: raw materials, power, labour, transportation facilities, market, and capital. An understanding of these location factors will help you to appreciate the distribution of the manufacturing industries in Canada.

RAW MATERIALS

If raw materials are very bulky and expensive to move, or are highly perishable, then the manufacturing plant is usually located close to them. Sawmills are scattered throughout all commercial timber areas because logs are bulky and expensive to transport unless they can be floated down streams. Fresh fruits and vegetables are highly perishable and damage very easily. For this reason, most fruit and vegetable canning plants are located near agricultural areas specializing in those crops. Thus, fruit and vegetable canneries are important in Vancouver, in the Okanagan Valley, in southwestern Ontario, around Montreal, and in the Annapolis Valley.

POWER

Manufacturing cannot operate without power. In the early days, the manufacturing industries located near waterfalls so that they could use water power. Later, steam power and hydroelectric power made it possible for factories to locate away from the water-power sites. Today, steam-generated power, using coal (thermal power) or uranium (nuclear power), has made it possible to generate electricity wherever it is needed. Also, electricity can now be transmitted up to 1600 km, so that the location of manufacturing is no longer restricted by the availability of local power.

LABOUR

Despite the increased use of automatic machinery, labour is still an important factor in the manufacturing industry. Skilled labourers are particularly valuable and are usually less inclined to move from one part of the country to another. Sometimes manufacturers will establish their plant in a large city because they know that there they will have a better chance of hiring workers with special skills. If skilled labour is not really important, manufacturing firms sometimes locate in small towns because there they can find workers at lower rates of pay.

TRANSPORTATION

Transportation, of course, is a vital factor in the location of manufacturing industries. Water transportation is the cheapest method for moving bulky cargo like coal, ore, and grain. Where water transportation is not available, railroads are used for long-distance hauling of bulky cargo. Trucks are used primarily for short-run hauls, although huge trailer-trucks now travel from coast to coast. Increasing use is being made of airplanes to transport high-value products. Pipelines are another form of transportation used to move oil and gas. Many modern industries want to locate where several of these forms of transportation are available.

MARKET

Secondary manufacturing is attracted to densely populated areas that provide a market for its products. Being close to the largest markets minimizes the cost of transporting finished products.

CAPITAL

Financial capital moves readily to any area where a good return on investment is assured. However, it is usually more difficult to convince money lenders that more remote areas are good places in which to make investments. Therefore areas in Canada far removed from the industrial manufacturing belt of central Canada have more difficulty in attracting the capital necessary to develop manufacturing.

In addition to these location factors, industrialists will consider other things before choosing a town or city for their plant. They will want to know what other manufacturing plants exist in the place because they may require goods produced by them or may produce products used by other industries. Thus manufacturing industry tends to attract more manufacturing industry. Manufacturers also want to be assured that the municipality will have an adequate supply of water. They are interested in locating in a well-planned municipality that has adequate streets, shopping facilities, schools, and parks. Finally, they want a flat industrial site, well-serviced with roads, sewers, water, and electricity.

LOCATION OF MANUFACTURING IN CANADA

The Atlantic Provinces

Most of the manufacturing industry in the Atlantic Provinces is primary. Products such as canned fish, meat, cheese, pulp, and lumber, as well as iron and steel, are produced from raw materials found in the region. The ports of Halifax and Saint John, however, have major oil refineries based on imported crude oil.

Secondary manufacturing is not of major importance in the Atlantic Provinces because there is not a large enough population to provide an adequate market. Most of the newsprint, which is made from local raw materials, is exported from the region. Several other secondary manufacturing industries are related to transportation. Wooden fishing boats are built in several towns along the south shore of Nova Scotia, including Lunenburg, where the famous *Bluenose* was built. This famous vessel is depicted on the Canadian ten-cent piece (except in 1967).

Steel shipbuilding is important at Saint John and Halifax. The shipbuilding industry at Halifax has increased its activity greatly as a result of the demand for special boats and equipment needed by the companies exploring for petroleum off the east coast. A modern railway-car plant is located at Trenton, Nova Scotia. Several foreign automobile assembly plants are also located in the province. Two large French tire factories were attracted to Nova Scotia by generous financial grants from the federal and provincial governments.

Port Hawkesbury on the Strait of Canso has become a boom town. The building of the Canso Causeway connected Cape Breton Island to the mainland by highway and rail and created an ice-free "bay" which made it possible to develop excellent port facilities at Port Hawkesbury. Its industries include a pulp and paper mill, an oil refinery, a heavy water plant and a ship-repair dock.

In New Brunswick most of the manufacturing is concentrated in Saint John, the largest city, with a smaller amount in Moncton. Both Saint John and Moncton have been chosen by the federal and provincial governments' regional development plan as major growth centres of the province. There is currently underway a scheme that would bring a very important group of metal-working industries to Saint John.

Despite recent manufacturing developments in the Atlantic Provinces, growth in manufacturing has not kept pace with that of the manufacturing belt of central Canada. Also, new manufacturing has not been able to create enough new jobs to alleviate the high unemployment rates that have plagued the region.

Shipbuilding and ship repair are important industries at Saint John N.B. Huge cranes (A and B) are used to lift heavy steel beams and plates into position. At C you see a large ocean freighter nearing completion. Smaller ships are being assembled in the dry docks. A thermal electric plant (D) supplies electricity to this shipbuilding yard, other nearby industries, and part of the city of Saint John.

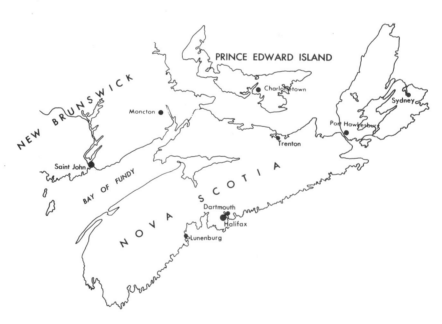

Ontario and Quebec

Ontario and Quebec are the most important manufacturing provinces in Canada, producing more manufactured goods than all of the other provinces combined. These two account for seventy percent or more of the Canadian production of pulp and paper, iron and steel, petroleum products, metal products, automobiles, aircraft, and farm machinery, as well as a host of manufactured items such as electrical appliances, clothing, textiles, soap and detergents, books, and miscellaneous hardware. The importance of the manufacturing industry to Ontario and Quebec is shown in Figure 11.2.

Most of the manufacturing of Ontario and Quebec is carried on in the Great Lakes-St. Lawrence Lowlands. This region is known as the *manufacturing belt*, or the *industrial heartland* of Canada.

The Great Lakes-St. Lawrence Lowlands has many advantages for the manufacturing industry (Figure 11.3):

(1) It has productive farmland that forms the basis of the food-processing industries.
(2) It is close to the Canadian Shield with its storehouse of minerals to be refined and processed, and with its abundance of forests suitable for pulp and paper.
(3) Power is readily available. Hydroelectric power is available from the many falls and rapids in the Canadian Shield, from Niagara Falls, and from the rapids of the St. Lawrence River. Coal can be readily imported from the United States Appalachian region to provide the fuel for the production of thermal electric power. Nuclear power plants can use the Great lakes for cooling. Oil and natural gas are piped to the region from Alberta.

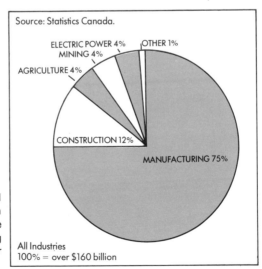

Figure 11.2 TYPES OF INDUSTRIES IN ONTARIO AND QUEBEC. This graph shows that in Ontario and Quebec the value of production of the manufacturing industry is greater than that of all other industries combined.

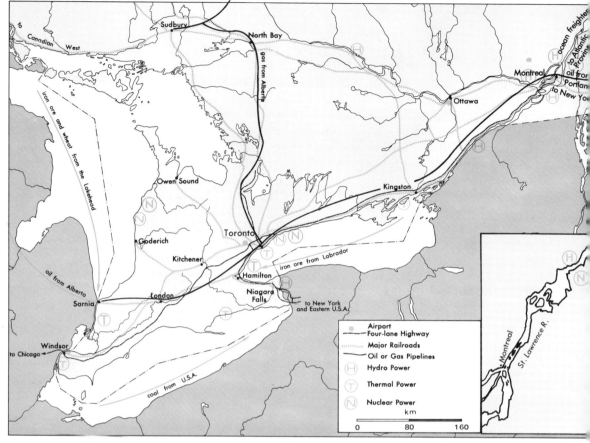

Figure 11.3 TRANSPORTATION AND POWER IN THE GREAT LAKES-ST. LAWRENCE LOWLANDS. This region has the advantage of excellent port, railroad, airport, and highway facilities. It also has hydroelectric, thermal electric, and nuclear power, and oil and gas piped from western Canada.

At Beauharnois southwest of Montreal, hydroelectricity has been developed from the rapids in the St. Lawrence River. This is one of the largest power plants in Canada. It produces about 1 700 000 kW. Much of the electricity generated at Beauharnois helps to operate manufacturing industries in the Montreal area including a large aluminum smelter nearby.

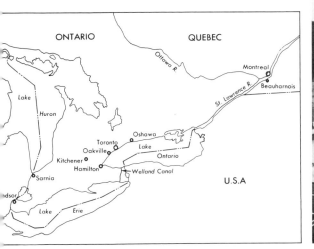

Ships going through the locks in the Welland Canal. The ships in the middleground of the photo have just entered the left lock. Water will be released and the ships will be lowered until they are level with the water in the next lower lock. Then the gates will be opened and the ships will move into the next lock where the same process will be repeated.

Ships must go through a total of eight locks in moving through the Welland Canal.

The Welland Canal is only one part of the St. Lawrence Seaway. The sketch below shows the various levels of the Great Lakes and the location of the locks. How many metres does a ship have to climb in going from Montreal to Lake Superior? How much of this is accounted for by the St. Lawrence and Welland locks? How many metres per kilometre does a boat climb in going from Lake Ontario to Lake Erie?

From information in this book and other geography reference books answer the following questions about the St. Lawrence Seaway:

1. Why did the completion of the St. Lawrence section of the Seaway take some port business away from Montreal and Halifax?
2. In what way has the St. Lawrence Seaway added to the importance of Toronto as a port?
3. Why is the St. Lawrence Seaway very important to Hamilton?
4. In what ways has the St. Lawrence Seaway added to the development of the manufacturing industry in the Great Lakes-St. Lawrence Lowlands?
5. Why are most of the ships using the St. Lawrence Seaway lake freighters instead of ocean freighters?
6. In what way has the St. Lawrence Seaway reduced the cost of moving wheat to overseas markets?
7. In what way did the St. Lawrence Seaway help to make it possible to develop the Quebec-Labrador iron deposits on a large scale?

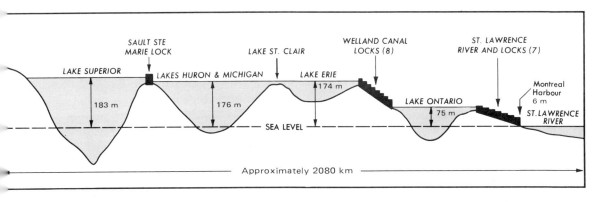

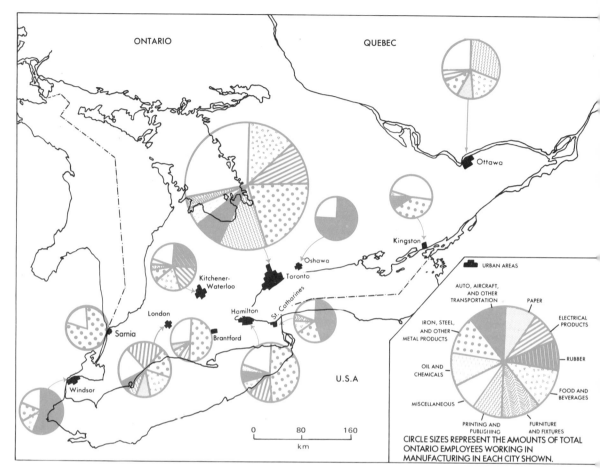

Figure 11.4 TYPES OF MANUFACTURING IN SELECTED CITIES IN SOUTHERN ONTARIO. In common with the rest of the Great Lakes-St. Lawrence Lowlands, Southern Ontario has a combination of all the important locational factors for the manufacturing industry. Therefore, most cities have a wide variety of manufacturing industries.

(4) The large population provides both a market for products and a supply of labour.
(5) All forms of transportation are available: the St.Lawrence Seaway, which connects the area with Europe and with the industrial heart of North America; a network of railroads and excellent highways; and airports, both large and small.
(6) There is an abundant supply of fresh water in the Great Lakes. Some industries, such as the iron and steel industry, require vast amounts of water.
(7) The primary industries provide many of the products required by secondary manufacturing. For example, the iron and steel industry at Hamilton provides the steel needed for the important automobile and farm machinery industries.

(8) The Great Lakes-St. Lawrence Lowlands is a part of the major manufacturing belt of North America. This closeness to the United States has encouraged American companies to establish branch plants in the area.

The two largest cities, Montreal and Toronto, are the most important manufacturing centres of the Great Lakes-St. Lawrence Lowlands. Together, they account for over one-third of all Canadian manufacturing. However, manufacturing is important in all the cities scattered throughout the region. Most cities have such a wide variety of manufacturing that it is difficult to remember what is produced where (Figure 11.4). A few places are well known for their specialization in certain products.

Montreal produces most of Canada's clothing. Toronto is the major publishing centre of Canada. Hamilton produces most of Canada's iron and steel. Sarnia is important for oil refining and for the chemical industry based on petroleum piped from the West. Kitchener is well known for its rubber companies and its meat-packing industries. Windsor, Oshawa, and Oakville are known as automobile cities (see sketch map on page 401).

The Prairie Provinces

As in the Atlantic Provinces, the most important manufacturing industries in the Prairie cities are based upon the processing of raw materials found in the region. The leading industries are oil refineries, meat-packing plants, butter and cheese factories, and flour mills. A number of petrochemical industries have developed in Alberta based on the petroleum resources there.

A large oil refinery beside the North Saskatchewan River at Edmonton, Alberta. Petroleum supplies and equipment are manufactured in or distributed from this city. Why is Edmonton the logical centre for the oil and gas industry in the Prairie Provinces?

Winnipeg is western Canada's most important meat-packing centre. Plants of two of the leading meat-packing companies in Canada are shown in this photograph. What use is made of the long buildings and pens behind the packing plants? What factors have led to the importance of the meat-packing industry in Winnipeg? What trend in Prairie agriculture might lead to further growth of the meat-packing industry in Winnipeg?

Secondary manufacturing in the Prairies is not very important, compared to that of Ontario and Quebec. However, all of the larger cities have diversified manufacturing that includes candy factories, bakeries, dairies, and sheet-metal, woodworking and other miscellaneous manufacturing plants.

Because of its location as "Gateway to the West" and its excellent transportation facilities, Winnipeg is the most important manufacturing city in the Prairies. With a population of half a million, it has more diversified manufacturing than the other Prairie cities. Winnipeg has a number of large flour mills and is the most important meat-packing centre in the West. Besides the primary manufacturing products, Winnipeg makes railway cars, buses, textiles, clothing, and books, as well as various wood and metal products.

British Columbia

Manufacturing in British Columbia is dominated by the forest industry. Lumber, pulp, paper, plywood, and other wood products make up about forty percent of the province's manufacturing production. Lumber, as well as plywood made from the giant red cedars and Douglas firs, is shipped

A newsprint mill in the town of Powell River, B.C. Pulp logs come from the nearby forested slopes of the Coast Mountains and are fed into the mill from booms (A). Fresh water needed in the processing of the newsprint is piped from the reservoir behind a dam (B) across the Powell River. Chemical wastes are dumped into the mouth of the river (C) so that the current will carry them out to sea. From the warehouses (D), the newsprint can be transported by ship to local or foreign markets. A section of this town is partially shrouded in smog produced from the smoking chimneys of the mill.

all across Canada, to the United States, and to many other parts of the world.

Processing plants based on local agricultural production supply a large part of the food consumed in British Columbia. Fish-processing plants, meat-packing plants, cheese and butter factories, as well as fruit and vegetable canneries, are all important.

The refining of minerals and petroleum is another important industry. The petroleum is piped from the Edmonton and Peace River districts, whereas the minerals are found in various parts of the mountainous terrain of the province. Trail, on the Columbia River, has one of the largest lead and zinc smelters in the world. Aluminum smelting at Kitimat is based upon the availability of a vast quantity of cheap hydroelectric power.

Because the local market is so small, there is very little secondary manufacturing other than the manufacture of wood products, some iron and steel products, and clothing. Most of the secondary manufactured products must be imported from Ontario and Quebec, or other countries.

SOME SELECTED INDUSTRIES

There are so many manufacturing industries in Canada that it is impossible to describe all of them in one chapter. The following accounts provide some details about a few selected industries. We suggest that you make similar studies of the important manufacturing industries in your region.

The Iron and Steel Industry

Canada ranks twelfth among the steel-producing countries of the world. Canadian iron and steel mills are now producing most of the steel used by the Canadian manufacturers. Only a few specialized steel products that cannot be produced economically in Canada are imported from the United States. A tariff on imports of steel has made it possible for Canadian steelmakers to compete with the giant steel industries of the United States.

The iron and steel industry requires vast quantities of iron ore, coal, and limestone, as well as scrap iron and steel. It also uses huge amounts

Figure 11.5 MAJOR IRON AND STEEL CENTRES. This map shows only the major integrated iron and steelmaking places. Many other centres make steel from iron produced in these places. Nanticoke and Contrecoeur are two of the newest centres in Canada.

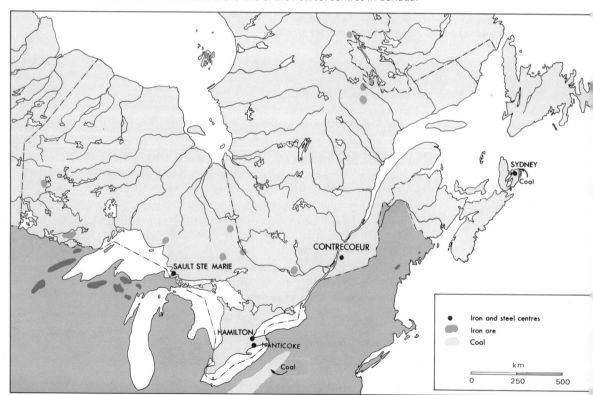

of water for cooling purposes. For this reason, most iron and steel industries are located on navigable waterways in order to bring the raw materials together as cheaply as possible. It is also important for the iron and steel industry to be located close to the large urban centres where the steel products are used (Figure 11.5).

How Iron and Steel are Made

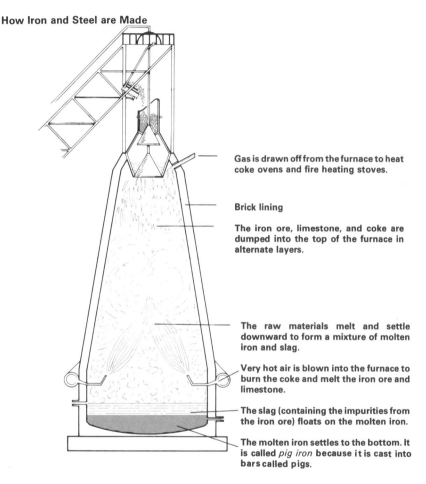

Gas is drawn off from the furnace to heat coke ovens and fire heating stoves.

Brick lining

The iron ore, limestone, and coke are dumped into the top of the furnace in alternate layers.

The raw materials melt and settle downward to form a mixture of molten iron and slag.

Very hot air is blown into the furnace to burn the coke and melt the iron ore and limestone.

The slag (containing the impurities from the iron ore) floats on the molten iron.

The molten iron settles to the bottom. It is called *pig iron* because it is cast into bars called pigs.

A blast furnace is a huge steel cylinder lined with heat resistant bricks. Some blast furnaces are as tall as a 15-storey apartment building. The furnace gets its name from the steady blast of very hot air (1925°C) that is forced into the bottom of the furnace.

The raw materials needed for making iron are *iron ore, limestone,* and *coal.* These raw materials are processed before they are put into the blast furnace. The iron ore is crushed and the rock waste is removed to make the ore concentrated. The limestone is crushed also. The coal is converted into *coke* by heating it in a tall coke oven. It takes approximately 1.8 t of iron ore, 0.7 t of coke, and 0.5 t of limestone to produce one tonne of iron.

The iron ore, coke and limestone are dumped into the top of the blast furnace. The blast of hot air causes the coke to burn. The heat produced by the burning coke melts the iron ore and limestone. The impurities in the iron ore mix with the melted limestone. This mixture of melted limestone and the iron ore impurities is called *slag* and floats on top of the melted iron. The molten iron then trickles down to the lowest part of the furnace. Every four or five hours, molten iron is drawn off or *tapped*. When the furnace is tapped, a white hot stream of iron pours out of the furnace with a shower of sparks. The iron made in blast furnaces is called pig iron because the iron is cast into bars called *pigs* if it is to be shipped any distance. Some pig iron is sent to foundries where cast iron is made. However, most of the iron produced today is not cast into pigs, but is carried in its molten form to a nearby steelmaking furnace.

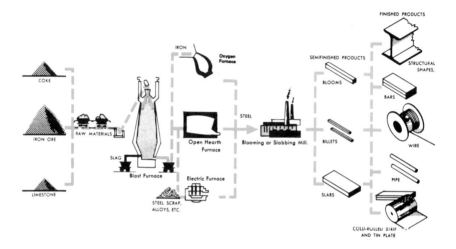

Oxygen Furnace

Steel is a mixture of iron and small amounts of other minerals (known as *ferroalloys*) such as nickel, chromium, or manganese. There is a different recipe for every kind of steel that is wanted.

Although there are several kinds of steelmaking furnaces, the *basic-oxygen* furnace is most commonly used by Canada's largest steel companies. A steel plant usually has a number of basic-oxygen furnaces standing side by side in one long building. Each basic-oxygen furnace is a hollow, jug-shaped structure which can be tilted on an axle. It gets its name from the high-speed injection of almost pure oxygen at the top of the furnace.

Molten pig iron, scrap iron and steel, and high grade iron ore are placed in the furnace. Flames from the burning of the oxygen raise the temperature in the furnace to around 1650°C. In less than half an hour, the furnace is tipped to first remove the slag and then the molten steel, which pours into a huge bucket called a ladle. Alloy materials (nickel, chromium, manganese, and others) are added at this time.

The ladle is moved by a crane and the steel is poured into molds to harden into *ingots*. These ingots vary in size from 3.6 to 16.4 t.

The ingots are taken to *soaking pits* where they are heated until they are white hot. They then go to the *rolling mills* where the ingots are rolled into different shaped bars or plates of steel.

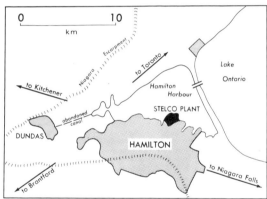

The Steel Company of Canada (STELCO) plant at Hamilton occupies a largely artificial peninsula that juts into Hamilton Harbour. An elevated expressway has been built on the sandbar that separates Hamilton Harbour from Lake Ontario. Ships that deliver iron ore, scrap iron, coal, and limestone to the plant must pass under this span. These raw materials are stockpiled (A) where they are readily available for use in the coking ovens (B) and blast furnaces (C). Because the steel company is hampered from expanding on three sides by other industries and by the city of Hamilton, land is being reclaimed from Hamilton Harbour. What locational advantages do the iron and steel mills at Hamilton have in terms of raw materials and market?

HAMILTON

Hamilton is the most important iron and steel centre in Canada. Two large companies at Hamilton, the Steel Company of Canada (STELCO) and the Dominion Foundry and Steel Company (DOFASCO), produce approximately two-thirds of the steel made in Canada.

Hamilton lies almost midway between the required coal and iron resources. The iron ore used at Hamilton comes from iron mines in the United States just south of Lake Superior, from Wabush, Labrador City, and Bristol in Quebec, and from Temagami in Northern Ontario. For many years the Hamilton iron and steel industry obtained most of its iron ore from the Lake Superior district in the United States, but in the 1960's it began turning to Canadian ore, until in the early 1970's most of the ore came from Canadian sources. (See Figure 10.8.)

Much of the coal used in Hamilton's steel industry is also imported. It is transported from the Appalachian region of the United States to Lake

Erie ports by rail and then shipped to Hamilton by boat. Recently, the Hamilton steel mills have begun increasing their use of coal from western Canada, despite the higher cost of transportation.

Scrap iron and steel are collected for the Hamilton steel industry in the many cities and towns of Southern Ontario. Limestone is obtained from quarries nearby, and unlimited supplies of water are available from Lake Ontario.

Hamilton steel mills are conveniently located in the heart of the Canadian manufacturing belt. Over eighty percent of the steel is consumed in the Great Lakes-St. Lawrence Lowlands, with large amounts of it being used by the automobile and farm machinery industries of Southern Ontario.

STELCO has built a new steel plant at Nanticoke on the Lake Erie shore just south of Hamilton. Nanticoke has locational advantages similar to those of Hamilton. It has the added advantage of being one of the most modern and efficient steel plants in the world.

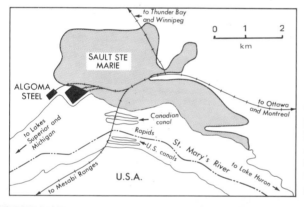

Winter at the Algoma Steel Corporation plant at Sault Ste Marie. What landform region do you see in the background? How can the steel company operate all winter when ice conditions do not allow ships to bring raw materials? List locational advantages and disadvantages of the iron and steel mills at Sault Ste Marie.

SAULT STE MARIE

Sault Ste Marie, home of the Algoma Steel Corporation, is the second largest iron and steel centre in Canada. In the early days, Algoma obtained its iron ore from Michipicoten (northwest of Sault Ste Marie, on the Lake Superior shore). Later, Steep Rock Lake (near Atikokan) provided much of its iron ore. More recently, most of Algoma's requirements have been filled from the iron ore ranges in Minnesota. Coal, too, is imported from the United States.

Sault Ste Marie supplies part of the demand for steel in the Canadian West. It also makes rails for railroads and other special products not manufactured in Hamilton. In this way, Sault Ste Marie is able to sell its steel in Southern Ontario and Quebec. It also exports large amounts of pig iron to the United States.

SYDNEY

For many years the iron and steel mills of Sydney, Nova Scotia, used raw materials close at hand. Coal came from mines a few kilometres away. Iron ore came from Bell Island, Newfoundland, and limestone from the west coast of Newfoundland, both of which were transported cheaply by ocean freighters. Unfortunately, the poor quality of both the iron ore and the coal resulted in high costs of steel production, which along with high transportation costs, made it difficult to compete with Hamilton for markets in the manufacturing belt.

Steel that has been shaped into rails and rods is stored outside a Sydney steel mill. When these products are ready to be shipped to Ontario and Quebec markets, a huge, electric crane can move back and forth over the piles to select and load the steel into railway cars. Why does the Sydney steel mill ship steel wires to Ontario instead of finished products such as nails?

Inside a large steel plant where molten steel is being poured into ingot molds.

Inside a rolling mill. A hot bar of steel passes back and forth over water-cooled rollers while it is shaped into rods and rails.

Major changes occurred in the Sydney iron and steel industry in the 1960's. The iron ore mines on Bell Island, Newfoundland, were closed down, forcing the Sydney iron and steel company to use Quebec-Labrador and other imported iron ore. Shortly after, the iron and steel company decided to shut down its plant because the operation was considered to be uneconomic. The Nova Scotia Government decided to purchase the industry in order to keep the workers employed. Under government ownership, the Sydney Steel Company (SYSCO) has been modernized and made more efficient. The same workforce of 3000 men was able to increase steel production by about forty percent in the early 1970's. The success of the steel industry provided a continuing market for the Sydney coal industry so that it too could remain in operation. These two industries provide employment for about 6000 men in an area where it is difficult to attract other manufacturing industry.

The major handicap of the Sydney iron and steel industry continues to be its distance from market. It is trying to compensate for this by specializing in certain steel products. For example, it is the only steel mill in North America producing double-length rails for railroads. It also concentrates on producing semiprocessed steel—blooms, billets and slabs—which can be transported more cheaply than more finished products. In addition it is strongly promoting export markets so that it can capitalize on its seaport location.

The Aluminum Industry

Canada smelts about one-fifth of the world's aluminum. The only country producing more aluminum than Canada is the United States.

The mass of aluminum makes it a particularly valuable metal. An object made of aluminum has a mass only one-third as much as one of the same size made from steel. Although pure aluminum is soft, it can be made as hard and strong as steel by alloying (mixing) other metals with it. Alloyed aluminum used in airplane wings can withstand pressures of more than 620 100 kPa.

Aluminum has many uses. Because it has a low mass, it is used in airplanes, ships, and trains. Because it is a good conductor of heat, it is used for cooking utensils, and for electric cable. Aluminum sheets make excellent siding and roofing for buildings because they do not rust. Paper-thin sheets of aluminum are used as insulation in the walls of houses because the shiny surface reflects heat well.

The location of aluminum smelters does not depend on the availability of raw materials and closeness to markets, as in the case of iron and steel mills. The most important location factor for the making of aluminum is a supply of vast amounts of cheap electric power.

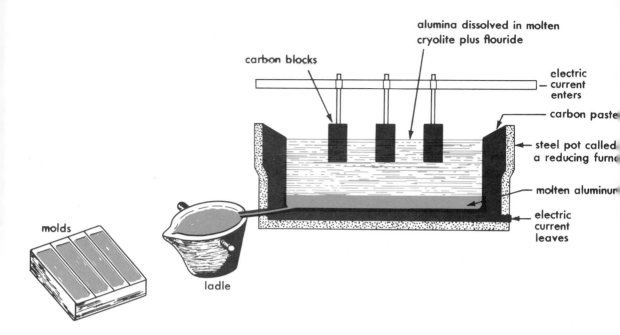

How Aluminum is Made

A reddish coloured ore called bauxite is first ground up and chemically treated to produce a chalky white powder called alumina. The alumina is then dissolved in a huge reducing furnace containing molten cryolite with fluoride added. The reduction furnace is lined with a paste of petroleum coke. Electricity enters the molten cryolite through carbon blocks and leaves via the coke lining. The electric current causes red-hot aluminum to sink to the bottom. The molten aluminum is siphoned into ladles and poured into molds to harden into aluminum ingots. The aluminum ingots are shipped to fabricating plants that produce aluminum sheets, rods, and other forms used by manufacturers.

Long, thin sheets of aluminum are produced from hot ingots that are passed through heavy rollers. Coils of sheet aluminum can then be shipped as easily as ingots.

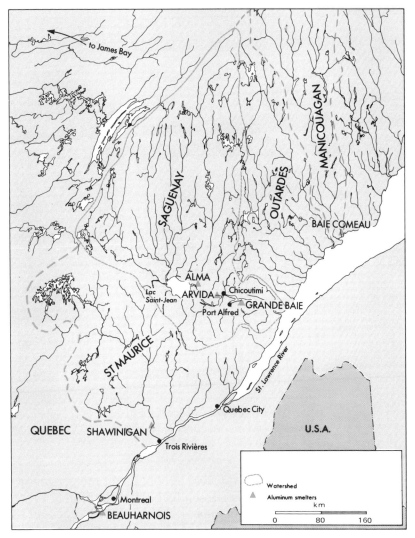

Figure 11.6 THE ALUMINUM INDUSTRY IN QUEBEC. The availability of hydroelectric power led to the establishment of large aluminum smelters in Quebec. The imported raw materials (bauxite, petroleum coke, fluorspar, and cryolite) are transported by ocean freighters.

ALUMINUM PRODUCTION IN QUEBEC

It was the potential hydroelectric power that led to the establishment of Canada's first aluminum plant in the St Maurice Valley. The St Maurice River plunges down 400 m in a series of waterfalls and rapids from the Canadian Shield to the St. Lawrence River. In 1900, the first plant was located beside the falls at Shawinigan, about 50 km north of Trois Rivières (Figure 11.6).

The cheap hydroelectric power of the St Maurice Valley soon attracted many other industries. By about 1925, the St Maurice hydroelectric power could no longer supply the demands of both the aluminum smelters and all the other industry. The aluminum industry began to look for other undeveloped areas where large amounts of electric power could be produced.

This search led to hydroelectric development of the upper Saguenay Valley, which was ideal for aluminum smelting. Dams have been built to

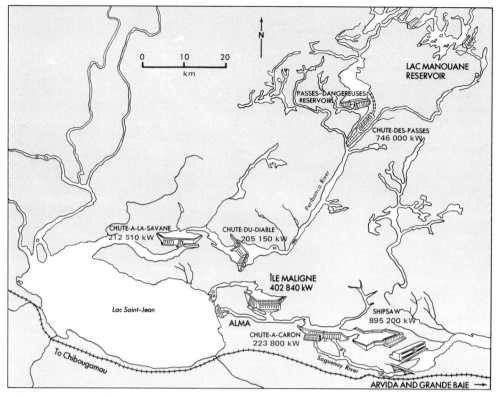

Figure 11.7 A STAIRWAY OF WATERS. Dams and powerhouses on the Saguenay River and its tributaries provide many thousands of kilowatts of hydroelectric power to run the aluminum smelters located at Alma, Arvida, and Grande Baie.

create a giant stairway of waters, commencing at the upstream reservoirs some 425 m above sea level. The Peribonca River has three hydroelectric plants producing nearly 1 200 000 kW. Lac Saint-Jean itself acts as a giant reservoir to regulate the flow of water. From Lac Saint-Jean to Shipsaw are another three hydroelectric plants producing over 1 500 000 kW (Figure 11.7).

It is important that the lower part of the Saguenay is navigable, for this permits ocean freighters to dock at Port Alfred (Grande Baie) only 32 km from Arvida. Thus, the raw materials and the aluminum ingots can be transported cheaply by water. Aluminum smelters have been built at Arvida, Alma, and most recently at Grande Baie. These three smelters produce almost 500 000 t of aluminum a year.

The smelting of aluminum requires that raw materials be brought together from different parts of the world. Aluminum ore, a clay material called *bauxite,* is imported from Guyana (formerly British Guiana). Petroleum coke is obtained mostly from the United States. Fluorspar (from which fluoride is made) comes from Newfoundland, and cryolite comes from Greenland.

The development of the aluminum industry in the upper Saguenay Valley has led to an urban population of over 250 000 in the area. The abundance of hydroelectric power and the timber resources of the area

The Shipsaw power plant is a complex of dams, river channels, hydro towers and roads. Much of the hydroelectricity produced here is used to make aluminum in Saguenay Valley smelters.

The modern aluminum smelter at Grande Baie on the Saguenay River that was opened in 1981. The 170 000 t smelter is located about 30 km from the Arvida smelter and is powered by hydroelectricity from the Lac Saint-Jean network. Bauxite from Jamaica and Australia, coke, cryolite, and fluorspar are shipped directly to Grande Baie; the aluminum is transported away more easily than from any other smelter in the area.

Harbour facilities of the Aluminum Company of Canada at Port Alfred. At the wharf on the right, cranes are used to lift heavy piles of aluminum ingots into freighters for shipment. On the left, bauxite is unloaded from smaller ships and conveyors take it to waiting railway cars that carry it to smelters. Across what landform region does the Saguenay River flow? In what way does the nature of the landform region play a role in the production of aluminum?

have attracted a number of pulp-and-paper mills, sawmills, and other wood-processing plants.

Other large aluminum smelters in Quebec are found at Baie Comeau and Beauharnois. The Baie Comeau smelter makes use of the huge hydroelectric developments on the Manicouagan and Outardes watersheds which by the early 1970's were producing even more kilowatts than the Saguenay region. The Beauharnois smelter, constructed during World War II, was idle in the first few postwar years. In 1951, when more hydroelectric power became available at Beauharnois, it was again put into operation to meet the increasing world demand for aluminum.

ALUMINUM PRODUCTION IN BRITISH COLUMBIA

In 1951, work was begun on a new aluminum smelter on the coast of British Columbia, 640 km north of Vancouver. As in Quebec, the location of the smelter was determined by the potential hydroelectric power.

The Nechako River, which drained eastward into the Fraser River, was dammed to create a reservoir 240 km long. A 16 km long tunnel was then bored through solid rock to let the water from the reservoir fall 790 m to generators housed in a cave blasted from the base of the mountains at

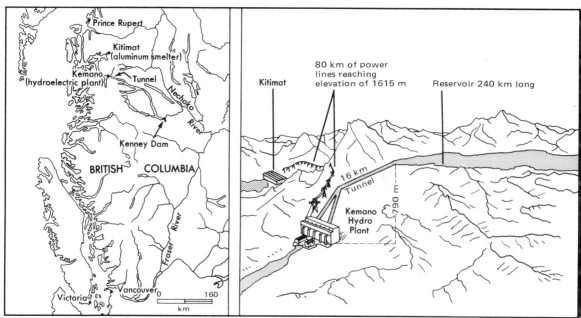

Figure 11.8 KITIMAT — THE ALUMINUM INDUSTRY IN BRITISH COLUMBIA. What are the locational advantages of Kitimat for an aluminum smelter?

The smelter at Kitimat, B.C. is located at the head of a fiord along the Pacific coast. The plant was built on a small, relatively flat piece of land between the steep, wooded slopes in the background and the water of the fiord. Docking facilities allow boats to bring in the alumina from Jamaica and to ship out the finished aluminum ingots. A road at the base of the slope behind the smelter leads to the town of Kitimat to the right of the photograph.

Kemano. Power lines were constructed 80 km over mountainous terrain to Kitimat where the smelter was built. This site permits ocean boats to dock right beside the smelters (Figure 11.8).

The choice of Kitimat for the smelter site also provided flat land for the development of the town to house the workers nearby. At Kitimat, town planners attempted to design a model community. Kitimat has paved streets, well-designed housing subdivisions, modern shopping plazas, and plenty of parks. The subdivisions are arranged so that people do not have to cross major streets to go to schools, parks, or shopping centres. The town is growing in an orderly fashion, according to a plan. When a new subdivision is built, it is serviced from the very start with paved streets, sidewalks, water, sewers, street lights, and playgrounds.

Kitimat obtains its ore from Jamaica, where the bauxite is processed into *alumina* before being shipped via the Panama Canal to British Columbia.

The Automobile Industry

It has been said that automobiles dominate our way of life in Canada. In the more densely populated areas, cars jam our highways and streets. In our cities, a large amount of urban space is occupied by streets and parking lots. In the larger cities, underground parking lots and multistorey parking garages have been built to store cars. Every year, more people are killed by automobile accidents than by some of the most dreaded diseases. Garages and service stations are found along every highway and on as many as three corners at some of the street intersections in our cities and towns. Giant billboards advertising automobiles are in evidence almost everywhere. Thus, it should not be surprising that the making of motor vehicles is one of Canada's leading secondary manufacturing industries.

The Canadian automobile industry produces over one million vehicles a year and employs over 50 000 people. In addition, hundreds of thousands of workers are employed in auto-parts factories and in the selling, servicing, and repairing of new and used cars and trucks. Since 80 percent of a motor vehicle is composed of iron and steel, the automobile industry provides the largest market for Canada's iron and steel industry. In addition, over 60 percent of the production of Canada's rubber industry goes into tires and other rubber parts for motor vehicles. The prosperity of the petroleum and oil refining industry depends largely on the consumption of fuel by automobiles as well as on the manufacture of products used in building highways, expressways, streets, and bridges for the use of automobiles.

DEVELOPMENT OF THE AUTOMOBILE INDUSTRY IN CANADA

The first important automobile factory in Canada was established by the Ford Motor Company at Windsor, in 1904. A group of Canadian businessmen convinced Henry Ford that he should establish a Canadian company in Canada that would have exclusive rights to sell motor vehicles to all parts of the British Empire outside of the British Isles. Because Canada was a member of the British Empire, it enjoyed a tariff advantage in exporting to other Empire countries. This tariff advantage was known as the Empire (and later, the Commonwealth) trade preference.

The McLaughlin Motor Car Company, created in Oshawa in 1907, first used Buick engines in McLaughlin cars, but by 1915 acquired Canadian rights to the Chevrolet. In 1918, the McLaughlin Company became General Motors of Canada.

To protect the infant automobile industry from cheaper imports from the United States, the Canadian Government imposed a 35 percent tariff on motor vehicles (except those from Britain) and on auto parts that could be produced in Canada. This tariff has been adjusted over the years, but

The motor vehicle industry provides a good example of mass production. In this plant, component parts are stored beside an assembly line. As the truck frames stop at points along the assembly line, parts are added by machines or by workers before the vehicle moves farther down the line. In this highly efficient production line, over fifty vehicles an hour can be completed.

was never reduced below 20 percent until the auto trade pact was made with the United States in 1965.

The Empire (Commonwealth) trade preference made it possible for Canada to be an important exporter of motor vehicles. Until 1938, Canada exported more motor vehicles to the Empire and Commonwealth than it imported from the United States. However, after World War II, Canada began importing far more automobiles than it exported. Most of the imports were small automobiles from the United Kingdom, Western Europe and Japan.

Canada has never imported large numbers of motor vehicles from the United States because of the protective tariff. However, a large number of the parts for an automobile have always come from the United States. This situation is to be expected when the major Canadian automobile factories are branch plants of United States automobile companies. By the early 1960's, Canada was importing over a half billion dollars worth of automobile parts from the United States each year. Canadian auto-parts manufacturers could not compete with the low prices of American firms that resulted from mass production for the huge automobile industry of the United States.

In 1962, the Canadian Government started working towards a special trade agreement in the automobile industry between the United States and Canada. The aim was to open the large United States market to Canadian-made automobiles and auto parts. In 1965, the U.S.-Canada auto trade pact agreements were concluded. These agreements meant that both Canadian and American tariffs were abolished on motor vehicles and auto parts at the manufacturer's level. The tariff on United States automobiles entering Canada remained in force for Canadians wishing to buy a car directly from United States car dealers. The United States guaranteed that Canada would obtain a certain minimum portion of the

North American market. Canada agreed to attempt to reduce the price of its automobiles over a period of years to a level closer to that in the United States.

As a direct result of the auto trade pact, the Ford Motor Company built a large automobile plant at Talbotville (between London and St. Thomas) and many new auto parts factories have been built in cities throughout Southern Ontario. As an example of the importance of the auto parts industry to Southern Ontario cities, in the early 1980's Kitchener-Waterloo had twenty firms related to the automobile industry that together employed about 6000 people.

Although Canada has retained a favourable balance of trade with the United States in automobiles, there has been a large deficit of trade in automobile parts. The deficit has been so serious that Canada has considered renegotiating the auto trade pact.

LOCATION OF THE AUTOMOBILE INDUSTRY

Early in this century, the first automobile factory in Canada was built in Windsor as a branch plant just across the river from the parent Ford Company in Detroit. Later, the Chrysler Company also established an automobile factory in Windsor. Thus, for many years, Windsor was known as the automobile centre of Canada. Windsor has a convenient location for the automobile industry because it is so close to the parent companies that supply parts; yet it is not too far from the major markets in Southern Ontario and Quebec.

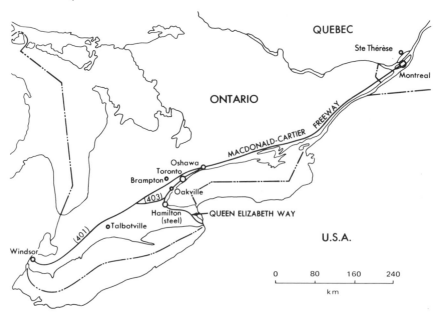

At Oakville, Ontario, the Ford Motor Company operates separate car and truck plants on the same site. The location of this site was determined by its accessibility to a four-lane highway, known as the Queen Elizabeth Way (at left), and to the Canadian National Railway (at right). Using these transportation facilities, Ford obtains the necessary steel parts and other materials to make the vehicles, and the finished products can be shipped easily to nearby markets. Double-decked trucks called auto carriers can carry up to eight cars each. Open two- and three-level railway cars can transport up to fifteen new cars.

Although Windsor is still an important auto centre, it no longer produces the largest percentage of Canadian-made motor vehicles. General Motors has for many years had its major Canadian plant at Oshawa, and in the 1950's, the Ford Motor Company established a large factory at Oakville. Both of these plants have the advantage of being close to the huge Toronto market and to the steel mills at Hamilton. Both of them have expanded their production greatly in recent years. As was mentioned earlier, Ford has also built a huge factory at Talbotville between London and St. Thomas.

General Motors, Ford, and Chrysler, often called "the Big Three", produce over ninety percent of Canada's motor vehicles. Another automobile manufacturer, American Motors, is located at Brampton, Ontario.

Ontario accounts for about ninety percent of Canadian-made motor vehicles. There are several European assembly plants in Nova Scotia and another in Montreal. In both of these cases, all of the parts are imported from Europe and the automobile is merely put together or assembled in Canada. General Motors also has an automobile assembly plant at Ste Thérèse in the Montreal area.

The Furniture Industry

In the early pioneer days, the settlers made most of their own furniture. However, as the people became more prosperous they were no longer satisfied with homemade articles. Those who had been cabinetmakers before coming to Canada began to make furniture for sale. At first, these

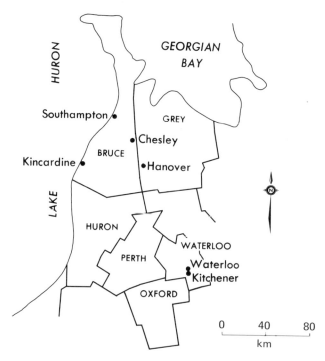

people made the furniture in their own homes, but as the demand for furniture increased, they set up small factories and employed other people. Some of these family firms continued to grow until they became important furniture manufacturers with a national market.

The first furniture centres in Ontario were established in the counties of Waterloo, Oxford, Perth, Huron, Bruce, and Grey. A typical story from this part of Ontario tells of a young German immigrant who walked through the woods with his carpenter's tools to find a job splitting rails at fifty cents a hundred. With the money he earned, he bought some lumber from a local sawmill to make articles of furniture in his spare time. He sold this furniture to the farmers for a little money and some farm produce that he then sold to the local grocery store. With this money, he bought more lumber. He continued this system until he had saved enough money to start a furniture factory and to employ other cabinetmakers. The firm grew quickly and some of his cabinetmakers decided to set up factories of their own. In this way, a furniture-factory town developed. Some of these towns grew into cities like Kitchener and Waterloo. Others, such as Hanover, Chesley, Kincardine, and Southampton, remained as small towns. In these towns, the making of furniture is still the most important industry.

By 1900, furniture manufacturing had spread to a number of towns and cities in different parts of Southern Ontario. The province of Ontario then accounted for 80 percent of the total Canadian furniture production. Quebec came second with 13 percent of the total.

From the early days of the industry, Ontario furniture companies have used skilled craftsmen to produce high-quality furniture. This trend has resulted in a low volume of production and high-priced items. The modern trend is towards mass-produced, lower quality furniture. Some of the historic Ontario firms are modernizing, but many still emphasize the production of quality furniture that requires skilled labour. A number of new furniture factories have been established in Ontario recently, particularly in the Metropolitan Toronto area.

Ontario is losing its predominance in furniture manufacturing to the province of Quebec, where many modern automated plants are opening. By the 1980's, Ontario's share of Canadian furniture production had dropped to below 50 percent, while Quebec's share had climbed to over 30 percent.

The most important furniture centres in Quebec are: Montreal, Ste Thérèse, Beauharnois, Daveluyville, Victoriaville, Arthabaska, Sherbrooke, Kénogami, and Chicoutimi. Because many of these places are small, furniture making is often one of their more important industries.

In the early days, the local trees such as walnut, cherry, chestnut, birch, and maple supplied the furniture industry with wood. As the mixed forests were cleared for farmland, the furniture factories were forced to obtain their lumber from places farther away. Today, Canadian furniture makers import large amounts of oak, mahogany, and walnut lumber from the United States and other countries. They also use a variety of other materials such as plexiglass, plastic, glass, aluminum and other metals in the manufacture of some types of furniture.

PROBLEMS AND PROJECTS

1. Write an essay about a manufacturing industry not described in detail in this chapter. For detailed information and illustrations, write to the headquarters of one or more companies in the industry you have chosen. Do not forget that in geography we are mostly interested in the location factors of capital, raw materials, power, labour, market, and transportation.

2. For some years, plans have been discussed for a large iron and steel plant at Bécancour, on the St. Lawrence River between Montreal and Quebec City. From where would you expect Bécancour to obtain its iron ore and coal? What market advantages would it have?

3. List all the "makes" of automobiles that you find in your area. Find out what companies manufacture these "makes" and where these companies are located. How are the automobiles transported to your community? Local car salesmen will have brochures that will help you. You may also write for information from the major Canadian automobile companies: General Motors at Oshawa, Ontario; Ford at Windsor or Oakville, Ontario; Chrysler in Windsor, Ontario; and American Motors in Brampton, Ontario.

4. From Census data, calculate the percentage of the workforce that is employed in the manufacturing industry in each province. Explain the differences from province to province.

5. The following is a list of abbreviated names or trade names for some important Canadian manufacturing firms. From advertisements in magazines, newspapers, and billboards, find out the full name of the company, the place where it is located, and what it produces. See how many more names your class can add to the list.

STELCO	DOMTAR	GSW	CIL
INCO	3M	GE	SKF
DOFASCO	ALCAN	CAN-CAR	SYSCO

6. For some additional information about manufacturing industries, consult the following sources:

W.B. Braund and W.C. Blake, *Studies in Canadian Economic Geography* (Toronto: McGraw-Hill Ryerson, 1982).
P.G. Burpee, *Aluminum and Power and the Saguenay Valley* (Toronto: Ginn, 1969).
L.R. Carson, *Assembling Automobiles at Oakville* (Toronto: Ginn, 1968).
J. Forrester, *Making Steel in Hamilton* (Toronto: Ginn, 1967).
P. Koch and E. Moore, *Kitchener: A Meat-Packing Centre* (Toronto: Holt, Rinehart and Winston, 1971).
J. Lavallée and E. Moore, *Granby: A Manufacturing Centre* (Toronto: Holt, Rinehart and Winston, 1968).

12 CONSERVATION AND RESOURCE MANAGEMENT

Canada is a vast country with a small population and a huge supply of resources. Because of these features, Canada has not been as much concerned about resource conservation as have countries where the population density is greater and the supply of resources smaller. However, we Canadians are beginning to realize that our resources are not unlimited and that we are rapidly destroying some of our most valuable assets and are ruining much of our environment.

We have polluted streams and lakes, thus destroying important fishing resources and making the water unfit for drinking or recreation purposes. We have destroyed the natural habitat of much wildlife, exterminated some wildlife species, and placed others on the verge of extinction. Indiscriminate use of insecticides has had a chain reaction affecting many species, and perhaps ultimately man. In the North we have carried on mineral exploration and development without much concern for the impact on a fragile environment nor the effect on the native population. We have dammed rivers and diverted waters from one drainage area to another without realizing what ecological damage might result. Much of our good topsoil has been eroded, and many hectares of our best farmland are being covered with highways and sprawling urban development. In our cities we have often created an unpleasant living environment with inadequate open space and recreational facilities so that people want either to live out in the country or at least to flee to recreational resorts on weekends. As a result, urban pressures are being felt for hundreds of kilometres around our major cities. Inadequate planning of recreational resources has led to overcrowded conditions and environmental pollution in our recreational resorts that sometimes is worse than that found in cities. In our large cities we are creating mountains of garbage that are difficult to dispose of, and we are even polluting the very air we breathe with smoke and auto exhaust fumes.

It is obvious that Canadians should be concerned about conservation and resource management. However, conservation does not mean that we should not use our resources; it means the wise and controlled use of our resources. Conservation does not mean hoarding resources for the future that can be used today; it means making the best use of resources in the present without ruining them for the future. Conservation does not mean that we leave nature completely alone; it means working with nature to help provide a better living and way of life for humans. It does not mean preserving all wilderness; it does mean preserving enough wilderness for people to see what our country is like in its natural state.

Because some people use the word conservation to describe the policy of not using resources, modern scientists and planners prefer to use the terms "resource management" or "resource planning." These terms mean planning the use of resources for the benefit of both present and future generations.

There are two basic types of resources: *renewable* and *nonrenewable*. The nonrenewable resources do not re-create themselves. Once we have used them up, they are gone. Minerals are good examples of nonrenewable resources.

Up to the 1970's, Canada did not have to worry about the supply of nonrenewable resources. In fact, Canada was not able to sell all the minerals it could mine, and vast areas had not been carefully explored for minerals. In addition, new technical developments had made some of our resources more useful. For example, the development of nuclear reactors to create electricity made Canada's uranium resources valuable. It was not until the energy crisis in the United States in the early 1970's that Canadians began to be concerned about the long-term supply of energy resources in Canada, particularly petroleum and natural gas. A serious shortage of petroleum is expected before the end of this century. This should force Canadians to cut back drastically on energy consumption; at the same time, Canadians should be developing energy from renewable sources such as the sun, wind, tides, hydrogen (found in water), or from the biomass (vegetation in the form of trees or cultivated crops).

The above Wizard of Id cartoon is reproduced by permission of John Hart and © Field Enterprises, Inc., 1972.

With mounting public pressures to conserve our renewable resources and preserve the quality of our environment, both federal and provincial governments have been making progress in enacting environmental legislation. In 1970, the federal government passed three important environmental bills: the Canada Water Act, the Arctic Waters Pollution Act, and the Northern Inland Waters Act. These three bills have given the federal government significant power over water pollution control in Canada, extending controls to 160 km from the coast in the Arctic. In 1971, the federal government passed a Clean Air Act which gives it similar powers over air pollution. Land use regulations have also been established for the North, which require petroleum companies to carry out exploration and development in ways that minimize ecological damage. Critics of the federal government's activities in the environmental field claim that the legislation is not adequately enforced. In some cases the government has insufficient information about the nature of the environment and the consequences of certain developments either to create adequate legislation or enforce it once it is enacted.

Under the Canadian Constitution, resources are under the jurisdiction of provincial governments. Some provinces have been more active in pollution control than others. The Province of Ontario has a program which forces every municipality to have both primary treatment (removing solids) and secondary treatment (breaking down remaining organic material) of its sewage. If the municipality does not comply, a government agency builds the sewage facility and the municipality pays for the cost over a period of years. Ontario is also beginning a program of tertiary treatment that will take nutrients out of the sewage effluent. Unfortunately, some provinces do not have such strict controls, and there are many municipalities across Canada which have only primary sewage treatment facilities, and many industries and a number of cities which dump their wastes directly into water bodies without any treatment at all. There are several major pollution problems for which Canada has not yet found adequate solutions:

(1) What should be done with chemical liquid wastes?
(2) How can wastes from nuclear thermal plants be stored safely for centuries to come?
(3) How can Canada and the United States co-operatively reduce the pollution that is creating acid rain?

Pollution control is only one aspect of conservation and resource management. The balance of this chapter considers some of the problems of, and strategies for, managing some specific renewable resources.

LOCAL STUDY

1. Are there any farm areas in your region that were abandoned because of soil erosion? Is gullying or wind erosion a problem to farmers? What are the farmers doing to halt soil erosion?
2. Have there been any recent floods that have damaged property? What is being done to prevent such floods?
3. From where do municipalities in your region obtain water? Have any municipalities had to restrict the use of water because of a low supply? Have any farm wells gone dry because of a lowering water table?
4. How far (in kilometres) do you have to travel to go for a hike in a woods? to go to a public beach for swimming? to go camping overnight? Does the public have access to your nearest lakefront?
5. Are there any natural beauty spots in your area that have been destroyed by the building of new highways or housing subdivisions?
6. Is the urban development in your area taking place on good or on poor agricultural land?
7. How pure is the air in your town or city? Is there any anti-air-pollution bylaw?

> 8. How is garbage in your community disposed of? Have there been any pollution problems related to garbage disposal?
> 9. Is the water in the streams and lakes in your area pure enough for swimming and fishing? If not, what is the source of pollution? Is anything being done about the pollution?
> 10. Does your municipality have primary and secondary sewage treatment? What proportion of the municipal sewage is given primary and secondary treatment?
> 11. Is there a conservation agency or environmental organization (e.g., Pollution Probe) in your area? If so, find out what it does.

AGRICULTURAL RESOURCE MANAGEMENT

For generations, Canadians have been told about Canada's vast areas of agricultural land. In the past, this idea of unlimited agricultural land has been supported by large surpluses of farm products. Thus, it comes as a shock to many Canadians that we should now be concerned about our supply of farmland.

There is little doubt that the world demand for food is going to skyrocket in the next few decades. The growth of the world's population has been so rapid that it has been referred to as a "population explosion." Each week, the number of people to be fed in the world increases by more than one million. At least one-third of the world's population is underfed, and these hungry people increasingly are going to look to Canada for food supplies. Will Canada be able to feed its own people and have surpluses for the hungry people of the world by the year 2000? No one has the answer to that question. It is clear, however, that what Canada does now with its farmland will be of great importance to Canadians and to the rest of the world in the future.

Most of Canada's best agricultural land has already been cleared for farming. In fact, many millions of hectares are being farmed that never should have been cleared. These areas, where either the soil is too thin or the slopes are too steep, are not providing a good living for the farmers and are not adding much to Canada's food supply. The greatest promise for increasing Canada's food production lies not in bringing more land into cultivation, but in using greater efficiency in farming methods. However, even good farming methods do not pay off unless there is good farmland available.

In heavy rainstorms, water draining across unprotected fields erodes small gullies. If gullying is allowed to go unchecked, a larger part of the landscape becomes ruined for farming. In this photograph, erosion has been so serious that there is no topsoil remaining and the land can no longer be farmed. Unless the bare slopes are protected, the gully sides will erode back toward the fence in the background.

Soil Erosion

Farmers must constantly work to prevent soil erosion. It takes hundreds of years for nature to create 2.5 cm of soil, but that same soil can be washed away in one rainstorm or blown away in one windstorm. Soil scientists estimate that 25 percent of Canada's farmland has been seriously eroded.

Farmers fight erosion in a number of ways. Steep slopes are taken out of production and are put back into trees or grass. Other slopes are ploughed, cultivated, and planted on the contour (that is, across the slope instead of up and down the slope). In this way, every furrow and crop row acts as a little dam that prevents the water from running down the hill and from creating gullies. The soil is made more porous by ploughing down barnyard manure, straw, and hay crops. The more porous soil lets rain soak in, instead of running off and carrying soil with it. Grain and row crops are often alternated with strips of hay to reduce the chances of gullies starting to form. In the chapter on agriculture, we discussed the way Prairie farmers use windbreaks, summerfallow, strip farming, and special methods of cultivation to combat wind erosion. These methods are available to farmers, but not all of them are used. When they are not, the production of the farmland decreases, and eventually, the land is ruined.

This photograph shows the result of serious wind erosion in the dry belt of Saskatchewan. Violent windstorms have blown away the unanchored topsoil, leaving a desolate landscape dotted with tufts of grass and sagebrush. Note the ripple pattern the wind made in the sand.

Furrows are ploughed on the contour during the reforestation of this sloping land. Small tree seedlings are spaced at intervals along each furrow. The trees have been planted in contour furrows so that any erosion of the slope would be prevented until the tree roots are large enough to anchor the soil themselves.

Strip farming is a method used by farmers to prevent soil erosion on sloping fields. In this photo, corn (light strips) is a row crop which is alternated with hay or grain (dark strips) on a hillside that slopes towards the right middleground. These crops have also been planted on the contour as well as in strips.

Urban Sprawl

In recent years, some Canadians have become concerned about the amount of agricultural land being ruined by urban expansion. Houses, factories, shopping plazas, parking lots, as well as streets and highways, are rapidly using up farmland around our urban centres. Because of the sprawl pattern of growth, cities ruin more land than they actually use. It has been estimated that for every 0.4 ha of land used by cities, another 0.8 ha are taken out of agricultural production.

At the Resources for Tomorrow Conference, in 1961, a geographer forecast that by the year 2000 there would be little agricultural production in the Lower Mainland of British Columbia (The Fraser Delta) and in most of the Great Lakes-St. Lawrence Lowlands. Although these areas represent only about five percent of Canada's total agricultural land, they are the most productive five percent. From the chapter on agriculture, you will recall that these are the areas with the longest growing season, and therefore, they have the greatest diversity of crops and the greatest value of production per hectare.

It is not necessary for urban growth to ruin so much land. Urban and regional planning make it possible to plan the land uses within a region so that there is room for both cities and farms. This kind of soil conservation requires the co-operation of both urban residents and farmers. The

Orchard land on this farm in Southern Ontario has been purchased to develop a new suburban housing project. All the fruit trees have been cut down; the stumps will be removed later. However, as you can see, no trees remain to provide the blossoms for Blossom Heights. The Ontario Government has a policy that requires municipalities to direct urban growth onto poorer agricultural lands where possible. However, the policy frequently is ignored. Both British Columbia and Quebec have set aside all good agricultural land for agricultural uses, but there is much pressure from developers for exceptions.

cities and the surrounding rural municipalities must join forces to create regional land-use plans that ensure that all land in the region will be put to its best use.

WATER RESOURCE MANAGEMENT

Water is one of our most valuable renewable resources. We use so much of it that we take for granted that we will always have an abundant supply. What a shock it would be if no water came out when we turned on the tap!

Water is valuable because it has so many uses. A modern household requires hundreds of litres a day to wash dishes, clothes, and people, to flush toilets, to quench thirst, to water lawns, and so on. In industry, water is used in huge quantities in many manufacturing processes. For instance, it may take as much as 454 600 l of water to make one tonne of paper or one tonne of steel. Water is used by farmers to water their livestock and to irrigate their fields. Falling water can be converted into hydroelectricity. Water channels provide cheap transportation by boat. When water is frozen into ice, it is used for refrigeration purposes. The availability of water determines the amount of wildlife and fish that can live in an area. Water is also much used for recreational activities such as swimming and boating.

On the whole, Canada is well endowed with water resources. A review of the chapter about climate will remind you that most of Canada has a humid climate with over 500 mm of precipitation a year. A glance at a map shows that Canada is laced with rivers and is studded with lakes. The Great Lakes provide one of the greatest sources of fresh water in the world. Canada's annual water runoff, or streamflow, is the third largest in the world, second only to those of the Soviet Union and Brazil. Our per capita annual streamflow is twenty times that of the United States.

Because of the above facts, Canadians are often misled into thinking that our country has unlimited supplies of usable water. This is not true. There are great variations in the amount of usable surface water from region to region. For example, the average annual streamflow in British Columbia is 2.5 m, but in the Prairies and the North, it is only 12.7 cm. Two-thirds of the runoff flows away from the populous regions towards the relatively empty North. In addition, there is great variation in runoff from one year to another, and the reliability of precipitation and resulting runoff is least in the driest part of the country. Thus the amount of water available for use is often much less than precipitation and streamflow records suggest.

As would be expected, the dry Prairies suffer from a lack of water supply. However, some of the most critical water shortages are now being faced by urban areas in regions with a humid climate.

In the spring, meltwater drains from the land so that rivers become swollen torrents and often overflow their banks. Spring flood waters cover farmland and urban centres are sometimes flooded. This photograph shows sandbagging in Winnipeg, Manitoba, under flood waters from the Red River, even though a floodway channel built around the city diverts much of the excess water (see Figure 8.11).

The reason for insufficient water supply in humid regions is poor management of the water resources. Before settlers cleared land in the humid areas of Canada, a large proportion of the rainfall soaked into the ground because the forest-covered land absorbed water like a sponge. The forest also slowed down the melting of snow in the spring. Large swamp areas acted as great reservoirs, storing water until later in the summer. As a result, the streams had an adequate flow of water all summer long. Because the streams and lakes had not been used as a dump for human and industrial wastes, the water was clean and pure enough to drink. In addition to an abundance of pure water in lakes and streams, the water table was high, making it easy to obtain water from shallow wells or natural springs.

The situation is very different today, particularly in the heavily populated areas of the country. Most of the forests have been cut, even from land that is too rocky or too steep for good farmland. Farmland has been drained with ditches and under-tiling, and many of the swamps have been drained. In the cities and towns, because large areas are covered by houses, streets, and sidewalks, most of the precipitation runs

In summer, rivers that had flooded in the spring are no more than mere trickles among the rocks in the riverbed. With so little water, this river cannot be used for water supply, recreation, or the diluting of sewage from cities.

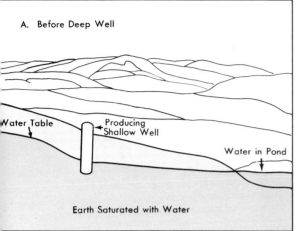 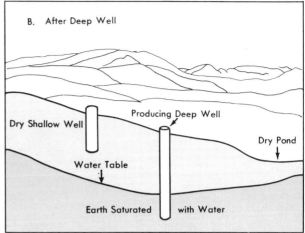

Figure 12.1 THE EFFECT OF A DEEP WELL ON THE WATER TABLE. In diagram A, the water table is high enough to have water in both the shallow well and the pond. The deep well in diagram B has pumped so much water that the water table has dropped. Both the shallow well and the pond are now dry. Who should own the water in the ground? Is it fair for a city to drill deep wells out in the country and make the farmers' wells go dry? What other things, besides deep wells, help to lower the water table? What conservation practices can be used to help keep the water table from dropping?

off into drains. As a result, much of the water from melting snow and spring rains runs rapidly off the land, into the streams and rivers. There are floods in the spring causing great damage, but in the middle of summer, most of the streams and even some of the larger rivers are dried up to trickles of water and a few stagnant pools.

Ground water supplies have also dwindled. Not only is there less water soaking into the ground, but there is more and more water being taken out of the ground. In some areas, farmers are using huge amounts of water for irrigation. In other areas, cities are drilling more and deeper wells farther and farther out into the country. The effect of all of this use of water is a lowering of the water table which, in turn, reduces the flow of spring-fed streams (Figure 12.1).

Even with the reduced recharging of ground water supplies and the increased use of water, our water supply problem would not be as great if we did not dump our untreated wastes into lakes and streams. As indicated at the beginning of this chapter, many municipalities and industries in Canada dump their wastes either partially or wholly untreated into bodies of water. As long as there was enough flow of water, the lakes and streams could sufficiently dilute the sewage. However, in recent years, the flow of water has been decreasing while the amount of sewage has been increasing.

Even farmers have unwittingly been responsible for water pollution. The heavy application of artificial fertilizer has resulted in much better crops, but it has also resulted in more nutrients being washed into the

Public access to many good beach areas in Canada is discouraged when cottage owners attempt to keep the beaches for their own use.

Industrial wastes dumped into the water at Thunder Bay have formed an unsightly scum along the harbour. Tugboats will have to churn through it in the hope that currents will take it farther out into Lake Superior.

Parts of the Lake Erie shoreline in Ontario have been affected by polluted water. In some beach areas, swimming is no longer enjoyable when masses of green algae are washed ashore. Algae is a water plant that thrives in polluted water and uses up so much of the oxygen that some species of fish will suffocate. Dead fish floating in the water or decaying on the beach are a common sight.

streams and lakes. These nutrients, added to those from urban sewage, promote the growth of algae. These single-celled plants sometimes become so thick that they form a slimy green mass on the surface of the water. They make the water unfit for swimming and they take oxygen out of the water, thereby ruining it for fish.

The net result of water mismanagement has been floods in spring, drought and inadequate water in summer, and polluted streams and lakes that are unsuitable for water supply and unfit for recreational purposes such as swimming and fishing. Even some of the Great Lakes are becoming polluted. Every summer, beaches along Lake Erie and Lake Ontario have signs that are posted to warn of polluted water. Canada has committed itself to reduce substantially the wastes it is adding to the Great Lakes, but without the active co-operation of the United States, it will not be possible to make much improvement of their water quality.

There are many approaches to water resource management. The following is only a partial list of specific conservation measures:

(1) Reforest areas of poor farmland.
(2) Encourage farmers to manage their land according to sound conservation principles.
(3) Build a number of small dams on the headwaters of river tributaries.
(4) Build major dams on the main river branches.
(5) Do not permit housing subdivisions to use septic tanks.
(6) Direct urban development away from major water recharge areas, and from swamp and forest land.
(7) Treat all city sewage and industrial wastes before they are dumped into the water. Primary and secondary treatment is essential; tertiary treatment is desirable.
(8) Encourage reduction of water use and water recycling by various means, including the raising of rates charged for water.
(9) Restrict further urban growth on watersheds in which the demand for water and the amount of effluent being discharged are too great for the volume of streamflow.

Various water resource management methods must be co-ordinated within each watershed. Certain standards of water quality and volume of streamflow have to be agreed upon and then all of the conservation practices must aim at those objectives. Dams and reservoirs are not a cure-all in themselves. If the volume of flow is too small, a dam and reservoir will increase the amount of evaporation (because of the larger water surface), will help to warm up the water (fish require relatively cool water), and will reduce the amount of downstream flow. In the case of large artificial reservoirs, a great deal of land including farmland, woods, and the habitat of wild animals is ruined by flooding. Also dams have a relatively short life-time; in areas where good soil conservation principles are not practised, a dam may silt up completely in twenty or thirty years.

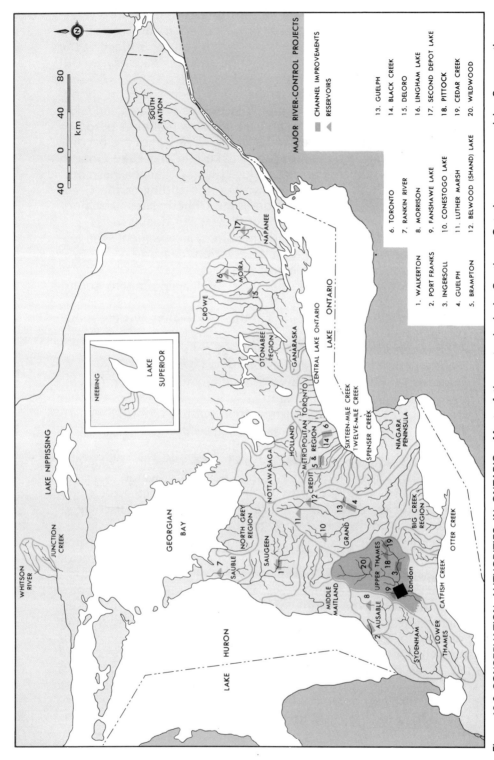

Figure 12.2 CONSERVATION AUTHORITIES IN ONTARIO. Most of the watersheds in Southern Ontario are managed by Conservation Authorities. These authorities build dams, improve channels, plant trees, preserve swampland and wilderness areas, and develop recreational parks.

In heavily populated areas water resources must be managed for multiple purposes. Water supply is required for industrial and domestic purposes in the cities and for irrigation and home and livestock on the farms. The volume of flow in the stream must be maintained at a certain level to dilute the sewage effluent of the cities. The crest of the spring runoff must be held back to avoid serious damage from flooding. The water recreational demands of the population should be met as close to the cities as possible.

It is not easy to meet all of these needs. For example, if a dam is built to regulate streamflow (prevent floods in spring and increase streamflow in summer), it is often also used as a recreational resource. Once beaches are established and cottages and docks are built along the shore, there is public pressure to keep the lake level relatively constant. However, to fulfill the purposes of streamflow regulation, the reservoir must be drained in the early spring to provide maximum water storage in order to prevent a flood, and must be lowered again gradually during the summer to increase the downstream flow of water.

As you can see, managing water resources is a complex business. Establishing water management regions and then getting the various users in the watersheds to agree on the priorities of water use is a difficult political process that usually takes considerable time. Some conservation measures, such as large dams, require large investments of capital; so much that local municipalities cannot afford it, but require assistance from both the federal and provincial governments. Other conservation practices are slow to show results. It takes many decades to grow a forest. It will take many years before all municipalities install adequate sewage treatment plants, and perhaps even longer before urban planning on a regional basis is effective across the whole country so that urban growth can be directed and controlled in harmony with water resource management objectives. However, if all levels of government put water conservation high on their list of priorities, we may still be able to manage our water resources in such a way that we avoid disaster.

The Upper Thames Valley Conservation Authority

Because it is densely populated, Southern Ontario has many problems with flooding, water supply, and water pollution. To help solve these problems, the provincial government has passed legislation that permits municipalities located within riversheds to organize a conservation authority. The conservation authority then receives financial assistance from the federal and provincial governments to carry out conservation projects (Figure 12.2).

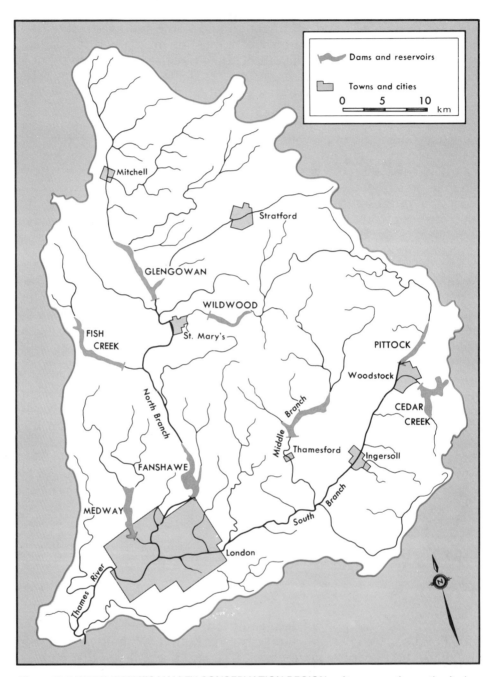

Figure 12.3 UPPER THAMES VALLEY CONSERVATION REGION. A conservation authority has made it possible for the municipalities in this region to manage their water on a watershed basis. The dams and resulting lakes prevent flooding in the spring and increase streamflow in the summer. The lakes also provide recreation facilities.

Water pours from the reservoir behind the flood control gates at Fanshawe Dam near London, Ontario. After heavy rains or when snow melts in the spring, the gates are lowered to reduce the flow of water downstream. At time of low water, the gates are opened (as shown in the photograph) to provide an increased flow of water.

The Upper Thames Valley Conservation Authority has been one of the most active of these conservation organizations. In the past, several towns and cities in the Thames watershed have suffered from spring floods. In 1937, the city of London had a devastating flood, with some parts of the city being under two metres of water. Therefore, as soon as the Conservation Authorities Act was passed, a group of citizens organized the Upper Thames Valley Conservation Authority.

The first task of the Authority was to conduct a thorough survey of the rivershed. Precipitation, runoff, streamflow, land use, water consumption, and water quality were all analysed, and a plan of action was drawn up. The second task was to inform the citizens of the region about the benefits of the various conservation proposals. Without public support, no planning or conservation program can succeed.

Since it was formed in 1946, the Upper Thames Valley Conservation Authority has undertaken many projects. Several large dams have been built to provide flood control and streamflow regulation (Figure 12.3). The lakes behind these dams have been developed for recreational purposes. For example, the Lake Fanshawe area near London has facilities for

The permanent lake created behind Fanshawe Dam has been developed into an important recreational area. Here you see part of the beach area at Fanshawe Lake with a picnicking area in the background. Other parts of the lake are suited to boating and fishing.

picnicking, swimming, boating, fishing, hiking, and golfing. Smaller dams have been built on the tributary creeks. At Ingersoll and Mitchell, the channel has been deepened to prevent flooding in those towns.

The Conservation Authority has been active in other conservation projects. It has planted millions of trees on both public and private land; it has helped farmers build ponds; and it has conducted education programs on soil conservation, woodlot improvement, pollution control, and wildlife management.

Despite these conservation measures, the city of London grew so rapidly that it still had major problems relating to water. There was not enough water in the Thames River to dilute the sewage of the city. The city had to improve old sewage treatment plants and had to build new ones, and factories were required to treat their wastes before dumping them into the sewers.

In the past, London depended on deep wells for its water supply. As the city grew, wells were drilled deeper and at some distance from the city. These new wells lowered the water table and the farmers complained that their wells were going dry.

London either had to find new sources of water or restrict its growth. It decided to ask the Ontario Water Resources Commission (a provincial government agency) to build a pipeline to bring water from Lake Huron. The citizens of London are to repay the Commission over a long period of time. Another pipeline was constructed from Lake Erie to Talbotville when the Ford plant moved in, and it has been suggested that the two pipelines be connected. These pipelines will assure the London area of an adequate supply of water for many years to come.

WATER TRANSFER

Because some parts of North America have apparent surpluses of water, whereas other parts suffer from a deficiency of water, a number of schemes have been proposed to transfer water from one drainage basin to another. The best measure of available fresh water resources is the volume of streamflow. Canada has some of the longest rivers in the world: the Mackenzie, the St. Lawrence, the Nelson, the Yukon, and the Columbia, some of which are shared with the United States. These and many other rivers have a high volume of streamflow. In fact, it is estimated that Canada has the third largest streamflow in the world. Canada's per capita supplies of available fresh water are twenty times those of the United States. However, a large amount of this available water flows through sparsely populated areas into the Arctic Ocean and Hudson Bay. Therefore, it is not surprising that there have been numerous proposals for the diversion of water from the Arctic and Hudson Bay river basins to regions of the continent that have water deficiencies.

Small flood control gates on a tributary river. The gates only close when water levels in the river threaten to flood land downstream.

River channel improvement. Along some streams, the channels have been deepened and widened to prevent flooding. The cement and grass-covered banks help prevent soil erosion.

A farm pond. Conservation authorities encourage farmers to build ponds to store water for dry periods. A good farm pond is lined with clay which prevents water from draining away.

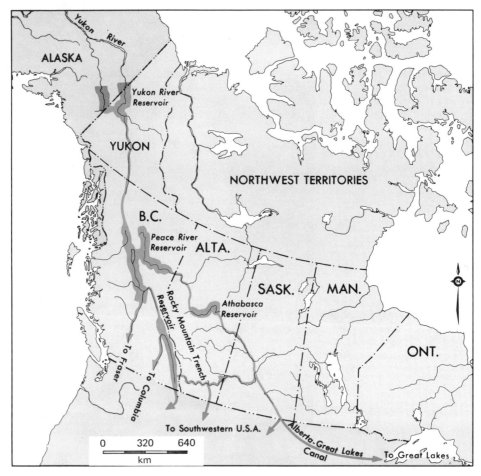

Figure 12.4 NAWAPA — AN ENGINEER'S DREAM, AN ECOLOGIST'S NIGHTMARE. A United States engineering firm has presented a plan for providing North America with water from the Arctic. The scheme is known as NAWAPA (North American Water and Power Alliance). It proposes to build a series of dams and reservoirs that would divert water from Alaska and the Yukon to the south. Some of the water would be released into the Fraser and Columbia River basins, some would be taken by a series of rivers and canals as far south as California, and some would be carried through an Alberta-Great Lakes Canal to the Great Lakes. From the Great Lakes, water could be diverted into the Mississippi River system.

This ambitious scheme would not only provide water for a large part of North America, but would also help to create vast amounts of hydroelectric power, and would enable ships to penetrate into the Interior Plains by way of the Alberta-Great Lakes Canal. This scheme, which has no official support from any government, is considered by Canada to be economically unfeasible and ecologically disastrous.

One scheme proposed at one time by Alberta, the Prairie River Improvement Management and Evaluation (PRIME), would transfer water from the headwaters of streams with water surpluses into river basins with water deficiencies in order to enhance the water supply and pollution dilution for cities in southern Alberta and Saskatchewan. The proposal would start out with small transfers but ultimately would involve massive transfers of water from the Peace and Athabasca Rivers into the Saskatchewan River system. Another scheme proposes diverting water into the

Great Lakes from several rivers now flowing into James Bay. In that way, the Great Lakes could act as huge reservoirs from which nearby municipalities could pipe water. It would also permit large amounts of water to be diverted to the Mississippi by way of Chicago. (By ruling of the International Joint Commission which has control over use of water in the Great Lakes, Chicago is permitted to draw only a limited amount of water from Lake Michigan to help dilute its sewage effluent which goes down the Mississippi by way of a connecting canal.)

The most ambitious scheme of all has been proposed by a United States engineering company. It suggests that water be brought all the way from Alaska and the Yukon to different parts of Canada and the United States. This scheme would reverse the flow of the Yukon River, turn the Rocky Mountain Trench into a 800 km long reservoir, and construct a canal from the Rockies to the Great Lakes. The whole project would take forty years and 100 billion dollars to build (Figure 12.4). The major benefits of this scheme would accrue to the United States. It makes no provision for paying Canada for any of the water, nor for compensating Canada for the vast areas of Canadian land that would be flooded. The huge reservoirs would ruin huge areas of timber land, mining country, and wildlife habitat and would disrupt a number of settlements. Canadian ecologists claim that ecological damage downstream from the giant dams would be disastrous. The scheme is so ill-conceived that it has no official support from the United States Government and Canada has completely rejected the whole idea.

Because of the many social, economic, and ecological implications of water diversion, only a few major water transfers from one drainage basin to another have taken place in Canada. Some minor diversions have been made in the headwaters of streams in Alberta, and in Ontario there have been diversions made from headwaters of streams flowing into Hudson Bay, into the Great Lakes drainage basin in order to increase the production of hydroelectric power.

As far as major diversions of Canadian water to the United States are concerned, the Canadian Government has made its position clear:

(1) Canadian waters are not a continental resource; they are as Canadian as any other resource found within our national boundaries.
(2) The United States has never officially requested Canada to consider selling water. The United States has its own surpluses of water which it could divert to deficient areas, and it could increase its usable supply of water by enforcing stricter conservation and pollution control measures.
(3) Canada will not sell any water to the United States until it is assured that its own future requirements can be met.
(4) Both the federal government and the provincial government concerned must agree before any negotiations are begun which might lead to the sale of water to the United States.

USING WATER FOR HYDROELECTRIC POWER

Canada has many rivers with year-round large volumes of water with many falls and rapids and steep-sided valleys that make them ideal for developing hydroelectric power. Thus Canada produces vast amounts of electricity from water resources and has the potential to develop much more in the future. With new technology making it economic to transmit electricity many hundreds of kilometres, it is no longer necessary to build hydroelectric plants near the market and, so, huge new hydroelectric developments are taking place, such as at Churchill Falls in Labrador, on La Grande River in Quebec, and on the Nelson River in Manitoba. Although there are usually heavy capital costs in constructing hydroelectric plants, these are offset by very low maintenance and operation costs. Hydroelectric plants can better adjust to varying peak demands for electricity than can thermal electric plants. In some cases hydroelectricity can be developed in conjunction with dams that are built to facilitate transportation (e.g., St. Lawrence Seaway) or to prevent flooding and regulate streamflow (e.g., Columbia River Project). The production of hydroelectricity does not consume any of the water resource; it merely converts the energy of the falling water into electricity and then the water is released to continue downstream with the quality little affected. However, as is illustrated in some of the following pages, there are certain negative environmental effects resulting from impounding huge volumes of water for hydroelectric purposes.

The Columbia River Project

The Columbia River has tremendous hydroelectric power potential. Its headwaters are in the Rocky Mountain Trench, only a few kilometres from the source of the Kootenay River (Figure 12.5). It first flows northwest, for about 320 km, until it is able to make its way around the northern end of the Purcell and Selkirk Mountains from where it flows south into the Arrow Lakes. Just before it reaches the forty-ninth parallel, the Columbia is joined by the Kootenay River. Not far below the United States border is the Grand Coulee Dam that backs the water of the Columbia into a lake 240 km long. At this dam is located the first of a series of hydroelectric plants that have been built on the Columbia River between the Canadian border and the Pacific Ocean. The total fall of the Columbia River from its source to its mouth is 808 m — half of which occurs in British Columbia. The full power potential of the Columbia and its tributaries is estimated at 40 million kilowatts — about as much as was being produced in all of Canada in 1970.

Two major problems led the United States to seek an agreement with Canada for control of the upper part of the river. During the spring and

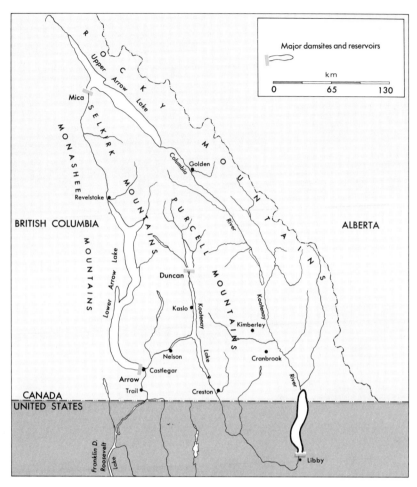

Figure 12.5 COLUMBIA RIVER PROJECT. Several large dams and numerous smaller ones provide flood control and additional hydroelectric power in the United States. Canada receives cash payments and the opportunity of developing additional power in the future. The Libby dam backs up water a few kilometres into Canada, but Canada gets no benefits from it.

summer, there was too much water for the generators to handle, and large volumes of water poured over the spillways, unused. During the fall and winter, however, when power demands were at their greatest, the river's flow was insufficient to drive all the generators. The spring floodwaters also caused a great deal of damage. Canadian and American engineers agreed that the best answer to both the flood and the power problem was the building of a series of dams on the Canadian section of the river, but how the costs and benefits would be shared between the two countries was not easily settled.

After many years of negotiations, finally in 1964 the Canadian and British Columbia governments signed the Columbia River Treaty with the United States. Under the terms of the treaty, Canada agreed to build three major dams in Canada and permit the United States to build the Libby

Above: Duncan Dam, the first of the Columbia River Treaty dams to be constructed.
Left: A close-up view of Duncan Dam, which is 40 m high and creates a reservoir which extends 8 km upstream.

Dam on the Kootenay River which would back up water a few kilometres onto Canadian soil. These dams hold back the spring floodwaters, thus controlling flooding downstream, and by gradually releasing the stored water, substantially increase the hydroelectric power developed at the power stations in the United States. The agreement gave British Columbia a cash payment for the American flood control benefits resulting from the Canadian dams, and half of the additional power generated in the United States. Because B.C. did not need additional electricity at that time (British Columbia already was planning a large hydro development on the Peace River) it opted to take cash in advance instead of its share of the electric power. It was hoped that the several hundred millions of dollars would be enough to pay for the cost of the dams. In addition, the existence of the dams and reservoirs would make it possible to develop more low-cost hydroelectric power if needed in the future.

At the time of the treaty, it appeared that Canada had made a good deal. However, with rapidly increasing costs, Canada discovered that by the time the dams were built, they cost more than the cash payments that had been made. In addition, no compensation was paid to Canada for the damages caused by the raised waters behind the dams. The reservoirs behind Mica and Duncan Dams have flooded large areas of forest land and destroyed the wildlife habitat. The raising of Lower Arrow Lake by some 24 m has meant that about 2000 people have had to relocate their homes. The whole Columbia River project was conceived in terms of hydroelectric power and flood control, and no thought was given to environmental and social consequences.

The Peace River Project

In the late 1960's, British Columbia constructed the Bennett Dam on the Peace River. The huge reservoir, Lake Williston, which formed behind it, provides water storage for producing hydroelectric power. Before the development, the spring flow was about twenty times as great as the winter flow. The dam and reservoir now regulate the seasonal flows and reduce the spring floods.

The reduction of the spring floods has had unexpected ecological results in the Peace-Athabasca Delta area. As a result of the dam, many of the shallow lakes in the delta area have dried up and others have been reduced in area. Much of the marsh has also dried up because the delta lakes such as Lake Claire and Mamawi Lake do not have their levels raised in spring flood and therefore do not have enough water storage to continue supplying the marsh with water through the summer (Figure 12.6). As a result of these lower water levels, the populations of pike, pickerel, beaver and muskrat have been greatly diminished.

The whole Peace-Athabasca Delta area forms one of the most important habitats for waterfowl in western North America, both as a breeding ground and a staging post for migration. Birds from all four major North

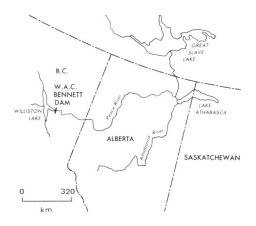

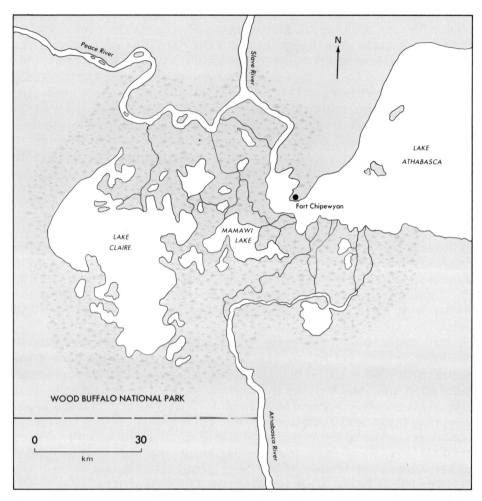

Figure 12.6 THE PEACE-ATHABASCA DELTA. The Peace-Athabasca Delta area is a maze of river channels, tributaries, lakes and marshes. The Bennett Dam, hundreds of kilometres upstream in British Columbia, reduced the level of the lakes and drained much of the swampland, thus seriously damaging the area as a habitat for fish, beaver, muskrats and wildfowl.

American flyways gather in this delta. The reduction of lakes and elimination of marsh is damaging this important bird sanctuary in Wood Buffalo National Park. The long-run impact on the birds has not yet been precisely determined, but certainly bird numbers will decline and a few species which nest exclusively in this area may become extinct.

Some engineering solution to again raise the water levels in the Peace River Delta may be possible. Nevertheless, what has happened to date illustrates the way in which interfering with the flow of a major river can cause ecological damage downstream. In this case, the downstream effects were not even anticipated because no ecological studies had been conducted ahead of time. Accounts of the effects of the Bennett Dam give us some indications of the disastrous ecological damage that would result from massive diversions of water from our Arctic rivers.

The W.A.C. Bennett Dam stretches 2 km across the Peace River Valley. The 183 m high earth structure stores water in the Lake Williston Reservoir. Ecological effects of this dam have been felt hundreds of kilometres downstream.

The James Bay Project

In 1971, the Quebec Government announced plans for a giant hydroelectric project in the James Bay drainage basin in Quebec to meet the growing need for power and to stimulate the economy of the province. It was announced that the project, which would involve the Eastmain, Broadback, Nottaway and La Grande rivers and La Grande Rivière de la Baleine, when completed would be the largest producer of hydroelectric power in Canada (Figure 12.8). The first phase of the development is on La Grande River (in many older atlases named Fort George River).

The project involves many huge dams and reservoirs and the diversion of water from one river to another. From the beginning, it was strenuously opposed by environmentalists who were concerned about the massive ecological damage that could result from the development. They claimed that fish and wildlife resources so important to the livelihood of the natives would be drastically reduced.

Natives living in the area also strongly opposed the development. They claimed that they legally owned the land concerned and that their ownership rights had never been altered by purchase, law, or treaty. After a number of legal battles, an agreement was negotiated between the Quebec Government and the native peoples.

The James Bay Agreement provides for $225 million in cash and royalties to be paid to the groups of native peoples (6500 Cree Indians and 5200 Inuit) over a twenty-year period. There is also a land claims settlement that divides the northern territory into three categories (see Figure 12.7). Category 1 includes approximately 13 000 km² set aside exclusively for the use of the two native groups. Category 2 comprises some

The James Bay Project: La Grande number 2 (LG2) is the first of four planned hydroelectric power sites to be developed on the La Grande River as part of this massive Hydro-Quebec project. A large rock-fill dam created a reservoir flooding thousands of square kilometres of native peoples' hunting and fishing grounds. Electricity is generated from an underground powerhouse blasted out of the solid rock below the dam.

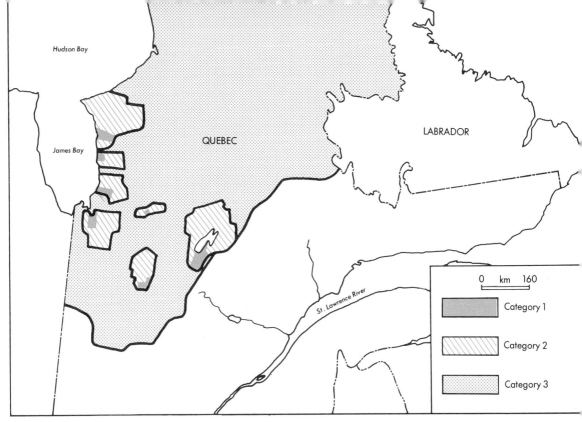

Figure 12.7 JAMES BAY AGREEMENT WITH NATIVE PEOPLES. Category 1 is for exclusive use of native peoples. Category 2 is for exclusive fishing and hunting rights to the native peoples. Category 3 gives native peoples certain privileges, but the public has access to the land.

Countryside in the James Bay area.

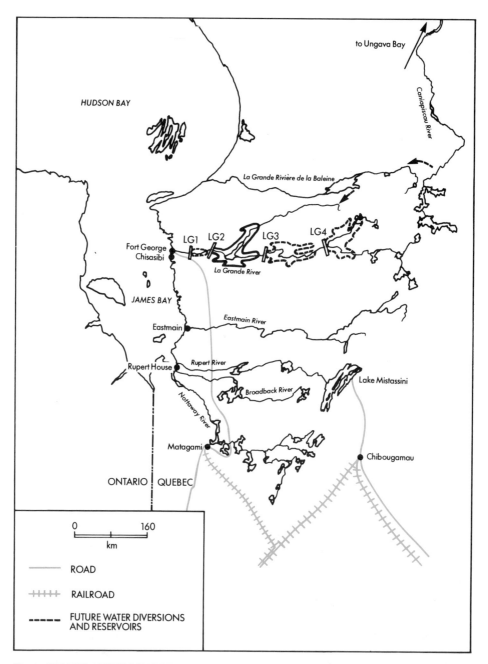

Figure 12.8 THE JAMES BAY PROJECT. The first phase of La Grande No. 2 (LG2) was completed in 1979. There are future plans for LG3 and LG4 upstream and LG1 downstream. The Caniapiscau River will be diverted into La Grande Rivière de la Baleine (Great Whale River), which in turn will be diverted into La Grande River to create still more power. At a later stage the Eastmain, Rupert, Broadback, and Nottaway rivers may also be developed.

The above photos were taken at Fort George (now Chisasibi) at the mouth of the La Grande River. Some of the Indians live in crudely built houses and obtain their water from the river. However, even before the James Bay project had been announced, some of the Indians at Fort George had begun to live in modern homes and to drive snowmobiles.

150 000 km² in which the natives have exclusive hunting, fishing, and trapping rights. Mining and other activities are permitted within Category 2 provided they do not conflict "unreasonably" with native uses. The balance of the territory falls into Category 3 where certain animals are reserved for the use of natives, and forest products will be free for their use; the general public will have access to this land. (See Figure 12.7.)

Concerning the natural environment, the James Bay Corporation is required to make environmental studies before any development proceeds. A committee of natives has been established to advise on environmental issues. A number of modifications in the hydro development plans have been made already to lessen the environmental impact.

The James Bay project has proved to be a boom to the Quebec economy. The construction phase created many jobs and the abundant supply of hydroelectricity has made the province less dependent on increasingly expensive petroleum products.

PROBLEMS AND PROJECTS

1. One chapter in a book cannot adequately cover such broad topics as pollution, conservation, and resource management. More information and points of view are found in the following books:

    R.G. Adamson, *Pollution: An Ecological Approach* (Scarborough: Bellhaven House, 1971).

    T.R. Berger, *Northern Frontier, Northern Homeland* (Ottawa: Supply and Services Canada, 1977).

    R. Bryan, *Much is Taken, Much Remains* (North Scituate, Mass.: Duxbury, 1973).

    O.P. Dwivedi (ed.), *Protecting the Environment: Issues and Choices — Canadian Perspectives* (Toronto: Copp Clark, 1974).

    F.D. Foster and W.R.D. Sewall, *Water: The Emerging Crisis in Canada* (Toronto: James Lorimer, 1981).

    Human Activity and the Environment (Ottawa: Statistics Canada, 1978).

    R.M. Irving (ed.), *Readings in Canadian Geography*, 3rd ed. (Toronto: Holt, Rinehart and Winston, 1978).

    R.R. Krueger and R.C. Bryfogle (eds.), *Urban Problems: A Canadian Reader* (Toronto: Holt, Rinehart and Winston, 1971). See sections on "Pollution" and "Urban Conservation".

    R.R. Krueger and B. Mitchell (eds.), *Managing Canada's Renewable Resources* (Toronto: Methuen, 1977).

    G.R. McBoyle and E. Somerville (eds.), *Canada's Natural Environment* (Toronto: Methuen, 1976).

    F. Morgan, *Pollution: Canada's Critical Challenge* (Toronto: McGraw-Hill Ryerson, 1970).

    W.P. Neimanis, *Canada's Cities and their Surrounding Land Resource* (Ottawa: Environment Canada, 1979).

    N. Sheffe, *Environmental Quality* (Toronto: McGraw-Hill Ryerson, 1971).

    W. Simpson-Lewis *et al.*, *Canada's Special Resource Lands*, Map Folio 4 (Ottawa: Environment Canada, 1978).

    F.J. Taylor *et al.* (eds.), *Pollution: The Effluence of Affluence* (Toronto: Methuen, 1971).

    D.M. Welch, *for land's sake!* (Ottawa: Environment Canada, 1980).

2. The more energy we use, the more we pollute the environment. What are the environmental impacts of each of the following: hydroelectricity; thermal electricity from fuel oil; thermal electricity from coal; thermal electricity from nuclear fuel; natural gas; or burning wood? None of these sources of energy is as clean as solar energy. Why does Canada not use solar energy as its major heat source?

3. Ontario is increasingly turning to nuclear reactors for electrical power. Quebec is turning to hydroelectricity instead. Explain why these two provinces are taking different approaches to the energy issue. In the long run, which approach do you think will have the greatest environmental impact? Why?

4. Obtain Canada Land Inventory (C.L.I.) maps of your region for: (i) soil capability for agriculture; (ii) land capability for forestry; (iii) land capability for outdoor recreation; and (iv) land capability for wildlife. Compare actual land uses with these capability ratings. On the basis of this study, make some land-use planning recommendations. (The C.L.I. index maps and *A Guide for Resource Planning: The Canada Land Inventory* are available from: Lands Directorate, Environment Canada, Ottawa, Ontario, K1A 0E7.)

13 CANADIAN REGIONAL DEVELOPMENT PROBLEMS

REGIONAL DISPARITIES

Canada is a huge country with great regional differences in physical geography, resources, density of settlement, intensity of economic activities and history of development. Because of these regional differences, there are also great differences in prosperity from one region to another. These regional differences in prosperity are often termed *regional* or *economic disparities*. A number of Canadian political leaders have claimed that regional disparities are one of Canada's gravest problems, and are as serious a threat to Canadian unity as Quebec's separation or the oil-pricing controversies.

The degree of regional disparity is very hard to measure. On the basis of income per person, Ontario and Alberta are the richest provinces, while Newfoundland and Prince Edward Island are the poorest. However, the cost of living (particularly housing) may be much higher in one province than another. For example, because of the oil boom in Alberta, houses there cost two or three times as much as in the Atlantic Provinces.

Income data for provinces or larger regions must be interpreted carefully. Because they cover broad geographic areas, within each region there are in turn great differences in prosperity. In the Atlantic Region, which has the lowest average income and the greatest percentage of people living in poverty, there are areas of considerable prosperity as well as areas of almost unbelievable poverty. Similarly, in prosperous Ontario there are pockets of serious rural poverty, and there are large numbers of poor people living in the cities.

Income statistics indicate the magnitude of poverty only in terms of money. The poor also suffer social disadvantages and mental anguish, as is indicated in the following quotation from *The Real Poverty Report:*

> By the time you are a teenager you accept without question ... that you are not really good enough to go any further with your education. You know that it would be a waste of time even to think about it because your parents couldn't afford to send you anyway.
>
> From then on, as you go from one menial job to another, you come to know that machines are more important than you are ... During hard times when jobs are scarce, employers tell you that it is your fault that you don't have enough education, enough skills ...
>
> As you move through a succession of crummy apartments, where the rents are always just too high, your kids start growing up the same way you did—on the street. And you suddenly realize there is no way out, that there never was a way out, and that the years ahead will be nothing but another long piece of time, spent with an army of other sick, lonely and desperate old people.*

*I.A. Adams, et al., *The Real Poverty Report* (Edmonton: Hurtig, 1971).

Another measure of regional disparities is the percentage of the labour force that is unemployed. When the wage-earner of a family is unemployed the whole family suffers. When large numbers of people are unemployed, they are unable to buy things and so the economy of the whole region declines.

Statistics show that, over the years, the Atlantic Provinces and Quebec have suffered the highest unemployment rates, while Ontario and the Prairies have had the least unemployment. Unemployment in British Columbia fluctuates a great deal because of its dependence on primary resources for which the world demand often rises and falls quickly.

The regional data of Figure 13.1 masks differences from province to province. For example, in 1979, when the unemployment rate for the Atlantic Region was about 12 percent, the figure for Newfoundland was about 17 percent and, for Prince Edward Island, 10 percent. Even within one province, the unemployment rate can vary greatly from one district to another.

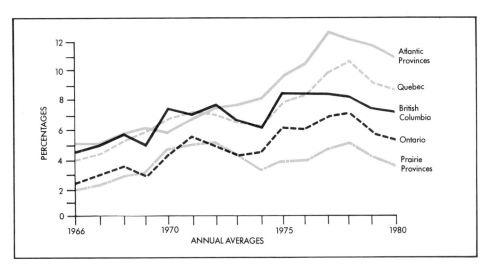

Figure 13.1 CHANGING REGIONAL UNEMPLOYMENT RATES. Which regions have the most favourable unemployment rates? Why?

CAUSES OF REGIONAL DISPARITIES

Regional disparities in Canada have deep historic and geographic roots. At one time, the colonies that now compose the Maritime Provinces were very prosperous. Rich forest and fishing resources and the British Empire trade preferences made it profitable to ship vast amounts of cod to the West Indies and lumber to Britain. The accessible timber resources led to an extensive shipbuilding industry and shipping business. However, in the latter part of the nineteenth century, technological change removed the region's geographic advantages. Wooden boats were replaced by steel ships and the Maritimes' shipbuilding and sea-carrying trade began to decline. In addition, Britain began abandoning Empire trade preferences, thus reducing the Maritimes' competitive advantage.

Confederation in 1867 did not bring the prosperity the Maritimes had anticipated. The National Policy of high tariffs on the import of manufactured goods benefited the manufacturers in the Great Lakes-St. Lawrence Lowlands. The result to the Maritimers was increased costs of consumers' goods imported from foreign countries.

The national tariff policies had a similar effect on the Prairie Provinces after they joined Confederation. Both the Atlantic and Prairie Provinces claim that high tariff protection has discriminated against them up to the present time. Both regions must pay higher prices for imported manufactured goods while their major products, mainly raw or semi-processed materials, are sold abroad at competitive world prices.

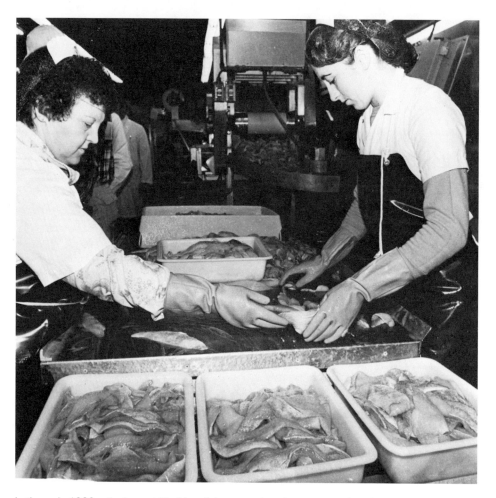

In the early 1980s, the largest Maritime fish processing plants teetered on the edge of bankruptcy due to national economic conditions. The federal government took steps to restructure the fish processing industry and pledged 138 million dollars to stabilize the fishery in Newfoundland, the Maritimes, and Quebec.

Other federal government policies and programs also seem to favour the already prosperous provinces. The building of canals in the Great Lakes-St. Lawrence Lowlands has been of primary benefit to Ontario and Quebec. At a time when there is rapid economic growth resulting in inflation, the federal government restricts the supply of money and raises interest rates. Although these policies are aimed at "cooling the overheated economy" of the more prosperous provinces, the greatest slowdown in growth and resulting unemployment occur in the slow-growth regions.

Economic and physical geographical factors also have contributed to regional disparities. Some areas are lacking in natural resources that are in demand. Others are so remote that they cannot compete in the major markets of the country. In still others, severe winters and short summers have discouraged economic development and population growth. Farm poverty is heavily concentrated in areas where climate and soils are not

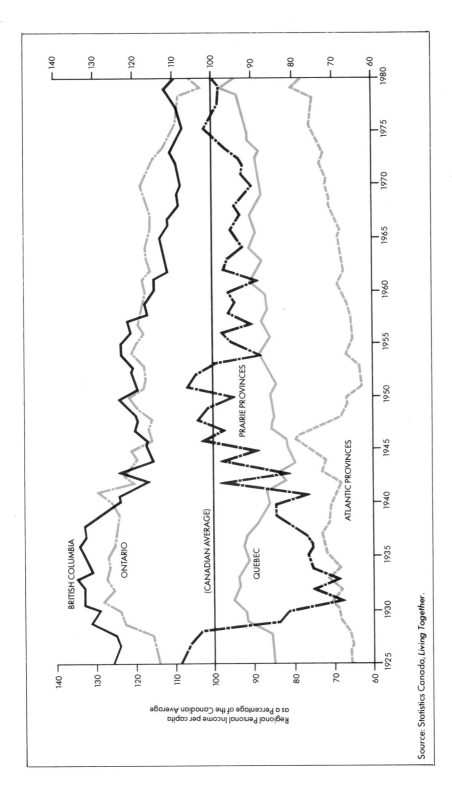

Figure 13.2 CHANGING REGIONAL INCOME DISPARITIES. Because of various regional development programs, income disparities in some regions of Canada are being reduced or improved (i.e. rising closer to the Canadian average).

Geographical factors can contribute to the uncertainties of the farmer's existence. Here, two farmers watch the rushing waters of the flooding Red River move over their land near the community of St. Jean Baptiste, Manitoba.

favourable for commercial agriculture. These poor farm areas include most of Newfoundland; large portions of the Maritime Provinces; the Gaspé Peninsula; the southern margin of the Canadian Shield in Quebec, Ontario, and Manitoba; the pioneer fringe north of the prime farming areas of Saskatchewan and Alberta; and scattered pockets in the interior of British Columbia (see Figure 9.2).

The gap between the richest and poorest regions is just as much a result of great prosperity in some parts of the country as it is of the poverty in others. The Great Lakes-St. Lawrence Lowlands, with its favoured geographic location near the great manufacturing belt of North America, its rich resources, and availability of water transportation and hydroelectricity, has attracted a great concentration of manufacturing industry. Industrialization turns to urbanization and dense populations; this, in turn, creates both a large market and a large labour pool that help to attract more industry. And so growth attracts growth.

Regional disparities have existed for a long time in Canada. In fact, it can be seen from Figure 13.2 that they were a much more serious problem in the 1930's. Can you explain the various fluctuations in the lines of the graph? For example, why did the Prairie income plunge so low in the early 1930's? Why did it rise so rapidly in 1940? Why has it shown such ups and downs since then? And why did it rise above the national average in 1975?

ATTEMPTS TO SOLVE THE PROBLEM OF REGIONAL DISPARITIES

Even in the early days of Confederation, regional disparities were recognized (although they were not known by that name) and policies and programs were initiated to help solve the problem. Transportation subsidies were established to help the grain farming economy of the Prairies and coal mining in the Maritimes. A special freight rate was introduced on all goods moved by rail in the Maritime Provinces. Old age pensions, family allowances and unemployment insurance are of more importance to a low income fisherman in Newfoundland or an Indian trapper in northern Quebec than to a well-paid tradesman in a large city in Ontario. Throughout the 1950's and 1960's, the national oil policy helped the economy of Alberta by excluding oil imports west of the Ottawa Valley and by assisting in the financing of oil pipelines. Then, in the late 1970's and early 1980's, the federal government suppressed Alberta's potential income by keeping Canadian oil prices below the world level. At the same time it assisted the economy of the Atlantic Provinces with subsidies to offset the high price of imported oil.

In addition to these various programs listed above, the federal government pays equalization grants to the "have-not" provinces. In some provinces these equalization grants form a significant proportion of their total revenue and help them provide services at a standard approaching the Canadian average.

REGIONAL DEVELOPMENT PROGRAMS

Despite numerous policies and programs, serious regional disparities persisted and, as a result, the federal government initiated a series of special regional development programs aimed at assisting the less prosperous regions in the country. A few of these are described here. Because of their long names, most of them are known by their acronyms, or initials.

PFRA (PRAIRIE FARM REHABILITATION ACT, 1935)

PFRA was intended to rehabilitate the Prairie land and farmers stricken by the combination of drought and the Great Depression in the 1930's. Federal and provincial funds were made available for projects such as:

(1) assembling land for community pastures,
(2) water supply projects for domestic uses,
(3) tree production and planting as shelter belts,
(4) encouragement of improved conservation practices by the farmers, and
(5) community development projects.

MMRA (MARITIME MARSHLAND RECLAMATION ACT, 1948)

MMRA was a means by which the federal government gave financial assistance to rural areas in the Maritimes similar to the aid it had been giving the Prairies. Many thousands of hectares of saltwater marshland bordering the Bay of Fundy were reclaimed for agriculture. By increasing the amount of productive land it was hoped that the incomes of the farmers would be increased. However, this program was not particularly successful because additional farmland was not needed at the time. Most of the reclaimed land has reverted to marshland.

ARDA (AGRICULTURAL AND RURAL DEVELOPMENT ACT, 1961)

The ARDA legislation was the first attempt to solve rural economic problems across the whole nation. It encouraged improved use of natural resources and efficiency of land use. One of ARDA's major accomplishments has been the Canada Land Inventory which has provided information on land capability for the settled areas of Canada. Maps are now available which show how good various areas of land are for agriculture, forestry, wildlife, sports fish production, and recreational uses. This information is useful in planning resource use on a broad regional basis.

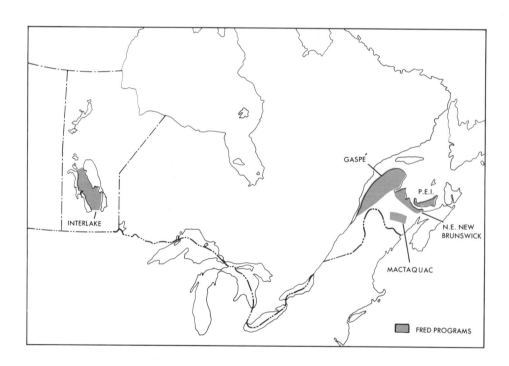

In the early stages of the ARDA program, the emphasis was on improving the efficiency of farming in economically depressed areas. Farm assembly schemes helped farmers obtain larger and more economic-sized farms. Community pastures helped farmers enlarge their cattle herds. Financial assistance was made available to drain land and improve the productivity of the land in other ways.

Although the program helped those farmers who increased the size of their farms, others were being forced to move off the land and were poorly trained to find other jobs. Also it became clear that rural poverty was not exclusive to agriculture. There were large numbers of non-farm rural poor who needed help just as desperately.

In the last half of the 1960's, the ARDA program was broadened to include the alleviation of rural poverty in general. This broadened scope included expenditures of funds for things such as employment training and assistance to move to where jobs were available.

Under ARDA, a special Fund for Rural Economic Development (FRED) was established to stimulate the economy in certain subprovincial regions with persistent and chronic poverty. FRED programs were established in five regions: Northeast New Brunswick, the Mactaquac Region in New Brunswick, all of Prince Edward Island, the Gaspé Peninsula of Quebec, and the Interlake Region of Manitoba. Intensive regional development programs in these special regions included assistance to modernize the primary resource industries, to upgrade the public education system, to provide job training and re-training for adults, and to help families re-locate in areas where jobs were available.

Large sums of money were spent on ARDA projects, but it is difficult to assess the results. The program undoubtedly helped many people and improved the economy in specific areas. Generally, however, the improvements were too small to combat widespread rural poverty or to narrow the regional income gap (Figure 13.2).

DREE (DEPARTMENT OF REGIONAL ECONOMIC EXPANSION, 1969)

The programs just described had a rural emphasis. Other government programs gave financial incentives to manufacturing industries to locate in cities and towns with high unemployment rates. Unfortunately, the urban and rural development programs were not always integrated. Unemployed rural people received training for jobs, but often no suitable jobs opened up in the cities and towns of their region.

Because of a lack of co-ordination of government programs and the persistence of regional disparities, in 1969 the federal government established the Department of Regional Economic Expansion. It took over all

Figure 13.3 REGIONAL INDUSTRIAL INCENTIVES. The above is a copy of an advertisement placed in major Canadian newspapers by the Department of Regional Economic Expansion.

the PFRA and ARDA programs as well as industrial incentive grants and other programs related to economic development. One of its major objectives is to co-ordinate all government programs aimed at fighting regional disparities. It integrates rural development and urban development. In each slow-growth region it has selected certain urban centres with growth potential. Development is encouraged in these *growth poles* or *growth centres* through industrial incentive grants and through assistance in providing the cities with basic services such as expressways, and water and sewage facilities. By attracting industry to these cities it is hoped that the rural unemployed can find jobs after they take employment training programs (Figure 13.3).

At the time of writing, it is too early to assess the effectiveness of DREE's regional development approaches. We should not expect a quick solution to a complex problem that has been developing for over a hundred years.*

The cost of the various regional development programs is great. But the cost is not too great if the result is the reduction of regional disparities, the improvement of the quality of life of a number of Canadians, and the preservation of Canada as a unified nation.

*In 1983 the Department of Regional Economic Expansion was reorganized into the Department of Regional and Industrial Expansion (DRIE).

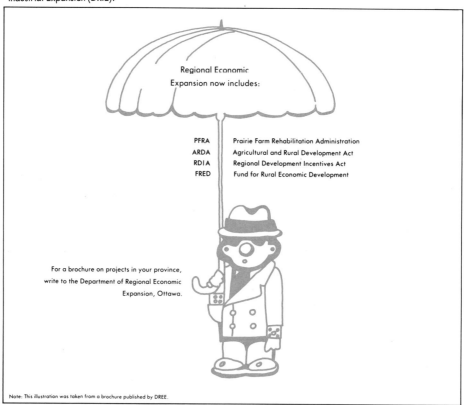

PROBLEMS AND PROJECTS

1. Debate the following controversial statements:

    If a region does not have the necessary geographical attributes (primarily resources and location) it may not be possible to improve the economy. In such regions it would be wiser to move the people out of the region to another, more prosperous region.

    Regional development programs destroy the way of life of the people of a region. It is not right for outsiders to impose their social, cultural, and economic values on the people of a particular region.

    The Atlantic Provinces are a depressed region because the people like it that way. They prefer to have less work and more relaxation.

2. For more information on regional disparities and regional development, consult the following:

    I. Adams, et al., *The Real Poverty Report* (Edmonton: Hurtig, 1971).

    H. Innis, *Regional Disparities* (Toronto: McGraw-Hill Ryerson, 1972).

    R.M. Irving (ed.), *Readings in Canadian Geography*, 3rd. ed. (Toronto: Holt, Rinehart and Winston, 1978), especially "Disparities Canada".

    R.R. Krueger, et al., *Regional Patterns: Disparities and Development* (McClelland and Stewart, 1975). This monograph provides a review of the nature and seriousness of regional disparities and an assessment of various regional development programs. It was written as an overview source book for the following three case studies published by McClelland and Stewart:

    R.C. Langman, *Poverty Pockets: The Limestone Plains of Southern Ontario*.

    R.R. Krueger and J. Koegler, *Regional Development in Northeast New Brunswick*.

    P. Sheehan, *Social Change in the Alberta Foothills*.

    Living Together, The Economic Council of Canada's annual report, 1977, provides a thorough appraisal of regional disparities and possible solutions.

    Natural Resources and Regional Disparities (Ottawa: Supply and Services Canada, 1979).

    Poverty in Canada: A Report of the Special Senate Committee on Poverty (Ottawa: Information Canada, 1971).

    B. Riddell, *Regional Disparity* (Toronto: Maclean Hunter, 1972). This small booklet gives a number of examples of poverty in different parts of the country and proposes a number of alternate solutions to regional disparities.

    P. Philips, *Regional Disparities* (Toronto: Lorimer, 1982).

14 REGIONS OF CANADA

In the preceding chapters, we have discussed the geography of Canada as it relates to a number of different topics. You will realize by now that there are many kinds of regions in Canada and that they all have their own boundaries. It is fairly easy to draw the boundaries of *single-purpose regions* such as climatic regions, landform regions, or soil regions. It is extremely difficult to draw the boundaries of *multiple-purpose* or *geographic regions*. A region is supposed to be an area of land that has characteristics common to the whole region and has boundaries that ensure that the region is quite different from surrounding regions. However, one might come up with different boundaries if one emphasizes physical factors instead of human factors. There would be different boundaries still if the regions were made up from political units such as provinces. Thus, it is not surprising that different geographers divide Canada into different geographic regions.

STUDY OF LOCAL REGION

Before you tackle a review of the geography of Canada on a regional basis, you should first gain some practice by writing a geography of your local region. First, review all of the local studies your class has done, as suggested in the different chapters. Then; decide on the boundaries of your local region. You may find that no one boundary is suitable for all purposes. In this case, you will have to decide which are the most important aspects of geography in the local area, and use these features to delimit the boundaries.

In writing your local study, remember that the geographer is interested not only in the facts, but also in how the facts are related. We want to know not only *where* things are, but also *why* they are there. In order to explain present geographic patterns, we often have to dig back into history.

In your local geography study, you should try to answer questions such as the following. You will be able to add further questions that are more pertinent for your particular local region.

QUESTIONS

1. Describe the location of your local region in relation to the rest of Canada. (A map showing large cities and major transportation routes will be of help.)

2. How many political units (municipalities) are in the region you have delimited? (A map is the easiest way to answer this question.) Are there any organizations that help the municipalities to co-operate with one another in planning land uses, in providing services, and in developing resources? How much have the municipal boundaries changed in the last ten years?

3. Describe and explain the physical geography of your local region. Explain the relationships among landforms, climate, natural vegetation, and soils. How has the physical geography affected the major industries of your region? Do not forget to include agriculture, transportation, and recreation as industries.

4. How does climate affect the costs of running a municipality? How does weather affect your daily activities?

5. Describe the population of your local region under the following headings: numbers, rate of growth, national origins, religion, and migration. Use a line graph to show the population growth over a long period of time. Try to explain the changes in rates of growth of population in your region.

6. Describe the rural settlement patterns. Explain how these patterns came to be.

7. Describe and explain the location of cities, towns, and villages in your local region. Draw generalized land-use maps of these settlements. Obtain copies of official maps showing land-use plans for these settlements. What are the major urban problems?

8. Analyse the economic geography of the primary industries in your region. Are they based on renewable or non-renewable resources? Trace the products from raw material to market. How many people in the region are employed in primary industries?

9. Analyse the economic geography of the secondary manufacturing industries in your region. In your answer consider the following factors: resources, transportation, labour, and markets. How many people are employed in these industries?

10. How important are the service industries (e.g., banks, law offices, insurance companies, finance companies, headquarters for major companies, educational institutions, wholesale and retail trade, personal services) in your region? How does the number of people employed in services compare with the number employed in primary and secondary industries?

11. What are the major environmental, conservation and resource management problems in your region? Describe any programs being used to combat these problems.

12. Are there any serious social or economic problems in your region? Describe the conditions existing in urban slums or areas of rural poverty. What social and economic development programs are being used to combat these problems?

GEOGRAPHIC REGIONS OF CANADA

Canada is sometimes divided into geographic regions by grouping provinces in the following way: Atlantic Provinces (Newfoundland, Prince Edward Island, Nova Scotia, New Brunswick), Central Canada (Ontario and Quebec), Prairie Provinces (Manitoba, Saskatchewan, Alberta), British Columbia, and the Canadian North. Such political geographic regions have certain advantages. The boundaries are precise, and many statistics are available by province (Figure 14.1).

However, these political geographic regions overlap important physical and economic geographic regions. For example, the Gaspé Peninsula of Quebec is in the same landform region as the Atlantic Provinces, and it shares landscape, natural vegetation, soils, and economic activities, as

Figure 14.1 POLITICAL-GEOGRAPHIC REGIONS OF CANADA

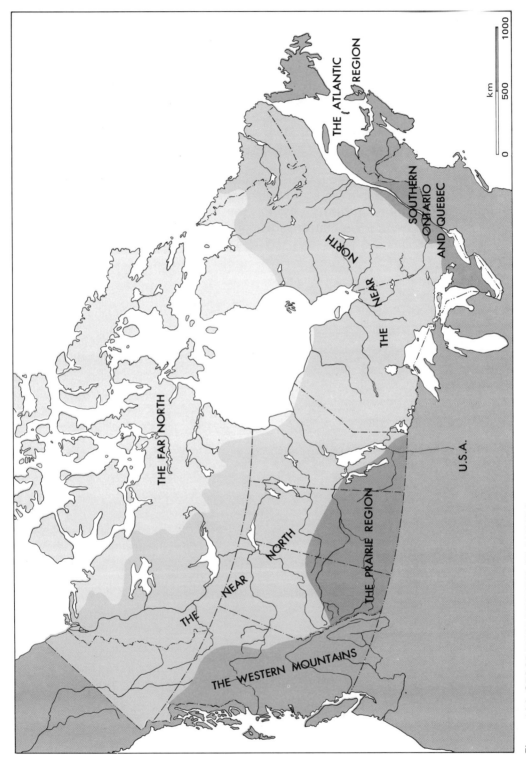

Figure 14.2 GEOGRAPHIC REGIONS OF CANADA.

well as common problems, with the Atlantic Provinces. As a further example, the southern and northern parts of Ontario have little physical or economic geography in common. Also, a large portion of the Prairie Provinces is not prairie at all but is in the northern Coniferous Forest region.

For the purpose of this review of Canadian geography, we are going to divide Canada into the regions shown in Figure 14.2. You will notice that these regions completely ignore political boundaries. We have attempted to delimit regions in such a way that each region has its own peculiar "personality," and the people within each region have common ways in which they develop the resources within their region. In class, you may wish to debate the choice of some of our regional boundaries.

You will be able to answer many of the questions asked in this chapter by using an atlas and by referring to earlier sections in this book. However, some of the questions will require that you consult encyclopedias, and government publications such as the *Canada Year Book* and the *Canada Handbook*. The *Census of Canada* has even more data broken down by census division so that geographic patterns within the major regions can be shown. Additional general information can be obtained from other geography textbooks such as:

H. Girt and W. Wiley, *Canada: This Land of Ours* (Toronto: Ginn, rev. ed. 1976).

C. Hannell and R. Harshman, *Across Canada: Resources and Regions* (Toronto: Wiley, 1980).

J. Molyneux and E. Jones, *Canada: Profile of a Nation* (Toronto: McGraw-Hill Ryerson, 1974).

D.M. Thompkins *et al.*, *Canada: The Land and its People* (Toronto: Gage, 1975).

Although they may be too difficult to read in their entirety, more advanced geography textbooks may also be used as a source of specific information and for statistical data, maps and photos. Some of the more recent of these are:

R.M. Irving (ed.), *Readings in Canadian Geography*, 3rd. Ed. (Toronto: Holt, Rinehart and Winston, 1978).

D.F. Putnam and R.G. Putnam, *Canada: A Regional Analysis* (Toronto: Dent, rev. ed. 1979).

L.D. McCann (ed.), *Heartland and Hinterland, A Geography of Canada* (Scarborough: Prentice-Hall, 1982).

After you have reviewed all of the regions, we suggest that you make a booklet on the region in which you are most interested. In it, include maps, diagrams, pictures, newspaper items, tables of information, as well as some notes you have written yourself. You will find that the departments of information of the various provincial governments will supply you with useful information. For information on the Yukon, Northwest Territories, and the Arctic Islands, you should write to Indian and Northern Affairs Canada, Ottawa.

The Atlantic Region

QUESTIONS

1. Discuss the statement: "The Atlantic Region is dominated by the sea."

2. What accounts for the excellent fishing grounds in the Atlantic Region?

3. Why are many Atlantic Region fishermen not very prosperous? What resource depletion problems are they facing? What is being done about these problems? How has the 320 km (200 mi) fishing zone helped the industry?

4. How have climate, landforms, and location of markets affected the agricultural industry?

5. In the past, what have been the major problems of the orchard industry in the Annapolis Valley? How were these problems solved? What does the Valley produce besides apples?

6. Where are the major potato growing districts and why are they located there?

7. List the locational advantages and disadvantages of the iron-and-steel industry at Sydney.

8. List the problems of the coal mining industry in Nova Scotia.

9. In which province does pulp and paper compose more than 50 percent of the value of all manufacturing production? Explain why.

10. Describe how the tide may be used to develop electric power at Passamaquoddy Bay.

11. Discuss the impact that the major oil discovery off the east coast is having on the economy of the Atlantic Region.

12. What is the most important factor limiting the secondary manufacturing industry in the Atlantic Region?

13. How has population growth in the Atlantic Region compared with other regions of Canada? How does its proportion of urban population compare with that of other regions?
14. Draw a line on a map dividing the area of French-speaking people from the area of those of primarily British descent.
15. Many residents of the Atlantic Region remain where they are because they like living there. What things do you think you would like about living in the Atlantic Region?
16. What advantages and disadvantages does the region have for attracting tourists?
17. What factors have caused the Atlantic Region to have the lowest per capita income in Canada?
18. Describe and assess the regional development programs in Northeast New Brunswick, Prince Edward Island, and the Gaspé Peninsula. If you had to decide on four major urban growth centres in the Atlantic Region, which would you choose and why?
19. Describe and discuss the Newfoundland resettlement program.
20. In Figure 14.2, Labrador—although politically part of Newfoundland—is not included in the Atlantic Region. Discuss the appropriateness of this regional boundary decision.
21. What advantages would accrue from union of the Atlantic Provinces? (Note that this would include Labrador and exclude the Gaspé Peninsula.)

Southern Ontario and Quebec

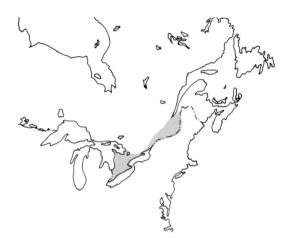

QUESTIONS

1. The most dominant physical geography features of this region are the Great Lakes and the St. Lawrence River. Discuss their importance to the region under the following headings:
   (a) modification of climate,
   (b) transportation of raw materials and manufactured goods,
   (c) water supply,
   (d) fishing,
   (e) hydroelectric power,
   (f) recreational activities,
   (g) dilution of municipal sewage effluent.

2. Draw a cross-section diagram illustrating the levels of the different Great Lakes. What problems have these differences created for navigation? How have these difficulties been overcome? What benefits have come from these differences in elevation?

3. Describe the use of the St. Lawrence River for navigation from the eighteenth century until today. What commodities form the bulk of the cargo carried on the St. Lawrence Seaway? What benefits did the St. Lawrence Seaway development bring besides better water transportation? What are some of the limitations of the St. Lawrence Seaway?

4. What are the major factors that have led to this region becoming the manufacturing belt of Canada?

5. What are the advantages of Hamilton and Nanticoke as sites for the iron and steel industry?

6. In what way does development of resources in the Canadian Shield create jobs in the Great Lakes-St. Lawrence Lowlands?

7. Why is this region the leading producer of special agricultural crops such as tobacco, soybeans, field beans, peanuts, and fruit crops?

8. What are the physical factors that make the Niagara Fruit Belt so suitable for growing tender fruit crops? What factors are leading to urbanization of the area? What steps are being taken to control the urban spread?

9. At what percentage rate per year has the region's population increased since 1951? Why has Quebec's rate of increase lagged behind Ontario's since 1971? Why do so many immigrants come to Ontario?

10. What proportion of this region's population was urban in 1951? 1961? 1971? 1981? What proportion of the census subdivisions were urban or semi-urban in 1971? (See Figure 8.12.)

11. Compare the density of population of this region with that of several western European countries.

12. What proportion of Canada's population lives in Greater Montreal and Metropolitan Toronto? Why has Toronto's population recently surpassed Montreal's? Describe the pattern of urban growth around these two cities. What problems does urban sprawl create for cities and rural municipalities?

13. In what way can it be said that Toronto and Montreal control the economy of the whole country?

14. Discuss the major problems of the fishing industry in the Great Lakes. How are these problems being solved?

15. Explain why the Great Lakes are badly polluted. Describe Canadian and American programs to solve this problem.

16. What are the major problems facing the largest cities in the region? Describe the urban planning and local government reorganization programs in Ontario and Quebec.

17. Culturally and politically the Great Lakes-St. Lawrence Lowlands are distinctly divided. If a stranger were to visit the region, what differences would he notice between Southern Ontario and the St. Lawrence Lowlands of Quebec? Consider land survey, fields, farmsteads, architecture, language, religion, entertainment, food, and general way of life.

18. In what ways is the St. Lawrence River of strategic importance to Canada? Do you think Canada could survive as a nation if Quebec became a separate country?

19. In what ways is the economy of this region linked to the economy of the United States? What are the advantages and disadvantages of these links?

20. Discuss: "The prosperity of the Great Lakes-St. Lawrence Lowlands is dependent on the exploitation of resources from outside the region."

The Prairie Region

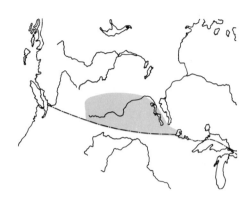

QUESTIONS

1. What factors were used in determining the boundaries of the Prairie Region?
2. Describe how and why precipitation, natural vegetation, and soils change from the Manitoba Lowland to the foothills of the Rockies.
3. Describe how the geological history helps to explain the mineral deposits found in the region.
4. The climate of the Prairies is truly continental. What does this term mean? Compare January and July average temperatures of Medicine Hat and Halifax.
5. In what way are the types of farming related to physical geography factors?
6. List some of the problems that face grain farmers.
7. In what ways does the rural settlement pattern of the Prairie Region differ from that of Southern Ontario?
8. In what ways has farming been changing in the Prairie Region?
9. Compare the population of the Prairie Region with that of Greater Montreal and Metropolitan Toronto combined.
10. What factors have led to the growth of Winnipeg, Calgary, and Edmonton as the major cities of the Prairie Region?
11. Describe the impact of the petroleum industry on the economy of the Prairie Provinces.
12. Which Prairie Province has had the most rapid rate of population increase since 1951? which the lowest? Explain why.
13. Which of the Prairie Provinces has the highest proportion of urban population? the highest percentage of employment in the secondary manufacturing and service industries?

14. Which province is most dependent on primary industry for employment?
15. Which of the provinces has the highest per capita income? the lowest? Why?
16. Which Prairie Province will likely benefit most from petroleum development in the Arctic? from mineral development in the Canadian Shield?
17. Describe and assess the regional development program in the Interlake Region of Manitoba.
18. Petroleum is a non-renewable resource. Explain why this fact has led Alberta to press for world prices for oil. What has Alberta been doing to prepare for the time when petroleum resources may be depleted?
19. The oil in the tar sands and the heavy oil of Alberta and Saskatchewan are enough to last Canada for centuries. Why have these resources not been used extensively to date?
20. Some people in the Prairie Provinces feel sufficiently alienated from eastern Canada that they have formed a separatist party. Give a historical account of why the Prairies (and often British Columbia) have these feelings of alienation.

The Western Mountains

QUESTIONS

1. Discuss the way in which mountains in this region have an effect upon the following:
   (a) precipitation,
   (b) vegetation,
   (c) the production of electricity,
   (d) the salmon industry,
   (e) the tourist industry,
   (f) the settlement pattern,
   (g) the highway and railway routes,
   (h) the cost of transportation.

2. Why is British Columbia able to lead all provinces in the production of lumber?

3. Describe the life cycle of the salmon. Why and how can the development of hydroelectric power interfere with the salmon industry? What other problems face the salmon fishermen?

4. Explain how control dams on the Columbia River in British Columbia help the United States. What benefits does Canada receive from the scheme? Why do some people question the wisdom of this scheme?

5. Describe the impact of the Peace River hydroelectric project on the Peace-Athabasca Delta.

6. What minerals are smelted at Trail? What conservation problem has the smelter caused in the past? How has this problem been solved?

7. Give reasons why a huge aluminum plant has been built at Kitimat.

8. Where are the most productive agricultural areas of the Western Mountain Region? What does each region produce?

9. What are some of the problems facing the orchardists of the Okanagan Valley? How does the climate compare with that of the Niagara Fruit Belt for fruit growing?
10. Describe cattle ranching in the Interior Plateaus. How does it differ from ranching in the Prairies?
11. What proportion of British Columbia's population lives in Greater Vancouver?
12. What is happening to much of the agricultural land in the Fraser Delta? Why is this problem particularly serious? What is the British Columbia Government doing about it?
13. In the last ten years why have Vancouver and Prince Rupert become more important as wheat ports?
14. Why is the coal industry booming in British Columbia? What conservation problems does open-pit mining pose?
15. Many Prairie farmers retire to a fruit farm in the Okanagan or move to Vancouver or Victoria. Why do you think they go to these places instead of living in a Prairie town after retirement?
16. The tourist industry is becoming very important to British Columbia. What natural attractions does the province have for tourists?
17. What is the total population of the Pacific Coast states of the United States? In what ways are the Western Mountain Region and the Atlantic Region in a similar position regarding their potential for attracting American tourists?
18. Where is most of British Columbia's natural gas found? How is it transported to the major market of Vancouver?
19. How can you defend the choice of the northern boundary for this area?
20. Debate the following statement: "On the basis of physical and economic geography, British Columbia should have been politically united with the states of Washington and Oregon."

The Near North

The Near North is that area between the other regions to the south and the true Arctic to the north. Its northern limit is usually considered to be the treeline.

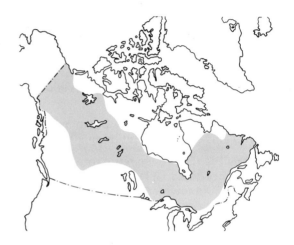

QUESTIONS

1. What three natural vegetation regions are included in the Near North?

2. How does the area of the Near North compare with the areas of the other geographic regions discussed? Is it as large as all of them combined? What is the approximate distance from the Mackenzie Delta to the eastern tip of Labrador?

3. What landform regions are included in the Near North?

4. Where are the coldest winters in the Near North? What is the effect of Hudson Bay on the climate of the region?

5. What kinds of soils are found in the Near North?

6. Why are vast areas of the Coniferous Forest region not yet used for the production of pulp and paper?

7. Since much of the Near North is in the Canadian Shield, it is important for the production of minerals. What minerals are mined at the following places?

Red Lake	Timmins
Flin Flon	Rouyn-Noranda
Thompson	Val d'Or
Wabush	Chibougamau
Baker Lake	Gagnon
Elliot Lake	Schefferville
Temagami	Labrador City
Sudbury	Allard Lake

8. Transportation is the key to developing the resources of the North. What forms of transportation have been most important in developing the following:
   (a) the fur trade in the nineteenth century,
   (b) the lead and zinc at Pine Point,
   (c) the iron ore at Lac Jeannine,
   (d) the gold at Yellowknife,
   (e) the silver near Dawson,
   (f) the oil at Norman Wells?

9. The Mackenzie River is used to transport supplies to northern communities such as Fort Norman and Inuvik. For how many months in the year would barges be able to use the Mackenzie? Why did an environmental report (Berger's) recommend against a major pipeline being built in the Mackenzie Valley?

10. Within the region of the Near North are several "islands" of agricultural land, including the Lac Saint-Jean area, the Great Clay Belt, and the Peace River district. Where are most of the products of Lac Saint-Jean marketed? In addition to climate, what is the major limitation to agricultural expansion in the Great Clay Belt? How do the length of the frost-free period and the amount of precipitation in the Peace River area compare with those for the major Prairie wheat belt?

11. How does the population of the Northwest Territories and the Yukon compare with the population of Prince Edward Island?

12. Describe the Churchill Falls, Nelson River, and James Bay hydroelectric projects under the following headings:
    (a) potential power,
    (b) market for power,
    (c) impact on environment and native people.

13. Trace the history of the development of several single resource-based towns in the region.

14. Give a historical account of the problems of the native peoples in the Near North. In what ways has the James Bay Agreement set a precedent for other groups? Why is it urgent for the federal government to settle native land claims in the Northwest Territories and the Yukon?

The Far North

QUESTIONS

1. The Far North, or Arctic, is usually defined as the area north of the treeline. How does the treeline boundary on the south compare with the location of the Arctic Circle and the southern limit of continuous permafrost?
2. Describe the landform regions found in the Far North.
3. Describe the tundra vegetation found in the region.
4. What ecological damage can result from exploring for, drilling for, and transporting petroleum?
5. How many months without frost would you expect to have at Resolute?
6. Why is it correct to call the Arctic a desert? Compare the precipitation of Resolute with that of Medicine Hat.
7. Compare the annual amount of snowfall of Resolute with that of Toronto.
8. How long would the days be at Resolute on June 21 and on December 21?
9. Discuss the problems facing the Inuit of the Canadian Arctic.
10. What are the problems of building towns and roads in the Far North? Look up the story of the building of the new village of Inuvik.
11. The Arctic is said to have a "fragile environment". What is meant by this?
12. What are the alternatives for transporting petroleum from the Arctic?

13. Describe Canadian Government legislation aimed at protecting the Arctic environment.
14. What impact may native land claims have on the development of Arctic resources?
15. What steps has Canada taken to show the rest of the world that it effectively occupies and controls all of this region?
16. Does Canada own the open water and channels in the Arctic Archipelago? Do other countries agree on Canada's stand on this question?
17. The United States disputes the position of the international boundary between Alaska and Canada where it extends into the Beaufort Sea. What is the basis for each country's position in this dispute? Why is this issue more important now than it was twenty-five years ago?
18. Compare the distance between Alert (the northern tip of Ellesmere Island) and Ottawa with Alert and Moscow.

GEOGRAPHERS AT WORK

In studying about the geography of Canada, you have learned that geographers are concerned with the natural environment and how people use, abuse, and modify that environment. To describe and analyse physical and human geographical patterns, geographers must combine the knowledge and research techniques of both the scientist and the social scientist. Because of their breadth of knowledge and their wide array of research methods, geographers are in demand by governments, industry, business, and consulting firms to do studies and make recommendations on a wide range of environmental, resource management, and economic development problems. Following is a short list of the kinds of jobs held by Canadian geographers:
— assessing the environmental impact of pipeline building in the Arctic,
— mapping and classifying agricultural land resources,
— managing renewable resources such as water, fisheries, and forests,
— developing and managing conservation programs,
— urban land-use analysis,
— urban and regional planning,
— regional economic development,
— location and market analysis for business and industry,
— recreational resource planning,
— planning for the location and use of parks,
— air photograph interpretation and remote sensing,
— computer analysis of statistical data,
— assessing natural hazard risks,
— cartography, including computer mapping,
— climate and landform analysis,
— determining the source, dispersal, and impact of air and water pollutants, and
— teaching, from elementary to university levels.

For a brochure describing careers in geography in more detail, write to:
    Secretary
    Canadian Association of Geographers
    Burnside Hall, McGill University
    Montreal, Quebec, H3A 2K6

To obtain jobs such as those described above, it is necessary to obtain a university degree in geography. Often, graduate study in geography or another field is also required. Information about university geography programs can be obtained from most Canadian universities.

GLOSSARY

alpine glacier *95* an ice mass that winds down valleys from ice centres among high mountain peaks.

Amerind *40, 209* the scientific contraction of American Indian, used to describe the people whose ancestors crossed from Asia between 11 000 and 40 000 years ago and who settled gradually as many different tribes in the southerly parts of North America.

arête *96* a jagged mountain ridge between two *cirques*.

atmosphere *133* the layer of gases, dust particles, and moisture surrounding the earth up to 100 km or more above the earth's surface.

badlands *see* **prairie**

bank *see* **fishing bank**

barrens *177* land areas north of the coniferous forest belt but south of the most severe Arctic conditions, consisting mainly of bare rock surfaces with little or no soil or vegetation. The barrens are dotted with swamps called *muskeg*.

break-of-bulk centre *242* a city in which a great many goods that are going elsewhere are moved from one method of transportation to another.

C.B.D. *253* the Central Business District of a city, where stores, offices, restaurants, apartments, and entertainment spots (often in tall buildings) are crowded closely together.

chernozem *193* the black soil belt in the Prairies, formed over thousands of years from the decay of tall grass vegetation. It produces very fertile land for grain production.

Chinook *147* a warm drying wind that blows down the eastern slopes of the Rocky Mountains. (See **rainshadow**.)

cirque *96* a circular basin at the source of a mountain valley in which alpine glaciers form.

climatologist *133* a person trained to study climate through long-term averages of weather conditions.

climatology, applied *151* the study of climate differences from place to place in order to help make living conditions more comfortable and land use more productive.

climograph *148* a combination graph that shows the climate conditions at one location for an entire year, consisting of a line graph for average temperature and a bar graph for average precipitation.

col *96* the gap between two *horns* in mountain ridges.

cold front *see* **frontal systems**

conifer *166* a cone-bearing tree with needle-like leaves that remain on the tree all year (except the tamarack); these trees have adapted to survive in regions with very cold winters. The softwood fibres of conifers are suitable for producing lumber, pulp, and paper.

conservation *428* the wise use and management of environmental resources, such as land, air, and water.

continental Arctic air mass *138* a cold, dry air mass that blows out of the northern polar regions of North America. As it moves southeast, the air is slowly warmed up, especially as it picks up moisture over large bays and lakes. This warmed-up air mass is called *modified continental Arctic*.

continental glacier *90* an ice mass that grew from separate ice centres to cover large areas of a continent. (*See* **alpine glacier**.)

continental shelf *381* a shallow, gently sloping extension of a continent out under the ocean to a depth of about

200 m; it is usually plentiful with fish.
conurbation *259* one continuous built-up region formed when several urban centres grow outward and merge.
coulee *see* **prairie**
cultural landscape *225* the land built on or modified for human living, including both urban and rural areas.

deciduous tree *166* a tree with broad leaves that are shed each fall when less moisture is available for food production. The wood from deciduous trees is used for special lumber products and in the manufacture of furniture.
degree-days *153* a system of rating daily temperature that measures how high and how long the temperature in an area remains above a given base temperature. It is calculated by subtracting the base figure from the average daily temperature and then adding together the day-to-day differences.
delta *86* the mouth of some rivers built up by sediment carried down the river and deposited in a roughly triangular shape. The soils in a delta are usually very fertile.
drainage basin *see* **watershed**
drift *see* **glacial drift**
drumlin *94* a hill or ridge formed from glacial drift. Generally long, with one steep slope and gentler slopes elsewhere, drumlins can vary in size. They are often found in groups, called a *drumlin field* (see Figure 4.20).
dryland farming *see* **fallow**

ecology, ecosystem *172* the study of how kinds of vegetation and animals, birds, and insects live interdependently with each other and the environment to form a community of living things, called an *ecosystem*.
ensilage *314* chopped corn or hay for cattle feed stored in metal or cement silos.
erosion *72, 86, 90, 304, 432* the process through which land is worn down and changed; water, ice, and wind are agents of erosion.

erratics *91* huge rock boulders carried by glaciers and left in random (erratic) places.
ethnic origins *204* the racial, religious, language, cultural, and geographic origins of people.
evapotranspiration *156* the amount of water that is evaporated from the soil and transpired (given off) by trees and other plants; measured daily, monthly, or yearly, these data are used to calculate how much moisture is needed in any one location.
extractive industries *339* economic activities based on the taking of minerals, wood, or fish from the natural environment.
fallow (dryland farming) *303* farm field cultivated but left unplanted in drier Prairie areas for one or two years to save up scarce soil moisture for the next grain crop.
fishing bank *381* a shallow, offshore sea-bottom site on raised parts of a *continental shelf* that provides ideal feeding and spawning grounds for many species of salt-water fish.
floodplain *87, 252* the area surrounding a river that is built up by sediment; this area floods with water when the river overflows its banks.
flume (agriculture) *331* a trough or channel built along the land surface to carry irrigation water from ponds (reservoirs) at higher elevations down to sprinklers or ditches in farm fields.
fossil fuels *see* **hydrocarbons**
frontal systems *138* the boundary areas between cooler and warmer air masses in a low pressure system. The leading edge of cool air is called a *cold front* and the leading edge of warm air is called a *warm front*.
frost-free season *153* the number of days between the average dates of the last frost in spring and the first frost in fall.
fry (salmon) *387* young salmon that live for a year in fresh-water rivers and lakes where they hatch before migrating to the sea to complete their growth.

gas field *353* a small land area in which many natural gas wells are drilled into underground petroleum deposits.

geographic region *473* a multi-purpose area that contains much the same kinds of human, physical, and economic conditions.

geological time scale *74* the estimated measurement of the age of the earth's rocks, landforms, plants, and animals.

glacial drift *or* **till** *91* the mixture of rock and clay picked up by moving glaciers and deposited as the glacial ice melts back.

glacier *89* a moving mass of ice. Yearly layers of unmelted snowfall become pressed into ice, then flow very slowly from the ice centre down the mountain valley or across the continent.

great circle route *9* the shortest shipping route between two points on the earth's surface; if followed in a line entirely around the earth, the arc of one such route would split the earth in half.

growth pole *or* **centre** *470* an urban centre identified by government programs that is to be the focal point for attracting additional business and industry.

gumbo *304* some Prairie soils developed from thick clay materials deposited by glaciers that often become wet and sticky after spring snowmelt or after heavy summer rains.

High *see* **pressure system**

hinterland *242* the land area surrounding a city in which urban goods and services are exchanged for raw materials.

hoodoo *see* **prairie**

horn *96* a pyramid-shaped mountain peak formed when glaciers erode several cirques from different slopes of the same mountain.

humus *168* decayed vegetative matter mixed into the top layer of soil that provides fertility.

hydrocarbons *82, 349* fuels formed when fossilized plant and animal matter trapped in sedimentary rock layers were compressed. Chemical changes produced coal, crude oil (petroleum), or natural gas.

ice age *88* one of four time periods when glacial ice covered much of the world's land surfaces.

igneous rock *70* rock formed underground by great heat and pressure, forced upward to cool at or just below the surface and to solidify into very hard rock types, such as basalt, granite, or pumice.

industrial park *255* a large parcel of land, usually near a city, that is provided with roads, sewers, and other services. Such parks are sold to companies to build factories, warehouses, and businesses.

inshore fishing *382* ocean fishing that occurs a few kilometres from shore; one or two fishermen use small motor boats to catch fish or other seafood with traps and hand lines.

Inuit *40, 213* formerly called Eskimos, they are descendants of Asian people who moved into North America and settled along the coastlines of Alaska and northern Canada.

irrigation *297, 331* drawing of water from lakes and reservoirs and its use on dry crop lands through sprinklers or a series of field ditches.

isobar *136* a line on a weather map joining all places with the same barometric pressure; all land area pressure readings are reduced to equivalent sea level readings.

isohyet *147* a line on a climate map joining all places with the same amount of precipitation; snowfall is converted to equivalent amount of rainfall.

isotherm *144* a line on a climate map joining all places with the same average monthly or yearly temperature.

latitude *and* **longitude** *5* both are sets of imaginary lines numbered in degrees and drawn in a grid pattern on globes and maps to locate points on the earth's surface. *Parallels* of latitude (parallel to the equator) are always the same distance apart, while *meridians* of longitude meet at the poles.

loam *170* a type of soil with both sand and clay particles; sandy loams usually warm up and cool down faster and hold less moisture than heavier clay loams.

Low see **pressure systems**

manufacturing (primary and secondary) *393* the processing of raw materials taken from land or water into semi-processed goods in mills or plants (primary manufacturing) before they are changed or assembled into finished products (secondary manufacturing). Steel-making and bauxite-refining for aluminum are examples of primary manufacturing, while the automobile and furniture industries are types of secondary manufacturing.

maritime Tropical air mass *138* a warm, wet (humid) mass of air that blows into North America from the Pacific Ocean or the Gulf of Mexico.

market gardening *332* the growing of vegetable crops in large amounts on farm fields and either sold fresh or sold for canning or freezing.

meltwater *94* water produced from the melting of surface ice at the top and edges of glaciers; this rapidly flowing water carries away rock materials once trapped in the ice.

metals *344* metallic minerals found in igneous rock ores, such as bauxite (aluminum ore), or magnetite (iron ore).

metamorphic rock *71* rock that was originally igneous or sedimentary, but under great heat or pressure from below the surface is changed into a new rock type. In this way, granite (an igneous rock) becomes gneiss, and shale (a sedimentary rock) becomes slate.

meteorologist *133* a person trained to study the day-to-day changes in the characteristics of the atmosphere; such technicians prepare weather maps, reports, and forecasts.

metropolis *250* a large urban centre with a business district, high population density, shopping, and industry servicing a wide tributary area.

mineral deposit *340* an accumulation of useful rock materials collected by natural forces. Deposits occur near the surface, often in thick even layers (beds), like salt, gypsum, or limestone, or in thin uneven ore veins, like lead, zinc, and copper.

mining *340* the locating, testing, and extracting of mineral deposits by digging open pits (strip mining) to remove deposits near the surface, or by drilling a network of shafts and tunnels (underground mining) to reach deeper deposits.

misfit stream see **spillway**

mixed farming *286, 294* the production of both field crops and livestock in many different combinations.

moraine *94, 96* a ridge formed at the edge of a glacier. In alpine glaciers, the ridges formed along the sides of the ice are called *lateral moraines*; those formed at the end of the ice are called *terminal moraines*.

mountain-building *72* the crumpling of rock layers into *folds*, the cracking and shifting of rock along *faults*, and *volcanic eruptions* of rock materials from underground pressures are three ways that high ridges and peaks of land can be formed.

muck *335* dark, spongy, very fertile soils often found in former glacial lake bottoms or in former bogs or swamps.

muskeg *177* swamps found in the *barrens*; these bogs are partly filled with dense masses of decaying vegetation, called *peat*, that freeze in the long cold winters and thaw in the short warm summers.

natural population increase *200* the increase of population in an area when the birthrate is greater than the deathrate.

natural soil *168* a combination of small rock particles, humus, air, moisture, and small living things; it is found at surface level and is generally undisturbed by human activity.

net migration *200* the difference between the number of people moving into a region (immigrants) and those leaving (emigrants); if immigrants outnumber emigrants, the population will increase.

nonmetallics *346* minerals found in some metamorphic rocks of sedimentary origin that contain no fuel or metallic traits, but have special uses in construction, such as sand or gravel, and in industry, such as asbestos or potash.

nonrenewable resources *see* **resources, renewable**

offshore fishing *381* ocean fishing that occurs many kilometres from home ports; a few fishermen working for a company on a large trawler ship usually spend several days at sea at a time.

oil field *353* a small land area in which many crude oil wells are drilled into underground petroleum deposits.

orographic precipitation *138* rain or snow resulting from moist air that is cooled by being forced up the wind-facing (windward) side of a mountain.

parent material (soil) *168, 170* loose, rocky, base material, such as clay, silt, sand, or gravel, that (with the addition of humus) develops into soil.

parkland *192* a natural vegetation region of long grasses and scattered clumps of trees, marking the change from dry, short grasslands in the southern Prairies to the more humid coniferous forests north of the central Prairie region of Canada.

paternalism *212* an attitude often applied to people who treat natives as if they were children dependent on other Canadians for their way of living.

physiographic region *84* a land area with similar types of landforms.

pioneer fringe (agriculture) *294* land cleared for farming in the mid-1890's along the northern edge of the major Prairie farm belt from Manitoba to British Columbia.

podzol *180* a type of less fertile soil, thin and more acidic, with a grey-coloured top layer, that is formed under coniferous trees.

pollution *429* waste gases, liquids, or solids dumped into the air (156), water (437), or soil (427) by industry, transportation, or human activity that partly or greatly affects the natural state of the environment.

prairie *122* the central region of Canada, mainly grasslands; the dry, semi-desert climate of southern Alberta-Saskatchewan has helped create some unusual features, such as wind- and water-carved slopes (*badlands*) and pillars (*hoodoos*), dry river valleys (*coulees*), and small water-storage ponds in glacial depressions (*sloughs*).

pressure systems *138* masses of air with high atmospheric pressure (a High) or low atmospheric pressure (a Low). Usually, a High will be clear, cold, and dry, while a Low generally will be cloudy, damp, and warm.

primary manufacturing *see* **manufacturing**

rainshadow *139, 147* the area that faces away from the direction of the wind (leeward side) of mountain slopes; the air flowing down the slopes warms up, producing dry winds for drier weather conditions. (*See* **orographic precipitation** *and* **Chinook**.)

regional disparities *63, 459* the differences in job opportunities, income levels, living standards, and general prosperity from one part of a country to another.

regions *see* **geographic region** *and* **physiographic region**

renewable resources *see* **resources, renewable**

resources, human *199* the intelli-

495

gence, skills, and interests of people needed as workers to make goods, as consumers to buy goods, and as designers to create better living conditions and services.

resources, natural *339* those materials in the natural environment, such as soils, trees, fish, minerals, and water, that are of importance as potential use for food or for industry.

resources, renewable *428* resources, such as water, soil, trees, and air, that can constantly re-create or re-generate (if people are careful to use them properly). Resources, such as minerals and fossil fuels, are called *nonrenewable*, since they cannot be replaced once they are used up.

rock *see* **igneous, metamorphic**, *and* **sedimentary rock**

runoff *156* extra water resulting from heavy precipitation or rapid snowmelt that cannot be absorbed by the soil; this water runs off the land and is collected by networks of streams. Additional rain can cause floods.

rural *225* describes the open countryside, mainly farm regions, or areas with few scattered homes, as opposed to more built-up areas that are called *urban*.

scale *6* the ratio of map size to the actual size of the area being mapped.

secondary manufacturing *see* **manufacturing**

sedimentary rock *70* rock formed when broken and ground up rock material from land surfaces is carried by rivers and laid down (sedimented) in layers on the bottom of water bodies; gradually the layers are pressed by their own mass into sedimentary rock, such as limestone, sandstone, and shale.

seismograph *351* the recording of time differences for seismic (shock) waves to travel down from and back to a surface; often, within sedimentary rock, seismographs can detect potential deposits of crude oil and natural gas.

service industries *241* business and community functions that do not make products, but offer or provide people with services, such as medical aid, education, fire and police protection, financial investment, goods for sale, and transport.

slough *see* **prairie**

slum *257* older, crowded housing and business area in bad repair, often in or near the downtown core.

soil *see* **chernozem, gumbo, humus, loam, muck, natural, parent material,** *and* **podzol**

soil profile *170* a vertical section dug down into soil showing different cross layers at various depths below the surface, identified as the topsoil, the subsoil, and the parent material.

soil texture *170* the size of the particles in a soil; small-sized particles result in a fine texture, and large-sized particles give a coarse texture.

spillway *94* a wide valley created by a former channel of glacial meltwater; such valleys are often filled now by small rivers called *misfit streams*.

steppe *189* a landscape in southern Alberta and Saskatchewan where the annual precipitation is under 300 mm; it consists of scattered tufts of grasses and semi-desert wildlife, such as antelope.

stereoscope *67* a magnifying instrument used on pairs of air photographs that allow a viewer to see the depth of land surface features on the photos.

storm centres and tracks *136* storm centres with moist, cloudy, unsettled air form in low pressure systems; these centres tend to follow predictable paths (tracks) from west to east across Canada.

tariff *34* a charge placed by some countries on imported goods, especially manufactured products, in order to give their own industries a better chance to sell more of their own goods within that country.

tarn *96* a lake that forms in a mountain *cirque* after glacial ice has melted.

thermal electricity *400* electrical current generated when water is heated (by burning of coal, oil, or natural gas) to produce steam that turns turbines; these turbines spin the generators that produce the electricity.

tidal bore *103* a ridge of water formed when the rising high tide water pushes river water back upstream; tidal bores occur in rivers flowing into tidal basins along shallow coastlines.

till *see* **glacial drift**

times zones *15* global areas, within which the time is the same, formed by north-to-south belts of longitude roughly fifteen degrees apart; any one time zone will be one hour behind the time zone to the east and one hour ahead of the time zone to the west.

trade balance *34* the difference between the value of imports and exports. If more goods are imported than exported, the country has a trade deficit (loss); if more goods are exported than imported, the country has a trade surplus.

trade bloc *34* an organization of several countries that group together to promote specific trade policies of interest to all the member countries.

transhumance *310* the herding of livestock to graze up to mountain pastures in summer and down to valley ranches in the winter.

treeline *172* the northern limit of coniferous tree growth beyond which the environment is too cold and dry to support tree growth. A treeline may also be seen on a mountainside.

tundra *175* the mixture of low shrubs, sparse grasses, mosses, and lichens that make up the vegetative ground cover in Arctic regions that are favourable to plant growth.

urban *225* describes central places of high population density, trade, transport, and industry; generally, to describe a built-up area.

urban planning *247* government programs for the orderly growth of a city based on where, how, and what kinds of stores, offices, homes, and factories, roads, parks, and other services are to be located.

urban renewal *247* the improvement of buildings and traffic movement in older, little-used downtown areas to attract people to live in a city core. Sometimes it involves the demolition of old buildings to make room for new ones.

urban sprawl *258, 434* scattered pockets and ribbons of homes, factories, and stores in the countryside beyond urban and suburban boundaries.

urban transition zone *255* a ring of older homes around a downtown area, some of which are being altered for small offices or stores. The land use is being changed from residential to commercial.

vegetation *see* **conifer, deciduous tree, parkland, steppe, treeline,** *and* **tundra**

warm front *see* **frontal systems**

watershed *or* **drainage basin** *439* the land area drained by a single river system, separated from other watersheds by a line of elevated land called a *drainage divide*.

water table *437* the uppermost level of water in the ground, below which the earth is saturated.

water transfer *444* the artificial movement of water from an area with a water surplus to an area with a deficit.

weather forecast *138* a report prepared from the most up-to-date analyses of weather data used to predict the conditions in an area soon afterward.

weathering *70* the process through which solid rock is broken into many small pieces.

weather map *137* a simplified map showing weather conditions in pressure systems at a given time.

ACKNOWLEDGEMENTS

The photographs in this book appear by kind permission of the following sources:

Front Cover   The rugged coastline of Cape Breton Island, Nova Scotia. Photo #E001328-M by E. Otto, courtesy of Miller Services.
Back Cover   A steel mill worker near a huge furnace. I-01114 © Toby Rankin/The Image Bank of Canada.
Chapter Openers   (18) Toronto International Airport, Photo #3042 by Mitchell/Miller Services; (132) Cloud Formations, Dr. Aubrey Diem, Dept. of Geography, University of Waterloo; (198) Crowd, Miller Services; (244) Don Mills and Eglinton, Northway Gestact Corporation Limited; (274) Combines harvesting wheat, Manitoba Government Photo Section; (338) Steep Rock Mine, Ontario, now closed and being reforested, Hunting Survey Corp.; (392) Boat construction, Digby, N.S., National Film Board of Canada; (472) Dawson City, Yukon, Nancy McMeekan.

Adair, Al; Peace River 265. **Air Canada** 61 (centre left). **Alberta Government Photographs** 83 (bottom right), 123 (bottom), 266, 282 (centre), 282 (bottom), 298 (bottom), 354, 403. **Alcan Smelters and Chemicals Limited** 417 (top), 417 (centre), 417 (bottom), 419. **The Algoma Steel Corporation Limited** 410 (bottom). **British Columbia Government Photographs** 96 (top right), 127 (top), 127 (bottom), 130 (top), 130 (bottom), 305 (top), 387. **British Columbia Hydro and Power Authority** 453. **Canadian Association in Support of Native Peoples** 210. **Canadian Government Travel Bureau** 103. **Canadian National** 60 (bottom), 105 (bottom), 106 (bottom), 121 (top), 121 (bottom left), 129 (top), 129 (bottom), 160 (bottom), 235, 241, 305 (third from top). **Canadian Pacific** 60 (top), 61 (centre right), 182 (top), 191 (top), 195 (top), 196 (top), 240 (top), 280 (bottom), 305 (second from top), 309 (bottom), 372 (top), 373 (top left), 373 (top right), 373 (bottom left). **Canapress Photo Service** 380, 462, 464. **Central Mortgage and Housing Corporation** 257 (top right), 257 (top left), 257 (bottom left), 257 (bottom right). **Convention and Tourist Bureau** 256 (centre). **Corder, R. G.** 93 (centre), 112 (top left), 348 (bottom left). **Department of Development Tourism Branch, St. John's, Newfoundland** 100 (top), 100 (centre). **Department of Regional Economic Expansion** 469. **Energy, Mines and Resources Canada** 68 (top left). **External Aid Office** 29. **Fisheries Canada** 141. **Ford of Canada** 421, 423. **Hampson, Dr. C. G.** 188 (centre), 189 (top). **The Image Bank of Canada** 28 (bottom), 1-12196 © Toby Rankin. **Imperial Oil Limited** 341 (bottom left), 341 (bottom right), 352, 356 (top right), 356 (bottom left), 356 (bottom right). **Inco Metals Company, Manitoba Division** 345 (bottom right). **Iron Ore Company of Canada** 362 (top). **Jackson, John N., Brock University, Ontario** 112 (bottom), 438 (top), 438 (bottom). **Johns-Manville Canada Ltd.** 347 (top left). **Kitchener-Waterloo Record Photos** 222, 445 (top). **Krueger, Dr. Ralph R.** 104 (top), 104 (bottom), 105 (top), 121 (bottom right), 125 (bottom left), 125 (bottom right), 128 (bottom), 180, 193 (bottom), 316 (centre left), 316 (bottom left), 316 (bottom right), 321 (bottom), 374 (bottom), 324 (bottom right), 328 (bottom left), 328 (bottom right), 329 (top left), 329 (top right), 329 (bottom left), 329 (bottom right), 331 (bottom), 345 (top right), 457. **Lakehead Harbour Commission** 305 (bottom). **Manitoba Department of Mines and Natural Resources** 93 (bottom right). **Manitoba Government Photo Section** 269 (top), 372 (bottom), 404. **Masterfile** 28 (top left) © Ly-1025 Gar Lunney, 108 (bottom) © HY-2034 Al Harvey, 109 © H-568 Sherman Hines, 209 (top) © KA-6891 Stephen J. Krasemann, 217 (top left) © V-2886 John de Visser, 217 (top right) © V-2871 John de Visser, 268 (bottom) © FS-2727 Douglas J. Fisher, 293 (bottom) © B-1568 Bill Brooks, 296 (top) © V-8723 John de Visser, 300 (top) HY-3526 Al Harvey, 364 (bottom) © V-8773 John de Visser. **McCallum, W. D.** 100 (bottom), 333 (top right). **McMeekan, Nancy** 125 (top), **Miller Services Ltd.** 188 (bottom left), 280 (third from top), 436 (top). **Montreal Transportation Commission** 256 (bottom). **National Film Board of Canada** 20, 25 (top), 28 (top right), 32 (top), 32 (bottom), 33, 61 (top), 67, 69 (top left), 69 (top right), 69 (centre), 69 (bottom left), 69 (bottom right), 96 (top left), 101 (bottom), 108 (top), 136, 160 (top), 173 (top), 173 (bottom), 174 (top), 174 (bottom), 176 (top), 185 (top), 189 (bottom), 208, 216, 217 (bottom), 264, 289 (bottom), 296 (centre), 300 (bottom), 324 (top right), 341 (top right), 357, 373 (bottom right), 388 (top), 388 (bottom), 405, 412 (top), 434, 438 (centre). **New Brunswick Museum** 45 (bottom). **Nova Scotia Government Services Photos** 47 (top), 101 (top right), 101 (centre), 102 (bottom), 257 (centre), 272, 280 (second from top), 289 (top), 321 (top), 347 (bottom right), 364 (centre), 383, 384 (top left), 384 (top right), 411, 412 (bottom). **Ontario Ministry of Agriculture and Food** 281 (bottom), 292, 293 (top), 293 (centre left), 293 (centre right), 313 (bottom), 333 (top left), 467. **Ontario Ministry of Industry and Tourism** 61 (bottom), 112 (centre), 113 (top), 113 (bottom), 116 (top), 116 (bottom), 117 (bottom), 181 (top), 184 (top), 184 (bottom), 209 (bottom), 220 (top), 200 (centre), 220 (bottom), 242 (top), 324 (top left), 348 (top left), 443 (top). **Ontario Ministry of Natural Resources** 88, 119 (top), 178, 432, 433 (centre), 433 (bottom), 436 (bottom), 443 (bottom), 445 (centre), 445 (bottom). **Petro-Canada** 356 (top left). **Prairie Farm Rehabilitation Administration** 298 (top). **Prince Edward Island Tourism** 290 (top). **Province of Quebec Film Bureau** 229, 230, 256 (top), 270 (bottom), 285, 361 (top), 400 (bottom). **Public Archives Canada** 45 (top) C-69745, 54 (top) C-14114, 54 (bottom) C-7658, 55 (top right) C-10380, 55 (bottom right) C-6413. **Quebec Ministry of Industry, Commerce and Tourism** 454. **Revell, Dr. A. M.** 188 (bottom right). **Royal Canadian Geographical Society** 46, 92 (top), 92 (bottom), 122 (bottom), 123 (top), 178 (top), 414 (bottom), 433 (top). **Royal Ontario Museum** 83 (top), 83 (centre). **The St. Lawrence Seaway Authority** 401 (top right). **Saskatchewan Tourism & Renewable Resources** 122 (top), 188 (top), 267 (top), 348 (centre left), 364 (top). **Stelco Inc.** 409 (top), 409 (bottom). **Toronto Star Syndicate** 158 (top). **Tourism New Brunswick Photos** 106 (top), 271 (top), 280 (top), 281 (top), 281 (second from top), 281 (third from top), 282 (top), 290 (bottom), 398 (top). **Transport Canada** 11, **USDA** 296 (bottom). **Warkentin, John** 237 (bottom). **Williams Bros. Photographers** 450. **Williams, Ghislain** 313 (top), 313 (centre left), 313 (centre right).

INDEX

This index is not a normal one in that it does not list individual places, products, terms, etc. Instead it lists themes and topics such as transportation, conservation, and the iron and steel industry. We feel that this kind of index is more useful for this textbook which has emphasized concepts and sample studies and has not attempted to be a comprehensive geography of the country.

A

agricultural regions
    Atlantic, 287-291
    British Columbia, 299-301
    Ontario and Quebec, 291-294
    Prairies, 294-299

agriculture
    dairying, 311-315
    farm poverty, 284-285
    importance, 275
    market gardening, 332-336
    marketing problems, 283-284
    natural hazards, 283
    orchard industry, 315-332
    part-time farming, 284
    ranching, 308-311
    statistics, 337
    trends in the industry, 276-282
    types of farming, 286
    wheat growing, 302-308

air photos, 67-68
aluminum industry
    importance and processing, 413-415
    Kitimat, 418-419
    Quebec production, 415-418

American Revolution, 51-52
Appalachian Mountain Region, landforms, 77-79, 98-108
Arctic Lowlands, 84
Atlantic Region
    agriculture, 287-290, 318-320
    coal mining, 364-365
    fishing, 381-385
    landforms, 77-78, 98-108
    manufacturing, 397-398, 411-413
    questions, 478-479
    regional disparities, 461, 464-466, 468

B

boundary disputes, 52, 58
British-French struggle, 49-51

C

Canadian Shield, landforms, 75-77, 118-119
cities (also, see urban geography)
    Churchill, 269
    Edmonton, 266
    Halifax, 272
    Kitchener, 244-248
    Medicine Hat, 266
    Montreal, 240-243, 251
    Peace River, 265
    Quebec City, 270
    Regina, 267
    Saint John, 271
    Thunder Bay, 240-243
    Toronto, 251-252, 255, 259
    Vancouver, 258, 264
    Winnipeg, 253, 268

climate
    and air pollution, 156, 158-159
    applied climatology, 151, 153-159
    climatic data, 162-163
    climographs, 148-150
    degree-days, 153, 155-156
    frost-free season, 153-154
    importance of, 159-160
    precipitation, 145-148
    regional differences, 150-152
    temperature, 140-145
    water deficit, 156-157

coal
    formation, 82, 349
    mining, 362, 365

colonization and settlement, 46-56
Columbia River project, 448-451
Commonwealth of Nations, 59
conservation and resource management
    agricultural resources, 431-435
    fisheries, 380, 383, 385-389
    forests, 371-373, 376
    fragile ecosystems, 172, 178-179
    hydroelectric power, 448-457
    Lower Mainland (Fraser Delta) farmland, 299-300
    meaning and importance of conservation, 427-430
    mining, 342, 345, 428
    Niagara Fruit Belt, 325-328
    Prairie farming, 303-304, 306, 308
    regional development programs, 465-470
    water resources, 435-444
    water transfer, 444-447
    wildlife, 168, 175-176, 183, 186-187, 190

D

defence
    air defence lines, 11, 25
    NATO, 24
    NORAD, 25

dinosaurs, 83

E

earth materials, 68-72
    (igneous, sedimentary, and metamorphic rocks)

erosion
    by glaciers, 90, 91
    by running water, 86-88
    by wind, 304
    soil erosion, 432

explorers of Canada, 42-46

F

Far North, the, 488-489

fishing
    Atlantic fisheries, 381-385
    importance, 377, 380
    Inland fisheries, 389
    lobster, 385-386
    methods, 378-379
    Pacific fisheries, 386-389
    salmon, 387-389

foreign aid, 26-29

forest industry
    importance and development, 368-370
    in British Columbia, 375-376
    location, 374-375
    resources base, 371-373

French Canadians, 62-63, 205-206
furniture industry, 424-425

G

geographic regions
    Atlantic, 478-479
    definition, 473-474
    Far North, 488-489
    Near North, 486-487
    Prairie, 482-483
    regions of Canada, 475-477
    Southern Ontario and Quebec, 480-481
    Western Mountains, 484-485

Geological Time Scale, 74, 75

glaciation
    alpine glaciers, 95-96
    glacial drift, 91
    glacial erosion, 90-91
    glacial landforms, 91-94
    Ice Age, 88-90

Great Lakes-St. Lawrence Lowlands
    (see Southern Ontario and Quebec)
    landforms, 81, 82, 109-117

H

Hudson Bay Lowlands, 84

I

igneous rocks, 70
Indians, 40-41, 209-213
Innuitian Mountains, 78, 79
Interior Plains, landforms, 80-84, 120-123
international organizations
    Commonwealth of Nations, 22-23, 59
    FAO, 26
    NATO, 24
    NORAD, 25
    OAS, 25-26
    UNESCO, 26
    UNICEF, 26
    United Nations, 26
    WHO, 26

international relations
    with the Commonwealth of Nations, 22-23
    with the Soviet Union, 21
    with the U.S.A., 20-21

Inuit, 40-41, 212-218

iron ore
    importance, 360-361
    Labrador Trough, 363
    mines, in eastern Canada, 360

iron and steel industry
    Hamilton, 409-410
    importance and processing, 406-408
    Sault Ste Marie, 411
    Sydney, 411-413

J

James Bay project, 454-457

L

land use problems
    agricultural, 431-435
    blight around the C.B.D., 255
    municipal government, 247-248
    regional disparities, 465-470
    renewing the C.B.D., 247, 254
    urban expansion in the Niagara Fruit Belt, 325-328
    urban sprawl, 258-259, 434
    water resource, 435-457

landform processes
    glaciation, 88-96
    mountain building, 72-73
    weathering and stream erosion, 72, 86-88

latitude and longitude, 12-15, 509

M

manufacturing

aluminum industry, 413-419
Atlantic Provinces, 397-398
automobile industry, 420-424
British Columbia, 404-406
furniture industry, 424-425
iron and steel industry, 406-413
location factors, 395-396
nature of Canadian manufacturing, 393-395
Ontario and Quebec, 399-403
Prairie Provinces, 403-404

Mennonites
of Manitoba, 236-238
Old Order
of the Waterloo area, 218-223

metamorphic rocks, 71
mining
coal, 365
importance and development, 339-342
iron ore, 360-362
location of minerals and mines, 343-350
metals, 343-346
mineral fuels, 349
nonmetallic minerals, 346-348
oil and gas, 350-359
statistics, 390-391
uranium, 366-367

N

national unity, 59-63
separation threats, 62-64
the role of transportation and communication, 60-62

nationhood
Canada comes of age, 58-59

natural vegetation
meaning, location, and relationships, 164-170
natural vegetation regions:
Coniferous Forest, 179
Deciduous Forest, 186-187
Grassland, 187-190
Interior Mountain, 193-194
Mixed Forest, 183
Parkland, 192
Subarctic, 177
Tundra, 172-175
West Coast Forest, 196-197

Near North, the, 486-487

O

oil and natural gas
Athabasca tar sands, 353-355
developing oil fields, 352-355
formation, 81-82, 350, 352-353
importance, 350-352
oil fields in Western Canada, 355-357
transporting oil and gas, 358-359

P

Peace River project, 451-453
physiographic (landform) regions, 75-85
political boundaries (1867 to today), 57-58
pollution
air, 156, 158-159
water, 375, 390, 429-432

population
ethnic origins, 204, 208
growth, distribution and density, 200-204
Indians, 209-213
Inuit, 212-218
Mennonites of the Waterloo area, 218-223
Mennonite settlements in Manitoba, 236-238

poverty
farm, 284-286
Indian, 211-213
Inuit, 216-218
regional, 459-461
urban slum, 255, 257

Prairie Region
agriculture, 294-299, 301-310
landforms, 80-84, 120-123
manufacturing, 403-404
natural vegetation, soils, and wildlife, 187-193
oil industry, 353-359
questions, 482-483
regional disparities, 461-468
settlement patterns, 233-238

R

regional development problems
development programs (PFRA, MMRA, ARDA, DREE), 465-470
Lake Ontario conurbation, 259-263
Lower Mainland, B.C., 299-300
Niagara Fruit Belt, 325-328
regional disparities, 63, 459-465
water resource management, 435-457

resource management
(see conservation)

rock types, 68-72

S

seasons, 12-13
sedimentary rocks, 70
scale, 6-9
settlement
early settlement, 46-49
English settlement, 46
French settlement, 47-49
later settlement, 52-56
rural settlement patterns, 225-238
settlement patterns in the Prairies, 233-238
settlement pattern in Quebec, 226-230
settlement patterns in Southern Ontario, 230-233

501

size
    area compared to other countries, 13
    distances across Canada, 14
    importance and problems of distance, 14-15
    population compared to other countries, 16
    time zones, 15

soil
    development and location, 168-170
    natural soil regions:
    Arctic, 177
    Coniferous Forest, 180
    Deciduous Forest, 187
    Grassland, 190-191
    Mixed Forest, 186
    Mountain, 194
    Parkland, 193
    Subarctic, 178
    profile, 170-171

Southern Ontario and Quebec
    agriculture, 291-294, 311-315, 320-328, 332-337
    fishing, 389-390
    landforms, 81, 109-117
    manufacturing, 399-403, 409-410, 415-425
    questions, 480-481
    regional disparities, 461-464
    settlement patterns in Quebec, 226-230
    settlement patterns in Southern Ontario, 230-233
    water management, 440-444

T

time zones, 15
topographic maps, 68, 131, 273, 353
trade
    importance to Canada, 30
    imports and exports, 31-33
    leading trading countries, 31
    trading blocs, 34-35
    trade in automobiles, 32
    where Canada trades, 33-34

transportation
    great circle routes, 9-11
    rail, road, air, seaway, 60-62
    Trans-Canada Highway, 62, 97-98
    transportation and location of cities, 242-250
    transporting oil and gas, 358-359
    transporting wheat, 307-308
    Welland Canal, the, 401

U

Upper Thames Valley Conservation Authority, 441-444
urban geography
    historical growth of cities in the Waterloo region, 244-248
    locations of cities, 248-252
    major metropolitan centres, 249-252
    origin and growth of cities, 241-243
    population of major Canadian cities, 249
    sites of towns and cities, 252
    some Canadian cities, 263-272
    urban growth patterns, 255, 258-259
    urban land uses, 253-255
    urban and regional planning, 259-263
    urbanization, 463-464

urban and regional planning
    at Kitimat, 419
    controlling urban sprawl, 434-435
    in the Niagara Fruit Belt, 325-328
    on the Fraser Delta (Lower Mainland), B.C., 299-300
    regional governments in Ontario, 247-248, 262
    Toronto-Centred Region Development Concept, 259-263

uranium industry, 366-367

W

water resources, 451-453
    Columbia River project, 448-451
    James Bay project, 454-457
    management problems, 435-440, 444
    Peace River project, 451-453
    PRIME and NAWAPA schemes, 446-447
    Upper Thames Valley Conservation Authority, 441-444

weather
    recording observations, 133-135, 140
    types of rain, 139-140
    weather maps, 136-137
    weather systems (Lows, Highs, Fronts), 136-138

Western Mountain (Cordillera) Region
    agriculture, 299-301, 310-311, 328-332
    fishing, 386-389
    forestry, 375-377
    landforms, 79, 80, 124-130
    manufacturing, 404-406, 418-420
    natural vegetation, wildlife, and soils, 193-197
    questions, 484-485
    water resources, 446-454

wildlife
    in these natural vegetation regions:
    Arctic, 175-176
    Coniferous Forest, 179-180
    Deciduous Forest, 187
    Grassland, 190
    Mixed Forest, 183-186
    Parkland, 192
    Subarctic, 177-178
    Western Mountain, 194-195